Regularity Techniques for Elliptic PDEs and the Fractional Laplacian

Pablo Raúl Stinga

CRC Press
Taylor & Francis Group
Boca Raton London New York

CRC Press is an imprint of the
Taylor & Francis Group, an **informa** business
A CHAPMAN & HALL BOOK

Designed cover image: ShutterStock Images

First edition published 2024
by CRC Press
2385 NW Executive Center Drive, Suite 320, Boca Raton FL 33431

and by CRC Press
4 Park Square, Milton Park, Abingdon, Oxon, OX14 4RN

CRC Press is an imprint of Taylor & Francis Group, LLC

Library of Congress Cataloging-in-Publication Data

Names: Stinga, Pablo Raúl, author.
Title: Regularity techniques for elliptic PDEs and the fractional Laplacian / Pablo Raúl Stinga.
Description: First edition. | Boca Raton : C&H/CRC Press, 2024. | Includes bibliographical references and index.
Identifiers: LCCN 2023057453 (print) | LCCN 2023057454 (ebook) | ISBN 9781032679440 (hardback) | ISBN 9781032679846 (paperback) | ISBN 9781032679853 (ebook)
Subjects: LCSH: Differential equations, Elliptic. | Laplacian operator. | Fractional differential equations.
Classification: LCC QA377 .S844 2024 (print) | LCC QA377 (ebook) | DDC 515/.3533--dc23/eng/20240201
LC record available at https://lccn.loc.gov/2023057453
LC ebook record available at https://lccn.loc.gov/2023057454

ISBN: 978-1-032-67944-0 (hbk)
ISBN: 978-1-032-67984-6 (pbk)
ISBN: 978-1-032-67985-3 (ebk)

DOI: 10.1201/ 9781032679853

Typeset in Latin Modern font
by KnowledgeWorks Global Ltd.

Publisher's note: This book has been prepared from camera-ready copy provided by the authors.

Dedicated to

Luis A. Caffarelli and José L. Torrea

my mentors and friends:
thank you for your wisdom

Contents

Foreword

Since the discovery of calculus, partial differential equations (PDEs) have been at the core of technological advances. The ability to represent the physical world by *equations* has allowed us to break through barriers we had not thought possible. There is a certain universality in mathematics in the sense that the same equations can model a wide range of phenomena observed in the physical world and in other, perhaps unexpected, contexts such as biology, sociology, chemistry and financial markets.

The appearance of computers signified a quantum leap for humankind. Solutions to equations can now be *seen* on the screen. Computer simulations and numerical implementations boosted an unprecedented technological revolution. Moreover, the world has become more connected thanks to new communication technologies, the internet and social media. Now we are more aware of the fact that we live in a global, nonlocal world. What happens far away has a direct impact in our daily lives. Nonlocal models that capture these phenomena are central in understanding this apparently new paradigm.

Questions of central importance arise. Up to what extent does the mathematical model describe the experiments and observations? Is the model stable under small perturbations or not? How can one be certain that what we see on the screen is actually what the solution looks like? Can we produce more efficient numerical schemes? The analysis of regularity estimates of solutions is of fundamental importance to answer these and other major questions.

This book presents a collection of powerful, robust and flexible analytic and geometric techniques to prove regularity estimates for solutions to second order elliptic PDEs, both in divergence and nondivergence form, and for the fractional Laplacian. The modern methods gathered here have played a fundamental role in breaking through major mathematical problems in areas like free boundary regularity, minimal surfaces, homogenization, fully nonlinear elliptic equations, nonlocal equations, harmonic analysis, the theory of functional spaces, etc. First, there is a family of elliptic regularity results with complete proofs, beginning with the Laplacian and developing divergence and nondivergence form equations. The second theme is the fractional Laplacian and includes everything from the basics to a family of the most relevant recent techniques.

Stinga has made a visible effort in presenting all the results and ideas in the cleanest possible ways, avoiding generalizations that would hinder the plain exposition of the main ideas. At the same time, all results are proved with full rigor and detail. It is clear that Stinga has kept the non expert reader in mind.

In addition to the more or less classical elliptic techniques such as the De Giorgi method, the Moser iteration and the Harnack inequality for viscosity solutions, Stinga

includes more recent methods that are not present in other books on the subject or are difficult to find with such clarity of exposition. Examples are Savin's method of sliding paraboloids, the Schauder estimates for divergence form equations by the method of compactness and the Sobolev estimates for divergence form equations by perturbation. The reader familiar with harmonic analysis will find known techniques in Chapter 4. However, Stinga's presentation is all about how those general methodologies are useful for elliptic regularity. In fact, that was the initial motivation of Calderón and Zygmund. Moreover, Stinga presents the a.e. pointwise formula for second derivatives, a result that is perhaps more known in the harmonic analysis community than in the elliptic PDE world.

Furthermore, Stinga gives a very detailed description of the fractional Laplacian and important regularity techniques. His approach stands on the *method of semigroups for fractional power operators* (that is also the underlying method for computing the fundamental solution of the Laplacian in Chapter 3). This is a powerful technique that has been very fruitful in many unexpected contexts. Hölder and Schauder estimates are proved with two methods: directly using pointwise formulas and with the characterization of Hölder spaces in terms of semigroups. The description of the extension problem for the fractional Laplacian follows again the lines of the semigroup methodology.

All throughout the text, Stinga maintains his philosophy of being accessible, explaining ideas and making the content readable to those having a solid background in analysis.

Anyone wanting to climb from the basis of classical elliptic PDEs to the summit of regularity theory for nonlinear elliptic equations and nonlocal equations of fractional order should pass through Stinga's book.

The contents, the way it is organized, and the teaching concepts behind it make the book unique in the field of theory of PDEs and fractional Laplacians. We see it becoming a classical reference in teaching and a most helpful companion in learning and research. We think that *Regularity Techniques for Elliptic PDEs and the Fractional Laplacian* should be on the shelves of any analyst and of any university library.

Luis A. Caffarelli and José L. Torrea

Preface

The purpose of this book is to present useful and important analytic and geometric techniques to prove regularity estimates for solutions to second-order elliptic partial differential equations (PDEs), both in divergence and nondivergence form, and to fractional nonlocal equations driven by the fractional Laplacian.

The emphasis is placed on ideas and the development of intuition, maintaining full rigor at the same time. The reader should keep in mind that this text is about *how analysis can be applied to regularity estimates.* Many methods are nonlinear in nature, but we focus on linear equations without lower order terms, thus avoiding bulky computations. Our philosophy is that ideas must be flushed out in the cleanest and simplest ways, showing all the details and preserving rigor. As such, the book is essentially self-contained. References and further notes are collected at the end of each chapter.

On the other hand, we intend to fill the gap between upper undergraduate-level textbooks like *Partial Differential Equations* by Evans [32] and *Partial Differential Equations in Action* by Salsa [57], and more advanced monographs such as *Fully Nonlinear Elliptic Equations* by Caffarelli and Cabré [7], *Elliptic Partial Differential Equations of Second Order* by Gilbarg and Trudinger [43] and *A Geometric Approach to Free Boundaries* by Caffarelli and Salsa [9].

Simultaneously, we complement and go beyond classical PDE texts by including recent regularity techniques for elliptic PDEs such as Savin's method of sliding paraboloids [58], and for the fractional Laplacian such as the Caffarelli–Silvestre extension problem [11], the Stinga–Torrea method of semigroups [69] and their applications.

Regularity Techniques for Elliptic PDEs and the Fractional Laplacian was conceived out of sets of lecture notes prepared by the author for the mini-course "Fractional Laplacians, semigroups and regularity" given at the 2014 Reunión Anual de la Unión Matemática Argentina (UMA) in San Luis, Argentina; for the "Topics in PDEs" graduate course he first taught in Spring 2015 at The University of Texas at Austin, USA; and for successive reading courses for graduate students and graduate PDE classes.

A typical one-semester course covers harmonic functions, Schauder and Calderón–Zygmund estimates for the Laplacian, the De Giorgi and Moser theorems, the ABP estimate and the Krylov–Safonov Harnack inequality for viscosity solutions, Savin's method of sliding paraboloids, the basics on the fractional Laplacian, including the Caffarelli–Silvestre extension problem and the Stinga–Torrea method of semigroups and their applications to regularity estimates.

Basic knowledge of elliptic PDEs is assumed, as well as a solid measure theory and functional analysis background. The reader should be familiar with PDE texts such as DiBenedetto [29], Evans [32], Folland [38] or Salsa [57], measure theory books such as Fava–Zó [35] or Wheeden–Zygmund [74], and real and functional analysis texts such as Folland [39] or Rudin [56].

If we were to give a word of advice to the reader, let it be this: when reading our book, keep in hand pen and paper and make the conscious effort to interpret what is going on *by drawing pictures*. Indeed, there is no better way than this to develop the geometric intuition needed to internalize and master the techniques presented in our book.

Pablo Raúl Stinga

Author Bio

Pablo Raúl Stinga earned his Licenciatura en Ciencias Matemáticas degree at Universidad Nacional de San Luis, in San Luis, Argentina (2005). He earned his Máster en Matemáticas y Aplicaciones (2007) and his Doctorado en Matemáticas *Doctor Europeus* under the direction of José L. Torrea at Universidad Autónoma de Madrid, Spain (2010). He held postdoctoral research positions at Universidad de Zaragoza, Spain (2010) and Universidad de La Rioja, Spain (2011–2012). During the period 2012–2015, he was the R.H. Bing Fellow in Mathematics No.1 Instructor at The University of Texas at Austin, USA, where he worked as a postdoctoral researcher under the supervision of Luis A. Caffarelli. He is currently Associate Professor at Iowa State University, USA. His research interests are in analysis, partial differential equations and nonlocal fractional equations.

CHAPTER 1

Introduction

In this book we present analytic and geometric techniques that are central in the regularity theory of elliptic partial differential equations (PDEs). Moreover, we introduce modern regularity techniques for nonlocal equations driven by the fractional Laplacian.

We begin with the study of harmonic functions, that is, functions u that satisfy the Laplace equation

$$\Delta u = 0$$

in a domain $\Omega \subseteq \mathbb{R}^n$, $n \geq 1$. It may come as a remarkable fact from complex analysis that the real part u of an analytic function, just because it is a harmonic function

$$u_{xx} + u_{yy} = 0$$

it becomes real-analytic. Furthermore, the oscillation of u in any given domain controls all the derivatives of u of any order in any subdomain. Then we consider the Poisson problem for the Laplacian

$$-\Delta u = f \qquad \text{in } \Omega$$

where f is some given datum.

There are several ways of studying the regularity for the Laplacian. The most powerful one is the **principle of superposition**. Here, the linearity of the equation and the symmetries of the Laplacian allow us to write down solutions and their derivatives using integral formulas, fundamental solutions, singular integrals, pseudodifferential operators, Fourier transform, etc. Then one can prove two of the most important regularity estimates.

- **Schauder estimates.** If $-\Delta u = f$ and $f \in C^{\alpha}$, $0 < \alpha < 1$, then $u \in C^{2,\alpha}$.
- **Calderón–Zygmund estimates.** If $-\Delta u = f$ and $f \in L^p$, $1 < p < \infty$, then $u \in W^{2,p}$.

These results for the Laplacian are the stepping stone for **perturbation methods**. The estimates above are the basis for the regularity theory for nondivergence form elliptic equations (we use the summation over repeated indices convention)

$$a^{ij}(x)\partial_{x_i x_j} u = f$$

DOI: 10.1201/9781032679853-1

in the case when the elliptic coefficients $a^{ij}(x)$ are a small perturbation of the identity matrix. Intuitively speaking, the equation is "close to the Laplace equation" at small scales. From here one can then derive similar regularity estimates by transferring the regularity from the Laplacian. As we said, the coefficients need to be good enough.

- **Schauder estimates for nondivergence form elliptic equations.** If $a^{ij}(x)\partial_{x_i x_j} u = f$, with $a^{ij}, f \in C^\alpha$, $0 < \alpha < 1$, then $u \in C^{2,\alpha}$.
- **Calderón–Zygmund estimates for nondivergence form elliptic equations.** If $a^{ij}(x)\partial_{x_i x_j} u = f$, with $a^{ij} \in C$ and $f \in L^p$, $1 < p < \infty$, then $u \in W^{2,p}$.

Going back to the Laplacian, let us observe that it shares both divergence and nondivergence structures.

- **Divergence structure.** This comes from the fact that harmonic functions are minimizers of the Dirichlet energy functional

$$E(u) = \int_\Omega |\nabla u|^2 \, dx.$$

We will come back to this in a moment. Notice also that

$$\Delta u = \operatorname{div}(\nabla u).$$

Hence, if $-\Delta u = f$ in Ω then, after multiplying the equation by a test function $\varphi \in C_c^\infty(\Omega)$ and integrating by parts, we arrive at the weak formulation

$$\int_\Omega \nabla u \nabla \varphi \, dx = \int_\Omega f \varphi \, dx.$$

- **Nondivergence structure.** The Laplacian can be written as

$$F(D^2 u) = \Delta u = \operatorname{trace}(D^2 u).$$

We see that $F(M) = \operatorname{trace}(M)$ is a monotone function of the symmetric matrix $M = D^2 u$. In other words, if $u_1 \geq u_2$ and they touch at x_0, $u_1(x_0) = u_2(x_0)$, then in some sense we must have $F(D^2 u_1) > F(D^2 u_2)$ at x_0. The Laplacian admits a comparison principle, which is the basic tool to prove regularity estimates for nondivergence form elliptic equations.

1.1 DIVERGENCE FORM EQUATIONS

Divergence form equations

$$\operatorname{div}(a(x)\nabla u) = 0$$

appear in conservation laws, continuum mechanics and the calculus of variations.

The calculus of variations deals with (local) minimizers of energy functionals like

$$E(u) = \int_\Omega F(\nabla u) \, dx.$$

Here $F = F(p_1, \ldots, p_n) : \mathbb{R}^n \to \mathbb{R}$ is a suitable function. We would like to minimize $E(u)$ among all admissible configurations u that, say, take a boundary datum $u = g$ on $\partial\Omega$. A typical example is minimal surfaces, where one wants to find the surface u_0 that minimizes the area functional

$$A(u) = \int_\Omega \sqrt{1 + |\nabla u|^2}\, dx$$

among all admissible configurations u such that $u = g$ on $\partial\Omega$. Here u can be thought of as a "soap surface" spanning the "wire" g.

Suppose that we are able to find a minimizer u_0 of $E(u)$. Then an equation for u_0 can be derived. Indeed, let $\varphi \in C_c^\infty(\Omega)$ and consider the small perturbation of u_0 given by

$$u_t = u_0 + t\varphi \qquad t \in \mathbb{R}.$$

This is again an admissible configuration for the minimization problem. Thus the function $i(t) = E(u_t)$ has a minimum at $t = 0$. Therefore, $i'(0) = 0$. We then compute

$$i'(t) = \sum_{i=1}^{n} \int_\Omega F_{p_i}(\nabla u_0 + t\nabla\varphi)\partial_{x_i}\varphi\, dx.$$

Letting $t = 0$, we find that u_0 satisfies

$$\int_\Omega F_{p_i}(\nabla u_0)\partial_{x_i}\varphi\, dx = 0 \qquad \text{for all } \varphi \in C_c^\infty(\Omega). \tag{1.1}$$

In other words, u_0 solves a first-order equation in integral form. As a matter of fact, as we will later see, the regularity methods in this case are integral methods, based on constructing appropriate test functions. If we now insist on integrating by parts in (1.1), we get

$$-\int_\Omega \partial_{x_i}(F_{p_i}(\nabla u_0))\varphi\, dx = 0 \qquad \text{for all } \varphi \in C_c^\infty(\Omega).$$

Since φ was arbitrary, we end up with the Euler–Lagrange equation

$$\partial_{x_i}(F_{p_i}(\nabla u_0)) = 0 \qquad \text{in } \Omega. \tag{1.2}$$

In general, the minimizer u_0 belongs to a quite weak functional space. Say, if F is coercive and has quadratic growth at infinity, then for $E(u)$ to be well defined, we require $\nabla u \in L^2$. If, in addition, F is convex, then the existence of a minimizer u_0 such that $\nabla u_0 \in L^2$ can be proved. Convexity guarantees that, in some sense, planes are minimizers. Indeed, it is easy to construct an example of a non convex F in dimension one such that $u = 0$ has 0 energy, but a sawtooth-like function with the same boundary values has negative energy. In addition, convexity of F ensures that the equation (1.2) is elliptic. The question is, can we prove further regularity of the minimizer u_0? For this we proceed as follows[1].

[1]Caffarelli teaches a lesson that, he says, learned from Serrin: "if an equation is translation invariant and has a comparison principle, then the derivatives of the solution must satisfy something." This is the basic idea for the explanations that follow.

(a) Pass the derivatives through. If we compute the derivatives in (1.2), we get

$$F_{p_ip_j}(\nabla u_0)\partial_{x_ix_j}u_0 = 0.$$

In the case when F is C^2 and convex, the equation is elliptic:

$$D^2F \geq 0, \qquad \text{that is,} \qquad F_{p_ip_j}\xi_i\xi_j \geq 0, \text{ for all } \xi \in \mathbb{R}^n.$$

If F is C^2 and uniformly convex, then there exists $\lambda > 0$ such that

$$D^2F \geq \lambda I, \qquad \text{that is,} \qquad F_{p_ip_j}\xi_i\xi_j \geq \lambda|\xi|^2, \text{ for all } \xi \in \mathbb{R}^n.$$

Suppose then that F is smooth and that, for some reason, we know that

$$\nabla u_0 \in C^\alpha.$$

Then $a^{ij}(x) := F_{p_ip_j}(\nabla u_0) \in C^\alpha$ and u_0 satisfies an elliptic equation in nondivergence form with C^α coefficients:

$$a^{ij}(x)\partial_{x_ix_j}u_0 = 0. \tag{1.3}$$

The perturbation theory mentioned above (Schauder estimates) can be applied to (1.3) to get $u_0 \in C^{2,\alpha}$. But then $\nabla u_0 \in C^{1,\alpha}$, which in turn gives that the coefficients $a^{ij}(x)$ in (1.3) are $C^{1,\alpha}$. Then (higher) Schauder estimates again apply and give $u_0 \in C^{3,\alpha}$. By continuing with this bootstrapping technique, we get $u \in C^\infty$.

However, the important point that was left open above is to prove that $\nabla u_0 \in C^\alpha$. The passage from $\nabla u_0 \in L^2$ to $\nabla u_0 \in C^\alpha$ is a nontrivial matter and is the celebrated result by De Giorgi, which later on evolved to be known as the De Giorgi–Nash–Moser theorem.

(b) The De Giorgi idea. We would like to prove that any solution u_0 to (1.2) has gradient in C^α. Let us choose a unit direction $e \in \mathbb{S}^{n-1}$ and compute the directional derivative ∂_e of the equation:

$$0 = \partial_e(\partial_{x_i}(F_{p_i}(\nabla u_0))) = \partial_{x_i}(F_{p_ip_j}(\nabla u_0)\partial_{x_j}(\partial_e u_0)).$$

If we call $v = \partial_e u_0$, then v satisfies the divergence form equation

$$\partial_{x_i}(a^{ij}(x)\partial_{x_j}v) = 0$$

where, as before, $a^{ij}(x) := F_{p_ip_j}(\nabla u_0)$. Recall that we want to prove that $v \in C^\alpha$. The only thing we know about ∇u_0 is that it is in L^2, so the coefficients $a^{ij}(x)$ above have no regularity at all and, they are just elliptic. The brilliant idea of De Giorgi was to forget about the dependence of the coefficients on ∇u_0, that is, to *decouple* the equation by simply looking at it as an equation in divergence form with *bounded measurable coefficients* and proving that solutions are Hölder continuous.

- **De Giorgi theorem.** Let v be a weak solution to $\operatorname{div}(a(x)\nabla v) = 0$. If $a(x)$ is uniformly elliptic and bounded, then $v \in C^\alpha$.

- **Moser Harnack inequality.** Let $v > 0$ be a weak solution to $\operatorname{div}(a(x)\nabla v) = 0$ in B_1. If $a(x)$ is uniformly elliptic and bounded, then there exists $C > 0$ depending only on ellipticity and dimension such that

$$\sup_{B_{1/2}} v \leq C \inf_{B_{1/2}} v.$$

As a consequence, weak solutions v to $\operatorname{div}(a(x)\nabla v) = 0$ are Hölder continuous.

(c) Perturbation methods. Suppose that, for some reason, we already have that the coefficients $a^{ij}(x)$ in the divergence form equation

$$\operatorname{div}(a(x)\nabla u) = f$$

have some regularity, say, C^α or continuous, and that the right hand side is in C^α or L^p, respectively. Then, at small scales, $a^{ij}(x) \sim \delta_{ij}$ (the identity matrix $I = (\delta_{ij})_{i,j=1}^n$) and the equation is going to be close to the Laplace equation $\Delta u = \operatorname{div}(I\nabla u) = f$. Perturbation techniques then give the following.

- **Schauder estimates for divergence form elliptic equations.** Let u be a weak solution to $\operatorname{div}(a(x)\nabla u) = f$. If $a(x), f \in C^\alpha$, $0 < \alpha < 1$, then $u \in C^{1,\alpha}$.

- **Calderón–Zygmund estimates for divergence form elliptic equations.** Let u be a weak solution to $\operatorname{div}(a(x)\nabla u) = f$. If $a(x) \in C$ and $f \in L^p$, $1 < p < \infty$, then $u \in W^{1,p}$.

Notice the difference between these perturbation results and the corresponding ones for nondivergence form equations. The gain in regularity is only of one derivative. This is because, strictly speaking, divergence form equations are not second-order equations, but first-order integral equations. In fact, the divergence form structure (1.2) arises after insisting on integrating by parts in (1.1). But the integration by parts is not allowed when F_{p_i} is not at least Lipschitz continuous. Also, we can see this gain of only one derivative in regularity from a simple observation in dimension one. Let $a(x) \in C^\alpha$ be a function uniformly bounded below away from zero and consider the equation

$$(a(x)u')' = 0.$$

Then $a(x)u' = c$ for some constant c, so u' cannot be better than C^α. As we mentioned before, $a(x)$ being continuous implies that at small scales the coefficients are close to constant, and the elliptic operator is close to the Laplacian. This is in high contrast with the De Giorgi theorem, where no matter how much the coefficients are rescaled, the equation remains within the same class. In this sense, the De Giorgi theorem makes a jump in invariance classes.

Finally, observe that it is not possible to exhibit an explicit solution to a divergence form equation, even if the equation has Hölder continuous coefficients. Indeed, given any explicit nonconstant u_0, the relation (1.1) would need to be verified for *every* test function φ.

1.2 NONDIVERGENCE FORM EQUATIONS

Nondivergence form equations

$$\text{trace}(a(x)D^2u) = 0$$

arise in optimal control and geometry.

These are the equations where we cannot multiply by a test function and integrate by parts. Instead, they satisfy a comparison principle. Two solutions u_1 and u_2 of the Laplace equation $\Delta u = 0$ cannot "touch without crossing," that is, if $u_1 - u_2$ is positive then it cannot become zero in some interior point of the domain. In other words, if $u_1 \geq u_2$ and they touch at x_0, that is, $u_1(x_0) = u_2(x_0)$, then in some sense we must have $\Delta u_1 > \Delta u_2$ at x_0.

The natural family of (nondivergence form) fully nonlinear elliptic equations to consider is

$$F(D^2u) = 0 \tag{1.4}$$

for $F = F(M)$ a function of symmetric matrices M of size $n \times n$. Of course, not every example of F can be covered by a reasonable theory. The natural assumption is for F to be elliptic, that is, a monotone function of symmetric matrices.

Standard examples of fully nonlinear elliptic equations are

$$\max_\alpha L_\alpha u = 0 \qquad \text{or} \qquad \min_\alpha L_\alpha u = 0 \qquad \text{(Bellman)}$$

where L_α is a family of uniformly elliptic linear operators $L_\alpha = a^{ij}_\alpha(x)\partial_{x_i x_j}$, and

$$\min_\alpha \max_\beta L_{\alpha\beta} u = 0 \qquad \text{or} \qquad \max_\alpha \min_\beta L_{\alpha\beta} u = 0 \qquad \text{(Isaacs)}$$

where $L_{\alpha\beta} = a^{ij}_{\alpha\beta}(x)\partial_{x_i x_j}$ and α, β belong to some family of indexes.

Important examples are the Pucci extremal operators. Say we consider *all* the equations of the form

$$\begin{cases} a^{ij}(x)\partial_{x_i x_j} u = 0 & \text{in } \Omega \\ u = g & \text{on } \partial\Omega \end{cases}$$

for which the matrix of coefficients $a(x) = (a^{ij}(x))^n_{i,j=1}$ satisfies

$$I \leq a(x) \leq 2I.$$

That is, we require all the eigenvalues of $a^{ij}(x)$ to be in between 1 and 2. We ask: is there an *envelope solution* u_0 that controls all of the other solutions? The answer is yes. One can find a function u_0 that is a supersolution to all such equations and that is itself a solution to an equation in the same class. In fact, one needs to find u_0 such that (say, in dimension 2)

$$2\lambda_{\max}(D^2u_0) + \lambda_{\min}(D^2u_0) = 0 \tag{1.5}$$

in Ω, where $\lambda_{\max}(D^2u_0)$ and $\lambda_{\min}(D^2u_0)$ denote the maximum and minimum eigenvalues of the Hessian matrix D^2u_0, respectively. Since D^2u_0 is symmetric, its eigenvalues

and eigenvectors are real. Moreover, if D^2u_0 is nonzero then, by using the equation in (1.5), $\lambda_{\min}(D^2u_0) < 0 < \lambda_{\max}(D^2u_0)$. Note that computing the eigenvalues of D^2u_0 is a highly nonlinear operation. If such u_0 were to exist, then u_0 would be a supersolution to all possible equations. Indeed, given any u we can compute the eigenvectors of $D^2u(x)$ and then write any of the equations $\mathrm{trace}(a(x)D^2u(x)) = 0$ (at each x) in the direction of the eigenvectors of $D^2u(x) = P^TDP$, where $D = \mathrm{diag}(\lambda_{\max}(D^2u(x)), \lambda_{\min}(D^2u(x)))$ to get

$$\begin{aligned} 0 &= \mathrm{trace}(a(x)P^TDP) = \mathrm{trace}(Pa(x)P^TD) = \mathrm{trace}(B(x)D) \\ &= \alpha(x)\lambda_{\max}(D^2u(x)) + \beta(x)\lambda_{\min}(D^2u(x)). \end{aligned}$$

The condition $I \le a(x) \le 2I$ is invariant under the orthogonal transformation P, so we still have $I \le B(x) \le 2I$, in which case the coefficients $1 \le \alpha(x), \beta(x) \le 2$. But then, since u_0 satisfies (1.5),

$$\alpha(x)\lambda_{\max}(D^2u_0(x)) + \beta(x)\lambda_{\min}(D^2u_0(x)) \le 0,$$

that is, u_0 is a supersolution to any of those equations. Additionally, if u_0 happens to be smooth then we can see that it is also a solution to one of the equations of the class. In fact, in the system of coordinates of the eigenvectors of $D^2u_0(x)$ one can write $a^{ij}(x)$ as $\mathrm{diag}(2,1)$, at each x. In conclusion, u_0 solves the (maximal) Pucci extremal operator equation

$$\sup_{I \le a^{ij}(x) \le 2I} a^{ij}(x)\partial_{x_ix_j}u_0 = 0.$$

The Monge–Ampère equation

$$\det(D^2u) = f$$

is another important example of a fully nonlinear equation. In fact, the Monge–Ampère equation is an extremal equation. If u is a convex C^2 function then

$$n[\det(D^2u)]^{1/n} = \inf\{\,\mathrm{trace}(A^2D^2u) : A = A^T > 0,\ \det(A) = 1\}.$$

To see this, notice that if $\lambda_1, \ldots, \lambda_n$ denote the eigenvalues of D^2u then, by the arithmetic mean-geometric mean inequality,

$$[\det(D^2u)]^{1/n} = (\lambda_1 \cdots \lambda_n)^{1/n} \le \frac{\lambda_1 + \cdots + \lambda_n}{n} = \frac{1}{n}\mathrm{trace}(I^2D^2u).$$

In addition, the infimum is attained at $A^2 = [\det(D^2u)]^{1/n}(D^2u)^{-1}$. The Monge–Ampère equation becomes elliptic when u is strictly convex, that is, when $D^2u > 0$, because in this case the infimum is achieved at a positive definite matrix.

In general, solutions to equations in nondivergence form and other fully nonlinear equations of the form (1.4) live in spaces of continuous functions and the equation is understood in the *viscosity* sense. This weak notion of solution is based on the comparison principle intuition explained above. Existence of continuous viscosity solutions can be shown by Perron's method.

Our goal is to understand how to prove regularity estimates. Recall that $F = F(M)$ is a monotone function of symmetric matrices $M = (m_{ij})_{i,j=1}^n$ of size $n \times n$. We use the notation

$$F_{ij}(M) = \partial_{m_{ij}} F(M) \qquad F_{ij,kl}(M) = \partial_{m_{ij} m_{kl}} F(M).$$

We proceed as before, by taking derivatives of the equation (1.4).

(a) Take a directional derivative ∂_e. We obtain

$$F_{ij}(D^2u)\partial_{x_i x_j}(\partial_e u) = 0. \tag{1.6}$$

We say that F is elliptic if $F_{ij} = F_{ij}(M)$ is a positive definite matrix at any symmetric matrix M, and that F is uniformly elliptic with ellipticity constants $0 < \lambda \le \Lambda$ if $\lambda|\xi|^2 \le F_{ij}(M)\xi_i\xi_j \le \Lambda|\xi|^2$ for every M and every $\xi \in \mathbb{R}^n$. We assume now that F is uniformly elliptic.

Even if F were smooth, we still know nothing about D^2u, so that the coefficients $a^{ij}(x) := F_{ij}(D^2u)$ in (1.6) have no regularity at all. Then, as De Giorgi did, one looks at $v = \partial_e u$ and realizes that v is a solution to an equation of the form

$$a^{ij}(x)\partial_{x_i x_j} v = 0$$

where $a^{ij}(x)$ are just uniformly elliptic and bounded. The analogue to the De Giorgi–Nash–Moser theorem in this case is the following result.

- **Krylov–Safonov Harnack inequality.** Let $v > 0$ be a viscosity solution to the uniformly elliptic equation $a^{ij}(x)\partial_{x_i x_j} v = 0$ in B_1, with $a^{ij}(x)$ bounded. Then there exists $C > 0$ depending only on ellipticity and dimension such that

  $$\sup_{B_{1/2}} v \le C \inf_{B_{1/2}} v.$$

  As a consequence, viscosity solutions v to $a^{ij}(x)\partial_{x_i x_j} v = 0$ are Hölder continuous.

From here, one can prove Caffarelli's $C^{1,\alpha}$ regularity result: any viscosity solution u to (1.4) is $C^{1,\alpha}$. This is the typical regularity for min-max or max-min Isaacs equations.

An important step toward proving the Harnack inequality for viscosity solutions is the so-called Alexandroff–Bakelman–Pucci (ABP) estimate, a tool that allows to pass local information from the equation to global information for the solution.

- **ABP estimate for viscosity solutions.** Let u be a viscosity solution to $a^{ij}(x)\partial_{x_i x_j} u = f$ in B_1, with $a^{ij}(x)$ uniformly elliptic and bounded, and $f \in L^n$. Then $u|_{B_{1/2}}$ is bounded and controlled by $\|f\|_{L^n(\Gamma_u = u)}$, where $\{\Gamma_u = u\}$ is a special *contact* set.

A recent technique to prove the Harnack inequality was devised by Savin, the so-called method of sliding paraboloids. Savin's methodology avoids considering Γ_u.

Now, $u \in C^{1,\alpha}$ is not enough to apply Schauder estimates in (1.6) to prove higher regularity of $\partial_e u$, as we need some control on D^2u. We differentiate the equation

once again. In general we do not expect second derivatives to satisfy an equation but, under an extra assumption, still some information can be extracted.

(b) Take second derivatives of the equation. By taking again ∂_e in (1.6),

$$F_{ij}(D^2u)\partial_{x_ix_j}(u_{ee}) + (\partial_{x_kx_l}u_e)F_{ij,kl}(D^2u)(\partial_{x_ix_j}u_e) = 0.$$

If F is convex then

$$(\partial_{x_kx_l}u_e)F_{ij,kl}(D^2u)(\partial_{x_ix_j}u_e) \geq 0.$$

Thus

$$F_{ij}(D^2u)\partial_{x_ix_j}(u_{ee}) \leq 0.$$

In other words, for convex (or concave) equations (like Bellman), second derivatives $v = u_{ee}$ are supersolutions to an equation of the form

$$a^{ij}(x)\partial_{x_ix_j}v \leq 0$$

where $a^{ij}(x) = F_{ij}(D^2u)$ are just uniformly elliptic and bounded. This fact and the original equation $F(D^2u) = 0$ can be combined to prove that $u \in C^{2,\alpha}$.

- **Evans–Krylov theorem.** If F is uniformly elliptic and convex then viscosity solutions u to $F(D^2u) = 0$ are $C^{2,\alpha}$.

(c) The Sobolev estimates. Regarding estimates in $W^{2,p}$, we have the following perturbation result.

- **Caffarelli $W^{2,p}$ estimates for viscosity solutions to fully nonlinear elliptic equations.** Let $F(M,x)$ be a fully nonlinear operator such that $F(0,\cdot) = 0$ and solutions w to $F(D^2w, x_0) = 0$ have $C^{1,1}$ estimates (like concave or convex equations) for any x_0. If u is a bounded viscosity solution to $F(D^2u, x) = f$ and $F(M,x)$ is close to $F(M,x_0)$ (in some suitable sense), with $f \in L^p$, $n < p < \infty$, then $u \in W^{2,p}$.

In this book we will not cover the proofs of the Evans–Krylov theorem or the Caffarelli $W^{2,p}$ estimate for fully nonlinear elliptic equations. These can be seen, for instance, in Caffarelli–Cabré [7].

1.3 NONLOCAL EQUATIONS: THE FRACTIONAL LAPLACIAN

At first sight, the Laplacian has a rather obscure formula:

$$\Delta u = \frac{\partial^2 u}{\partial x_1^2} + \cdots + \frac{\partial^2 u}{\partial x_n^2}.$$

Just by looking at this special linear combination of second derivatives of u, it is hard to recognize why this operator is so important. In particular, how is it that the Laplacian represents Fick and Fourier's notion of *diffusion*?

However, we may begin to uncover some insight if we write the Laplacian as

$$\Delta u(x) = \lim_{r\to 0^+} \frac{2(n+2)}{r^2} \cdot \frac{1}{|B_r|}\int_{B_r} (u(x+z) - u(x))\, dz$$

which can be checked by using a second-order Taylor expansion of u around x. In Caffarelli's interpretation, this identity is *an infinitesimal limit of a gain-loss of a density u at a point x*, that is, a "gain-loss" relation of *particles* arriving into position x minus those leaving x. In that sense, the heat equation $\partial_t u = \Delta u$ reflects the fact that the density u at the point x has compared itself with its infinitesimal neighborhood and is trying to revert to its surrounding average. Simply put, if u at x is below its surrounding average then $\Delta u \geq 0$ and $u_t \geq 0$, meaning that the density is growing at x to match its infinitesimal average. This is the idea of diffusion: the density flows from regions of higher to lower concentrations, and this is the ellipticity that forces the regularization effect. In other words, if the initial data in the heat equation has a "peak," then the peak instantaneously goes down and the solution becomes regular.

It is important to observe as well that the Laplacian is completely *local* in nature: at the end of the day, the only values of u that count in the computation of $\Delta u(x)$ are those *infinitesimally close* to x. Another way of evidencing the *locality* of the Laplacian is by observing that if u has compact support $E = \text{supp}(u)$, then Δu has support contained in E.

In the other, say, noninfinitesimal extreme, we have *nonlocal equations* of the form

$$\int_{\Omega} (u(z) - u(x))K(x,z)\, dz = f(x)$$

or

$$\int_{\Omega} (u(x+z) - u(x))K(x,z)\, dz = f(x)$$

in which the density is comparing itself at the point x with all its global values, not only those infinitesimally close to x as in the case of the Laplacian. The kernel $K(x,z)$ is assumed to be positive, so that it keeps count of those particles instantaneously reaching and leaving position x. For simplicity, we assume that K is symmetric in x and z. Thus, the first equation has a "divergence form structure" in that it has an associated symmetric bilinear form, while the second one is in "nondivergence form." In fact, when $K(x,z)$ becomes concentrated around x, properly scaled, we recover second order equations. From the first equation, in which x gives to z the same weight than z gives to x, we obtain the divergence form equation

$$\text{div}(a(x)\nabla u) = f.$$

If in the second equation we assume symmetry with respect to z, that is, $K(x,z) = K(x,-z)$, then $x+z$ and $x-z$ receive the same weight and in the appropriate limit we get the random process, nondivergence form equation

$$\text{trace}(a(x)D^2 u) = f.$$

Now, in the same way that there is no theory that englobes *all* PDEs at once, assumptions on the kernel K must be made in order to have a reasonable theory of nonlocal equations. Moreover, the regularity theory for second order elliptic PDEs hinges on a model operator, the Laplacian. Similarly, the regularity theory for nonlocal *elliptic* equations of *fractional order* will hinge on the adequate model operator: the *fractional Laplacian*.

The fractional Laplacian of a function $u : \mathbb{R}^n \to \mathbb{R}$ is defined as

$$(-\Delta)^s u(x) = c_{n,s}\,\mathrm{P.V.}\int_{\mathbb{R}^n} \frac{u(x)-u(z)}{|x-z|^{n+2s}}\,dz. \tag{1.7}$$

Here $0 < s < 1$, $c_{n,s} > 0$ is a normalizing constant behaving like $s(1-s)$ when $s \to 0$ and $s \to 1$, and P.V. means that the integral is understood in the principal value sense. This operator arises, for instance, in the theory of Lévy processes with jumps, and in models for anomalous diffusions and long-range interactions, where particles in a random walk can jump (and come from) far away with positive probability.

The name *fractional Laplacian* comes from its interpretation as the fractional power of the differential operator $-\Delta$. More precisely, for the Laplacian we have the Fourier transform identity

$$\widehat{-\Delta u}(\xi) = |\xi|^2\widehat{u}(\xi).$$

Using the heat semigroup generated by the Laplacian, we will see that, for the fractional Laplacian,

$$\widehat{(-\Delta)^s u}(\xi) = |\xi|^{2s}\widehat{u}(\xi).$$

From here, at least formally,

$$\lim_{s\to 0}(-\Delta)^s u = u \qquad \text{and} \qquad \lim_{s\to 1}(-\Delta)^s u = -\Delta u.$$

Therefore, the fractional Laplacian interpolates between the identity operator of differential order 0 and the Laplacian of differential order 2. One then interprets $(-\Delta)^s$ as a nonlocal (or *integro-differential*) operator of *fractional differential order* $0 < 2s < 2$. In fact, this has a completely rigorous sense at the scale of Hölder spaces.

- **Hölder estimates for the fractional Laplacian.** If $u \in C^\alpha$, for some $\alpha > 2s$, then $(-\Delta)^s u \in C^{\alpha-2s}$.

On the other hand, since the kernel $K(x,z) = c_{n,s}|x-z|^{-(n+2s)}$ of the fractional Laplacian in (1.7) has a nonintegrable singularity of order $n+2s$ for z close to x, a well-defined fractional Laplacian forces some regularity for u around x. For instance, $u \in C^{2s+\varepsilon}(x)$, $\varepsilon > 0$, and bounded in $\mathbb{R}^n$ suffices for $(-\Delta)^s u(x)$ to be finite. The gain in regularity of fractional order $0 < 2s < 2$ can be made precise.

- **Schauder estimates for the fractional Laplacian.** Let u be a bounded solution to $(-\Delta)^s u = f$. If $f \in C^\alpha$, $0 < \alpha < 1$, then $u \in C^{\alpha+2s}$.

For nonlocal equations with *elliptic* kernels, that is, kernels $K(x,z)$ behaving like that of the fractional Laplacian (1.7), a *fractional regularity theory* is expected. In this context, the fractional Laplacian turns out to be to nonlocal equations of fractional order what the Laplacian is to second order elliptic equations.

In this book we will focus on presenting analytic and geometric techniques that are useful for proving regularity estimates for the fractional Laplacian. The ideas presented here can also be applied, with nontrivial modifications, to more general nonlocal elliptic equations of fractional order and to problems involving other fractional power operators such as $(-\operatorname{div}(a(x)\nabla))^s u = f$ or $(-a^{ij}(x)\partial_{ij})^s u = f$.

We are then compelled to mimic a parallel theory to that of the Laplacian (or more general elliptic equations) for fractional Laplacians (or more general fractional nonlocal equations). However, several evident aspects prevent a direct application of local PDE methods to nonlocal equations. Let us mention a few of them.

(*a*) **Nonlocality and the Poisson–Dirichlet problem.** To solve the problem

$$(-\Delta)^s u = f \qquad \text{in } \Omega$$

it is not enough to prescribe u on $\partial\Omega$. Obviously, changing u on $\partial\Omega$ does not affect the value of its fractional Laplacian in Ω, because $\partial\Omega$ has Lebesgue measure zero. Moreover, to compute $(-\Delta)^s u(x)$ for $x \in \Omega$, we also need the values of u outside of Ω, see (1.7). Therefore, the appropriate *boundary* condition must be an *exterior* condition of the form

$$u = g \qquad \text{in } \mathbb{R}^n \setminus \Omega.$$

A perhaps surprising fact is that, even when $f = 1$ in $\Omega = B_1$ and $g = 0$ in $\mathbb{R}^n \setminus B_1$, the solution u is no better than C^s up to the boundary of B_1. This is in high contrast with the case of the Laplacian: the unique solution to $-\Delta u = 1$ in B_1 such that $u = 0$ on ∂B_1 is $u(x) = \frac{1}{2n}(1 - |x|^2)$, which is smooth up to the boundary of B_1.

(*b*) **Nonlocality and differentiating the equation.** It is not clear how one should differentiate the equation in the problem above, as one would need to differentiate u outside of Ω to compute $(-\Delta)^s(\partial_e u)$, and the exterior datum g may not necessarily be smooth. However, if f is smooth in Ω and g is merely bounded in $\mathbb{R}^n \setminus \Omega$, then u will still be smooth in the interior of Ω.

(*c*) **Nonlocality and variational techniques.** Localization techniques based on multiplication by a test function and integration by parts are not immediately available. First, if φ is a smooth, compactly supported function then, in general, $(-\Delta)^s\varphi$ has noncompact support. Second, if we apply the equation to a typical test function such as $\varphi^2 u$, the usual product rule fails as, for instance, $(-\Delta)^{1/2}(\varphi^2 u) \neq \varphi^2(-\Delta)^{1/2}u + u(-\Delta)^{1/2}(\varphi^2)$.

(*d*) **Nonlocality and nonvariational techniques.** From the very beginning, the usual notion of viscosity solution for second order elliptic equations does not directly apply to nonlocal equations. If we touch u by above or below with a quadratic polynomial P in a neighborhood of a point x_0, it is not clear how to compute the equation on P for two reasons: the test function needs to be globally defined and the fractional Laplacian of a quadratic polynomial on $\mathbb{R}^n$ is not well defined as the integral in (1.7) will diverge at infinity.

These issues, and many others, for the fractional Laplacian and other fractional nonlocal equations have been addressed in recent times. As we mentioned before, we will focus on developing useful techniques for the case of the fractional Laplacian.

Apart from integral and potential theory methods, there is a very powerful localization technique for the fractional Laplacian known as the *Caffarelli–Silvestre*

extension problem. The method allows to convert a nonlocal problem driven by the fractional Laplacian to a local PDE one, at the expense of adding an extra variable and dealing with a degenerate elliptic equation. Fix $0 < s < 1$. Given $u = u(x)$ defined on $\mathbb{R}^n$, extend it to the upper half space $\mathbb{R}^n \times [0, \infty)$ as a function $U = U(x, y) : \mathbb{R}^n \times [0, \infty) \to \mathbb{R}$ by solving the boundary value problem

$$\begin{cases} \Delta U + \frac{1-2s}{y}\partial_y U + \partial_{yy} U = 0 & \text{in } \mathbb{R}^n \times (0, \infty) \\ U(x, 0) = u(x) & \text{on } \mathbb{R}^n. \end{cases}$$

We call U the *extension* of u to the upper half space. It is a remarkable fact that

$$-\lim_{y \to 0} y^{1-2s}\partial_y U(x, y) = c_s(-\Delta)^s u(x) \qquad x \in \mathbb{R}^n \tag{1.8}$$

where $c_s > 0$ is an explicit constant depending only on $0 < s < 1$. One can interpret the role of the extra variable $y > 0$ as encoding the values of u at infinity in such a way that the local operation on U to the left hand side of (1.8) coincides with the nonlocal operation performed on its trace $U(x, 0) = u(x)$ in the right hand side. In other words, the fractional Laplacian is characterized as the Dirichlet-to-Neumann map for the extension equation: we begin with the Dirichlet datum $u(x)$, extend up to $y > 0$ by solving the extension equation for U, compute $-y^{1-2s}\partial_y U(x, y)$ and return back in the limit as $y \to 0$ to find that the Neumann condition is $c_s(-\Delta)^s u(x)$.

The simplest application of the Caffarelli–Silvestre extension problem is to the Poisson equation for the fractional Laplacian. Suppose that u is a solution to

$$(-\Delta)^s u = f \qquad \text{in } \mathbb{R}^n.$$

In view of the characterization of $(-\Delta)^s u$ in (1.8), to obtain regularity estimates for u it is enough to consider solutions U to the Neumann problem

$$\begin{cases} \Delta U + \frac{1-2s}{y}\partial_y U + \partial_{yy} U = 0 & \text{in } \mathbb{R}^n \times (0, \infty) \\ -\lim_{y \to 0} y^{1-2s}\partial_y U(x, y) = c_s f(x) & \text{on } \mathbb{R}^n \end{cases}$$

and show that U is regular *all the way up to* $y = 0$. Indeed,

$$U(x, 0) = u(x) = (-\Delta)^{-s} f(x).$$

The great advantage of this approach is that the equation for U is a purely local PDE, so many of the typical PDE techniques that we will see in this book are available. In particular, when $s = 1/2$, U is just a harmonic function satisfying a classical Neumann condition on a flat boundary $\{y = 0\}$.

Now, the extension equation for U can be written as

$$\operatorname{div}(y^{1-2s}\nabla U) = 0.$$

This is a degenerate elliptic equation because the coefficient y^{1-2s} blows up or vanishes as $y \to 0$ whenever $s \neq 1/2$. Nevertheless, it falls into a theory of elliptic PDEs with Muckenhoupt A_2 weight degeneracy developed by Fabes–Kenig–Serapioni. Many

regularity results such as the Harnack inequality and Hölder estimates can be proved for this class of equations and those estimates can be transferred back to u.

The Caffarelli–Silvestre extension problem is a particular case of a much more general theory of analysis and extension problem characterizations of fractional powers of linear operators known as the *Stinga–Torrea method of semigroups for fractional power operators.* The Stinga–Torrea general theory, which was generalized for fractional power operators on Banach spaces by Galé–Miana–Stinga, reaches far beyond regularity estimates. In particular, it has been crucial in developing numerical approximations for solutions to fractional nonlocal equations and to analyze nonlocal equations of fractional order in other geometric contexts involving fractional Laplacians on manifolds and graphs, and master equations. We will not go into these important developments in this book, but describe the semigroup methodology (in which the heat equation plays a pivotal role) only for the particular case of the fractional Laplacian on $\mathbb{R}^n$.

BASIC NOTATION

Along this book we use standard analysis and PDE notation.

Sets and measures. We denote by Ω a domain, that is, an open, connected subset of points $x = (x_1, \ldots, x_n)$ of $\mathbb{R}^n$, $n \geq 1$. Given a measurable set E, its Lebesgue measure is $|E|$, the surface area measure of its topological boundary ∂E is $|\partial E|$ and $\overline{E}$ is its closure. We use dS to denote the differential of surface area. The characteristic (or indicator) function of E is denoted by χ_E. We write $\Omega' \subset\subset \Omega$ whenever Ω' is a bounded domain such that $\overline{\Omega'} \subset \Omega$. The open ball of radius $R > 0$ centered at x is $B_R(x)$, with $B_R = B_R(0)$. We have $|B_R(x)| = \omega_n R^n$ and $|\partial B_R(x)| = n\omega_n R^{n-1}$, where $\omega_n = |B_1| = \frac{\pi^{n/2}}{\Gamma(n/2+1)}$ and Γ is the Gamma function. The open cube of side length $2R$ (or *radius* R) centered at x is $Q_R(x) = \prod_{i=1}^n (x_i - R, x_i + R)$, with $Q_R = Q_R(0) = (-R, R)^n$. Then $|Q_R(x)| = (2R)^n$ and $B_R(x) \subset Q_R(x) \subset B_{\sqrt{n}R}(x)$. For any cube Q, we denote by cQ, $c > 0$, the cube with the same center as Q and side length c times the side length of Q. In particular, $|cQ| = c^n|Q|$ and $cQ_r(x) = Q_{cr}(x)$.

Gamma function. The Gamma function $\Gamma(s)$ is defined for $s > 0$ by

$$\Gamma(s) = \int_0^\infty e^{-r} \frac{dr}{r^{1-s}}.$$

It satisfies the functional identity $s\Gamma(s) = \Gamma(s+1)$. In particular, $\Gamma(k) = (k-1)!$ for any integer $k \geq 1$. The Bohr–Mollerup theorem establishes that the Gamma function is the only positive, log-convex function $f(s)$, $s > 0$, such that $f(1) = 1$ and $sf(s) = f(s+1)$. The Gamma function can be extended to noninteger negative values via the functional identity above. In particular, using integration by parts,

$$\Gamma(-s) = \frac{\Gamma(s)}{-s} = \int_0^\infty (e^{-r} - 1) \frac{dr}{r^{1+s}} < 0 \qquad \text{for } 0 < s < 1.$$

Matrices. Let A and B be square matrices of size $n \times n$. Then A^T, $\det(A)$ and $\operatorname{trace}(A)$ denote the transpose, determinant and trace of A, respectively. We write $A \geq 0$ whenever A is nonnegative definite. Then $A \geq B$ means that $A - B \geq 0$. If A and B are symmetric, $A \geq B$ if and only if the eigenvalues of $A - B$ are nonnegative and, in this case, $\det(A) \geq \det(B)$. We denote by $I = (\delta_{ij})_{i,j=1}^n$ the identity matrix and by $\operatorname{diag}(\lambda_1, \dots, \lambda_n)$ a diagonal matrix with diagonal entries $\lambda_1, \dots, \lambda_n$. The norm of A we use is $\|A\| = (\operatorname{trace}(AA^T))^{1/2}$.

Functions. Derivatives of functions $u = u(x)$ are denoted by $\frac{\partial u}{\partial x_i} = \partial_{x_i} u = \partial_i u = u_{x_i}$, $\frac{\partial^2 u}{\partial x_i \partial x_j} = \partial_{x_i x_j} u = \partial_{ij} u = u_{x_i x_j}$, for $i, j = 1, \dots, n$; $D^\gamma u = \frac{\partial^{|\gamma|} u}{\partial_{x_1}^{\gamma_1} \cdots \partial_{x_n}^{\gamma_n}}$, where $\gamma = (\gamma_1, \dots, \gamma_n) \in \mathbb{N}_0^n$ is a multi-index with $|\gamma| = \gamma_1 + \cdots + \gamma_n$; and so on. The gradient and Hessian of u are $\nabla u = (u_{x_1}, \dots, u_{x_n})$ and $D^2 u = (\partial_{ij} u)_{i,j=1}^n$, respectively. The divergence of a vector field $F(x) = (F_1(x), \dots, F_n(x)) : \Omega \to \mathbb{R}^n$ is $\operatorname{div} F = \sum_{i=1}^n \partial_{x_i} F_i$ and the Laplacian of u is $\Delta u = \operatorname{div}(\nabla u) = \operatorname{trace}(D^2 u)$. The positive and negative parts of u are $u^+ = \max(u, 0) \geq 0$ and $u^- = -\min(u, 0) \geq 0$, respectively, so that $u = u^+ - u^-$ and $|u| = u^+ + u^-$. The oscillation of u over Ω is defined as $\operatorname{osc}_\Omega u = \sup_\Omega u - \inf_\Omega u$.

Summation convention. We use the standard summation over repeated indices convention. For example, if $a(x) = (a^{ij}(x))_{i,j=1}^n$ is a matrix of coefficients then we write $\operatorname{div}(a(x)\nabla u) = \sum_{i,j=1}^n \partial_j (a^{ij}(x) \partial_i u) = (a^{ij}(x) u_{x_i})_{x_j}$ for divergence form PDEs, and $\operatorname{trace}(a(x) D^2 u) = \sum_{i,j=1}^n a^{ij}(x) \partial_{x_i x_j} u = a^{ij}(x) \partial_{ij} u$ for nondivergence form PDEs.

Functional spaces. Classical notation for spaces of continuous, Hölder, differentiable, Lebesgue and Sobolev functions is used throughout, such as $C(E)$, $C^\alpha(E)$ (Hölder continuous), $C^{0,\alpha}(E)$ (Hölder continuous and bounded) for $0 < \alpha < 1$; $C^k(E)$, $C^{k,\alpha}(E)$ for $k \geq 1$; $\operatorname{Lip}(E) = C^{0,1}(E)$ (Lipschitz); $L^p(E)$, $L^p_{\mathrm{loc}}(E)$, $W^{k,p}(E)$, $W_0^{k,p}(E)$ for $1 \leq p \leq \infty$; $W^{k,2}(E) = H^k(E)$, $W_0^{k,2}(E) = H_0^k(E)$; etc. In particular, $C_c^\infty(\Omega) = \mathcal{D}(\Omega)$ is the set of smooth functions $\varphi \in C^\infty(\overline{\Omega})$ with compact support $\operatorname{supp}(\varphi)$ in Ω. The space of distributions is $\mathcal{D}'(\Omega)$. A function u is C^α at $x \in \Omega$, for $0 < \alpha < 1$, if $[u]_{C^\alpha(x)} = \sup_{z \in \Omega, z \neq x} \frac{|u(x) - u(z)|}{|x - z|^\alpha} < \infty$. The norm of $u \in C^{k,\alpha}(E)$ is $\|u\|_{C^{k,\alpha}(E)} = \sum_{j=0}^k \|D^j u\|_{L^\infty(E)} + [D^k u]_{C^\alpha(E)}$. A quadratic or second-order polynomial is a polynomial of the form $P(x) = \frac{1}{2} x^T A x + B \cdot x + C$, where $A = D^2 P$ is a symmetric matrix, B is a vector and C is a constant. A function u is $C^{1,1}$ at a point x if there are a concave quadratic polynomial P and a convex quadratic polynomial Q such that $P(x) = u(x) = Q(x)$ and $P \leq u \leq Q$ in a neighborhood of x. Caffarelli showed that if u is $C^{1,1}$ at every point of a bounded domain Ω in a uniform way (that is, there is $M > 0$ such that all the touching paraboloids P and Q satisfy $|D^2 P|, |D^2 Q| \leq M$) then $u \in C^{1,1}(\Omega)$ in the classical sense, that is $D^2 u \in L^\infty(\Omega)$, see [7].

Some inequalities. Let $a, b \geq 0$. The Cauchy inequality with $\varepsilon > 0$ says that $ab \leq \varepsilon a^2 + \frac{1}{4\varepsilon} b^2$. The Young inequality establishes that if $1 < q < \infty$ and $\frac{1}{q} + \frac{1}{q'} = 1$ then $ab \leq \frac{1}{q} a^q + \frac{1}{q'} b^{q'}$. For $1 \leq p, q, r \leq \infty$, Young's convolution inequality gives that if $f \in L^r$ and $g \in L^q$ then $f * g \in L^p$ for $\frac{1}{r} + \frac{1}{q} = 1 + \frac{1}{p}$ with $\|f * g\|_{L^p} \leq \|f\|_{L^r} \|g\|_{L^q}$. For a nonnegative measurable function $f(x, z) : E_1 \times E_2 \to \mathbb{R}$ and $1 \leq p < \infty$,

Minkowski's integral inequality establishes that

$$\left(\int_{E_2}\left(\int_{E_1} f(x,z)\,dx\right)^p dz\right)^{1/p} \leq \int_{E_1}\left(\int_{E_2} f(x,z)^p\,dz\right)^{1/p} dx.$$

The isoperimetric inequality says that if Ω is a sufficiently smooth bounded domain of $\mathbb{R}^n$ then $|\Omega| \leq C_n|\partial\Omega|^{n/(n-1)}$ for some $C_n > 0$ depending only on n.

I

The Laplacian

CHAPTER 2

Harmonic functions

In this chapter we study harmonic functions, that is, solutions u to

$$\Delta u = 0 \qquad \text{in } \Omega \subset \mathbb{R}^n.$$

The basic and most important property of harmonic functions is their characterization by the mean value property. The mean value property implies the interior Harnack inequality, the strong maximum principle, the Liouville theorem and the interior derivative estimates in terms of the oscillation. In particular, harmonic functions are real analytic. All these properties play a crucial role in the regularity theory of more general second order elliptic equations.

2.1 DEFINITION AND EXAMPLES

Recall that if a complex function f defined on an open subset D of the complex plane $\mathbb{C}$ is analytic (or holomorphic) and we write $f(x+iy) = u(x,y) + iv(x,y)$ then, by the Cauchy–Riemann equations, u and v are harmonic functions in D, that is,

$$u_{xx} + u_{yy} = 0 \qquad \text{and} \qquad v_{xx} + v_{yy} = 0.$$

Moreover, u and v are real analytic. This is one of the most striking features of harmonic functions. Indeed, this just happens because of a special cancelation of the pure second derivatives of u. Furthermore, not only all the derivatives of u are continuous, but they are all controlled in the interior by the oscillation of u.

Let Ω be a bounded domain of $\mathbb{R}^n$, $n \geq 1$. Recall that the **Laplace operator** acting on a function $u \in C^2(\Omega)$ at a point $x \in \Omega$, or just the **Laplacian** of u at x, is

$$\Delta u(x) = \sum_{i=1}^{n} \frac{\partial^2 u}{\partial x_i^2}(x) = \operatorname{trace}(D^2 u). \tag{2.1}$$

We can also write the Laplacian of u as

$$\Delta u(x) = \operatorname{div}(\nabla u(x)). \tag{2.2}$$

Therefore, the Laplacian shares both divergence form (2.2) and nondivergence form (2.1) structures.

DOI: 10.1201/9781032679853-2

The way of taking advantage of the divergence structure (2.2) is by means of the divergence theorem. Let Ω_0 be a domain in $\mathbb{R}^n$ whose boundary $\partial\Omega_0$ is C^1. Denote by $\nu = \nu(x)$ the exterior unit normal on $\partial\Omega_0$. Let F be a vector field in $C^1(\overline{\Omega}_0; \mathbb{R}^n)$. The divergence theorem establishes that

$$\int_{\Omega_0} \operatorname{div} F \, dx = \int_{\partial\Omega_0} F \cdot \nu \, dS.$$

An easy consequence for the Laplacian follows by taking $F = \nabla u$ for $u \in C^2(\overline{\Omega}_0)$:

$$\int_{\Omega_0} \Delta u \, dx = \int_{\partial\Omega_0} \nabla u \cdot \nu \, dS = \int_{\partial\Omega_0} \partial_\nu u \, dS. \tag{2.3}$$

We have the following classical definition. A function $u \in C^2(\Omega)$ is said to be

(1) **harmonic** in Ω if $\Delta u = 0$ in Ω;

(2) **subharmonic** in Ω if $\Delta u \geq 0$ in Ω;

(3) **superharmonic** in Ω if $\Delta u \leq 0$ in Ω; equivalently, if $-u$ is subharmonic in Ω.

Example 2.1. It is always convenient to have a set of examples of harmonic and subharmonic functions.

(i) Constants and linear functions are harmonic.

(ii) $u(x_1, x_2) = x_1^2 - x_2^2$ is harmonic on $\mathbb{R}^2$. More generally, in $\mathbb{R}^n$ one can construct families of harmonic polynomials.

(iii) Fundamental solution. The following function is harmonic everywhere in $\mathbb{R}^n$ except at the origin:

$$K(x) = \begin{cases} \dfrac{\Gamma(n/2 - 1)}{4\pi^{n/2}} \cdot \dfrac{1}{|x|^{n-2}} & \text{if } n \geq 3 \\ -\dfrac{1}{2\pi} \cdot \log |x| & \text{if } n = 2 \end{cases}$$

where Γ denotes the Gamma function. The constants chosen above will become evident in Chapter 3. Fundamental solutions are the building blocks of other harmonic functions.

(iv) Let u be a harmonic function with Taylor expansion

$$u(x) = \sum_{|\gamma| \geq 0} c_\gamma x^\gamma.$$

Then the homogeneous parts

$$p_m(x) = \sum_{|\gamma| = m} c_\gamma x^\gamma \qquad m = 0, 1, 2 \ldots$$

are also harmonic.

(v) $u(x) = \frac{1}{2n}|x|^2$ is subharmonic in $\mathbb{R}^n$:

$$\Delta u = 1 \geq 0$$

and $-u$ is superharmonic. The idea behind the name *subharmonic* is that a subharmonic function is going to be below any harmonic function having the same boundary values. The maximum principle will provide the precise statement.

In addition to these examples, it is important to know that we can always solve the Dirichlet problem for the Laplace equation in a ball.

Theorem 2.2 (Poisson formula for balls). *Let $B = B_R(0)$ and $g \in C(\partial B)$. Define*

$$u(x) = \begin{cases} \dfrac{R^2 - |x|^2}{n\omega_n R} \displaystyle\int_{\partial B} \frac{g(z)}{(1 - 2x \cdot z + |x|^2)^{n/2}} \, dS_z & \text{for } x \in B \\ g(x) & \text{for } x \in \partial B. \end{cases} \tag{2.4}$$

Then $u \in C^\infty(B) \cap C(\overline{B})$ is the unique solution to

$$\begin{cases} \Delta u = 0 & \text{in } B \\ u = g & \text{on } \partial B. \end{cases}$$

The classical formula (2.4) is a consequence of the very special invariances of the Laplacian and Green's identities. In dimension 2, it can be derived using complex analysis. The proof of Theorem 2.2 is left as an exercise.

2.2 THE MEAN VALUE PROPERTY AND SMOOTHNESS

The most important property of (sub/super)harmonic functions is the (sub/super)-mean value property. As it will be clear from the proof, the mean value property relies upon the divergence structure of the Laplacian.

Theorem 2.3 (Mean value property). *Let $u \in C^2(\Omega)$ be a subharmonic function in Ω, that is, $\Delta u \geq 0$ in Ω. Then*

$$u(x) \leq \frac{1}{|\partial B_R(x)|} \int_{\partial B_R(x)} u \, dS \qquad \text{for any ball } B_R(x) \subset\subset \Omega \tag{2.5}$$

and

$$u(x) \leq \frac{1}{|B_R(x)|} \int_{B_R(x)} u \, dz \qquad \text{for any ball } B_R(x) \subset\subset \Omega. \tag{2.6}$$

By considering $-u$, the reverse inequalities hold whenever u is superharmonic in Ω. In particular, if u is harmonic in Ω then, for any ball $B_R(x) \subset\subset \Omega$,

$$u(x) = \frac{1}{|\partial B_R(x)|} \int_{\partial B_R(x)} u \, dS = \frac{1}{|B_R(x)|} \int_{B_R(x)} u \, dz.$$

The value of a subharmonic function u at x is controlled by the integral mean of u around x, as it can be seen from the example $u(x) = |x|^2$, while the opposite is true for superharmonic functions (think of $u(x) = -|x|^2$). For harmonic functions, there is a perfect balance: near the origin there are directions where the harmonic function $u(x_1, x_2) = x_1^2 - x_2^2$ bends upward, while there are other directions where it bends downward, at the same rate.

Proof of Theorem 2.3. Let us begin by recalling that $\omega_n = |B_1| = \frac{\pi^{n/2}}{\Gamma(n/2+1)}$,

$$|B_R(x)| = \omega_n R^n \qquad \text{and} \qquad |\partial B_R(x)| = n\omega_n R^{n-1}. \tag{2.7}$$

Let u be subharmonic in Ω. We first show that (2.5) implies (2.6). Fix a ball $B_R(x) \subset\subset \Omega$. Since (2.5) holds true then, for any $0 < r < R$,

$$n\omega_n r^{n-1} u(x) \le \int_{\partial B_r(x)} u\, dS.$$

Integrating this inequality in r from 0 to R and using polar coordinates we get (2.6):

$$\omega_n R^n u(x) \le \int_0^R \int_{\partial B_r(x)} u\, dS\, dr = \int_{B_R(x)} u\, dz. \tag{2.8}$$

Therefore, we only need to prove (2.5). From (2.3), the fact that u is subharmonic and by using the change of variables $(z - x)/R = w$,

$$\begin{aligned} 0 \le \int_{\partial B_R(x)} \nabla u \cdot \nu\, dS &= \int_{|x-z|=R} \nabla u(z) \cdot \frac{(z-x)}{R}\, dS \\ &= R^{n-1} \int_{|w|=1} \nabla u(x + Rw) \cdot w\, dS \\ &= R^{n-1} \frac{\partial}{\partial R} \int_{|w|=1} u(x + Rw)\, dS \\ &= R^{n-1} \frac{\partial}{\partial R} \left[R^{1-n} \int_{|x-z|=R} u(z)\, dS \right] \end{aligned} \tag{2.9}$$

where in the last identity we changed variables back. This inequality says that the expression in brackets in (2.9) is increasing in R. Thus, for any $0 < r < R$, we have

$$r^{1-n} \int_{|x-z|=r} u(z)\, dS \le R^{1-n} \int_{|x-z|=R} u(z)\, dS. \tag{2.10}$$

But now, by the Lebesgue differentiation theorem,

$$\lim_{r \to 0} r^{1-n} \int_{|x-z|=r} u(z)\, dS = n\omega_n u(x).$$

Hence, after taking the limit as $r \to 0$ in (2.10) we arrive to (2.5). □

As a matter of fact, the mean value property characterizes harmonic functions.

Theorem 2.4 (Converse of the mean value property). *Let u be a continuous function in Ω. Suppose that u satisfies the mean value property everywhere in Ω, namely,*

$$u(x) = \frac{1}{|\partial B_R(x)|}\int_{\partial B_R(x)} u\,dS = \frac{1}{n\omega_n}\int_{|z|=1} u(x+Rz)\,dS$$

for every $B_R(x) \subset\subset \Omega$. Then $u \in C^\infty(\Omega)$ and $\Delta u = 0$ in Ω.

Proof. We use a mollifier. Take $\varphi \in C_c^\infty(B_1)$ of the form $\varphi(z) = \psi(|z|)$ and such that

$$\int_{B_1} \varphi\,dz = n\omega_n \int_0^1 \psi(r)r^{n-1}\,dr = 1. \tag{2.11}$$

Consider the rescaled dilations defined by $\varphi_\varepsilon(z) = \frac{1}{\varepsilon^n}\varphi(\frac{z}{\varepsilon})$ for any $\varepsilon > 0$. Observe that $\mathrm{supp}(\varphi_\varepsilon) \subset B_\varepsilon$.

Now let us fix $\varepsilon > 0$ small. We are going to prove that $u(x) = (\varphi_\varepsilon * u)(x)$ whenever $x \in \Omega$ is at distance at least ε from the boundary, namely, for $x \in \Omega$ such that $\mathrm{dist}(x, \partial\Omega) > \varepsilon$. This of course implies that $u \in C^\infty(\Omega)$, since u coincides with the smooth function $\varphi_\varepsilon * u$ in the open set $\Omega_\varepsilon = \{x \in \Omega : \mathrm{dist}(x, \partial\Omega) > \varepsilon\} \subset \Omega$, for any $\varepsilon > 0$, and $\Omega_\varepsilon \to \Omega$ as $\varepsilon \to 0$. Since φ is radially symmetric, by using the definition of φ_ε, the change of variables $z/\varepsilon = w$, polar coordinates, the mean value property of u and (2.11), we find

$$\begin{aligned}(\varphi_\varepsilon * u)(x) &= \frac{1}{\varepsilon^n}\int_{|z|<\varepsilon} \varphi(\tfrac{z}{\varepsilon})u(x+z)\,dz = \int_{|w|<1} \varphi(w)u(x+\varepsilon w)\,dw \\ &= \int_0^1 r^{n-1}\psi(r)\int_{\partial B_1} u(x+\varepsilon r\theta)\,dS_\theta\,dr \\ &= \int_0^1 r^{n-1}\psi(r)n\omega_n u(x)\,dr = u(x).\end{aligned}$$

Next let us see that u is harmonic in Ω. By (2.3) and the computation we did in the proof of Theorem 2.3, see (2.9), we have that, for every $B_R(x) \subset\subset \Omega$, by the mean value property of u,

$$\begin{aligned}\int_{B_R(x)} \Delta u\,dz &= R^{n-1}\frac{\partial}{\partial R}\int_{|w|=1} u(x+Rw)\,dS \\ &= R^{n-1}\frac{\partial}{\partial R}(n\omega_n u(x)) = 0.\end{aligned}$$

Since $B_R(x)$ was arbitrary, we get $\Delta u = 0$ in Ω. □

The interaction between the mean value property and the divergence theorem allows for more general definitions of harmonic functions, which we analyze next.

The mean value property permits to define the notion of *continuous* harmonic function. We say that $u \in C(\Omega)$ is **harmonic in the mean value sense** in Ω if it satisfies the mean value property everywhere in Ω. Theorems 2.3 and 2.4 show that this definition is equivalent to assuming that $u \in C^2(\Omega)$ and that $\Delta u = 0$. Moreover, in this case, $u \in C^\infty(\Omega)$. From now on, we will refer to continuous functions as harmonic without explicitly mentioning that they are smooth.

We can also define the notion of *continuous* subharmonic functions. We say that $u \in C(\Omega)$ is **subharmonic in the mean value sense** in Ω if it satisfies the sub-mean value property everywhere in Ω, that is, (2.5) or, equivalently, (2.6). Similarly, $u \in C(\Omega)$ is **superharmonic in the mean value sense** in Ω if $-u$ is subharmonic in Ω in the mean value sense in Ω.

In general, u subharmonic in the mean value sense does not imply that u is C^2, as the example $u(x) = |x|$ in $(-1, 1)$ shows. If we impose the extra assumption that $u \in C^2(\Omega)$ then the chain of identities (2.9) in the proof of Theorem 2.3 holds for any ball $B_R(x) \subset\subset \Omega$ and, by (2.3), we find that $\Delta u \geq 0$.

On the other hand, if $u \in C^2(\Omega)$ is subharmonic in Ω and $\varphi \in C_c^\infty(\Omega)$ is nonnegative then we can use integration by parts (or rather, Green's formula) to get

$$0 \leq \int_\Omega (\Delta u)\varphi \, dz = \int_\Omega u \Delta \varphi \, dz.$$

This motivates the notion of locally integrable subharmonic function.

We say that $u \in L^1_{\text{loc}}(\Omega)$ is **subharmonic in the distributional sense** in Ω if for any nonnegative $\varphi \in C_c^{1,1}(\Omega)$ (that is, $\varphi \in C_c^1(\Omega)$ and $\nabla \varphi \in \text{Lip}(\overline{\Omega})$; this last condition being equivalent to requiring that $\partial_{ij}\varphi$, defined in the distributional sense, are bounded functions in Ω), we have

$$\int_\Omega u \Delta \varphi \, dz \geq 0.$$

Similarly, $u \in L^1_{\text{loc}}(\Omega)$ is **superharmonic in the distributional sense** in Ω if $-u$ is subharmonic in the distributional sense in Ω. Finally $u \in L^1_{\text{loc}}(\Omega)$ is **harmonic in the distributional sense** in Ω if it is both sub- and super- harmonic in the distributional sense in Ω.

The distributional definition of subharmonic function above says that, as a distribution, Δu is a nonnegative measure. It turns out that subharmonic functions in the distributional sense are in fact subharmonic in the mean value sense, although they are not necessarily continuous.

Theorem 2.5 (Distributional subharmonic implies mean value subharmonic). *If $u \in L^1_{\text{loc}}(\Omega)$ is subharmonic in the distributional sense in Ω then its average over balls is an increasing function of the radius, that is, if $0 < r < R$ and $B_R(x) \subset\subset \Omega$, then*

$$\frac{1}{|B_r(x)|} \int_{B_r(x)} u \, dz \leq \frac{1}{|B_R(x)|} \int_{B_R(x)} u \, dz.$$

Consequently, u has a unique pointwisely defined representative

$$u(x) = \lim_{r \to 0} \frac{1}{|B_r(x)|} \int_{B_r(x)} u \, dz$$

which is upper semicontinuous and satisfies the sub-mean value property in Ω. In particular, if $u \in L^1_{\text{loc}}(\Omega)$ is harmonic in the distributional sense in Ω, then it has a unique continuous representative that is harmonic in the classical sense in Ω.

Proof (Caffarelli). For simplicity, assume that $x = 0$. Let $0 < r < R$ with $B_R \subset\subset \Omega$. The idea is to find a nonnegative test function $\varphi \in C_c^{1,1}(B_R)$ such that

$$\Delta\varphi = C\Big(\frac{1}{|B_R|}\chi_{B_R} - \frac{1}{|B_r|}\chi_{B_r}\Big) \qquad \text{in } \Omega \tag{2.12}$$

for some $C > 0$. Indeed, if such a φ exists, by the definition of subharmonic function in the distributional sense,

$$0 \leq \int_\Omega u\Delta\varphi\, dz = C\Big(\frac{1}{|B_R|}\int_{B_R} u\, dz - \frac{1}{|B_r|}\int_{B_r} u\, dz\Big)$$

as desired. The pointwise representative of u is then obtained by applying the Lebesgue differentiation theorem; the fact that it is upper semicontinuous follows because it is the pointwise decreasing limit of a sequence of continuous functions. The statement about harmonic functions is readily verified.

We are left to construct φ. Consider the fundamental solution $K(z)$ given in Example 2.1(*iii*). We do the proof of dimension $n \geq 3$, for which $K(z) = c_n|z|^{2-n}$; the case $n = 2$ goes along the same lines. We fit a paraboloid P_r of the form

$$P_r(z) = -c_1|z|^2 + c_2$$

from below that is tangent to K on $|z| = r$. To compute c_1 and c_2 explicitly, we use the facts that, when $|z| = r$, both $P_r(z)$ and its radial derivative $\partial_{\mathrm{rad}}P_r(z)$ satisfy

$$\begin{cases} P_r(z) = -c_1r^2 + c_2 = c_nr^{2-n} \\ \partial_{\mathrm{rad}}P_r(z) = -2c_1r = c_n(2-n)r^{1-n} \end{cases}$$

Therefore, the paraboloid is explicitly given by

$$P_r(z) = -\frac{c_n(n-2)}{2r^n}|z|^2 + \frac{c_nn}{2}r^{2-n}.$$

We define

$$V_r(z) = \begin{cases} K(z) - P_r(z) & \text{in } B_r \\ 0 & \text{in } \Omega \setminus B_r. \end{cases}$$

It is not difficult to check that, except at the origin, V_r is $C^{1,1}$ and has compact support. Moreover, since K is harmonic away from $x = 0$,

$$\Delta V_r(x) = C(n,r)\chi_{B_r} \qquad x \neq 0$$

where $C(n,r) = -\Delta P_r > 0$ depends only on n and r. In addition, notice that if $r < R$ then $V_r \leq V_R$. The nonnegative test function is defined as

$$\varphi = V_R - V_r.$$

Now $\varphi \in C_c^{1,1}(\Omega)$ because the singularity at the origin of K has been canceled and

$$\Delta\varphi = C(n,R)\chi_{B_R} - C(n,r)\chi_{B_r} \qquad \text{in } \Omega. \tag{2.13}$$

To compute $C(n,R)$ and $C(n,r)$ we use that both 1 and -1 are subharmonic:

$$0 \leq \int_\Omega (\pm 1)\Delta\varphi \, dx = \pm(C(n,R)|B_R| - C(n,r)|B_r|).$$

Therefore, $C(n,R)|B_R| = C(n,r)|B_r| = C$ for some constant $C > 0$, for all $R, r > 0$. Substituting back into (2.13), we finally obtain (2.12). □

If $u \in C^2(\Omega)$ is subharmonic in Ω and $\varphi \in C_c^\infty(\Omega)$ is nonnegative then, integrating by parts only once (instead of twice as before),

$$0 \leq \int_\Omega (\Delta u)\varphi \, dx = -\int_\Omega \nabla u \nabla \varphi \, dx.$$

This identity motivates the notion of subharmonic function for Sobolev functions.

We say that $u \in H^1(\Omega)$ is **subharmonic in the weak sense** in Ω if, for any nonnegative $\varphi \in H_0^1(\Omega)$ (that is, $\varphi \in H^1(\Omega)$ with $\varphi = 0$ on $\partial\Omega$ in the sense of traces),

$$\int_\Omega \nabla u \nabla \varphi \, dx \leq 0.$$

The notions of superharmonic and harmonic functions in the weak sense in Ω are defined in a similar way.

It is obvious that if u is subharmonic in the weak sense in Ω then, by definition of weak derivatives, u is subharmonic in the distributional sense in Ω. Thus, u (or, more precisely, its unique upper semicontinuous representative) satisfies the sub-mean value property everywhere in Ω. In particular, if u is weakly harmonic then its continuous representative is smooth and harmonic in the classical sense.

From now on, and for the rest of this chapter, to avoid cumbersome statements about (semi)continuous representatives, we will just work with continuous functions.

2.3 CONSEQUENCES OF THE MEAN VALUE PROPERTY

Two conclusions can be drawn from the previous section. First, harmonic functions are smooth and satisfy the mean value property. Second, and conversely, any continuous function satisfying the mean value property at every ball inside a domain is smooth and harmonic. In fact, any other weaker notion of harmonic function is equivalent to the classical one. We should remember that these statements are consequences of the divergence structure of the Laplacian (2.2).

In this section we take advantage of the mean value property to show many other important features of harmonic functions: maximum principles, ordering, uniqueness of the Dirichlet problem, derivative estimates, real analyticity and Harnack inequality.

2.3.1 The maximum principle and its consequences

Let us begin with the first main corollary of the mean value property, the strong maximum principle. Notice that if a function u is in $C^2(\Omega)$, then it does not necessarily attains a maximum or a minimum value in $\overline{\Omega}$. Indeed, an example is $u(x) = \frac{1}{x}$ in $\Omega = (0,1) \subset \mathbb{R}$. Of course, in order to ensure the existence of extreme values we have

to require $u \in C(\overline{\Omega})$. Recall that $u \in C(\Omega)$ is subharmonic in the mean value sense in Ω if it satisfies the sub-mean value property (2.5) or, equivalently, (2.6), in Ω.

Theorem 2.6 (Strong maximum principle for subharmonic functions). *Let Ω be a domain, not necessarily bounded. Let $u \in C(\Omega)$ be subharmonic in the mean value sense in Ω. Suppose that there is a point $x_0 \in \Omega$ such that*

$$u(x_0) = \sup_{\Omega} u.$$

Then u is constant in Ω.

In other words, the strong maximum principle says the *if* a subharmonic function in Ω *attains* a maximum value *in the interior* of Ω then it must be constant. Observe that it does not say anything about a possible minimum value of u. The reader should keep in mind examples such as a hyperplane and $u(x_1, x_2) = x_1^2 - x_2^2$, which are harmonic, and $u(x) = |x|^2$, which is subharmonic. Moreover, in the statement of Theorem 2.6 we do not need to assume that $u \in C^2(\Omega)$. Indeed, the maximum and comparison principles are consequences of the (sub)mean value property for continuous functions and do not depend explicitly on the Laplacian as a differential operator. The reader should recall the example $u(x) = |x|$, which is subharmonic in the mean value sense but not C^2.

Proof of Theorem 2.6. The idea is a nice topological argument. To simplify the notation let us call

$$M := u(x_0) = \sup_{\Omega} u.$$

Consider now the set of points in Ω where u coincides with its maximum:

$$\Omega_M := \{x \in \Omega : u(x) = M\}.$$

We want to show that $\Omega = \Omega_M$, or, which is the same, that $u(x) = M$ at every point $x \in \Omega$. As Ω is an open set, this will follow from the following three properties:

(*i*) Ω_M is nonempty (because $x_0 \in \Omega_M$);

(*ii*) Ω_M is relatively closed in Ω (because u is continuous on Ω);

(*iii*) Ω_M is relatively open in Ω.

For (*iii*), let $x \in \Omega_M$. We need to show that there is a ball $B_R(x)$ such that $B_R(x) \cap \Omega \subset \Omega_M$, or, which is the same, that $u \equiv M$ in $B_R(x)$ if $R > 0$ is sufficiently small. Let $R > 0$ be such that $B_R(x) \subset \Omega$. We obviously have $u \leq M$ in $B_R(x)$. By the sub-mean value property,

$$M = u(x) \leq \frac{1}{|B_R(x)|} \int_{B_R(x)} u \, dz.$$

By subtracting M we deduce that

$$0 \leq \frac{1}{|B_R(x)|} \int_{B_R(x)} (u - M) \, dz.$$

The integrand $u - M$ is nonpositive, so it follows that $u = M$ in $B_R(x)$. □

Let us now derive some direct crucial consequences of Theorem 2.6.

Corollary 2.7 (Direct consequences of the strong maximum principle). *Let Ω be a domain.*

(1) (Strong minimum principle for superharmonic functions). *Let $u \in C(\Omega)$ be superharmonic in the mean value sense in Ω. Suppose that there is a point $x_0 \in \Omega$ such that*

$$u(x_0) = \inf_{\Omega} u.$$

Then u is constant in Ω.

(2) (Strong maximum principle for harmonic functions). *If $u \in C(\Omega)$ is harmonic in Ω, then u cannot attain an interior maximum or minimum value in Ω unless it is constant.*

(3) (Weak maximum principle for subharmonic functions). *Suppose that Ω is bounded. Let $u \in C(\overline{\Omega})$ be subharmonic in the mean value sense in Ω. Then*

$$\sup_{\Omega} u = \sup_{\partial\Omega} u.$$

(4) (Weak minimum principle for superharmonic functions). *Suppose that Ω is bounded. Let $u \in C(\overline{\Omega})$ be superharmonic in the mean value sense in Ω. Then*

$$\inf_{\Omega} u = \inf_{\partial\Omega} u.$$

(5) (Weak maximum principle for harmonic functions). *Suppose that Ω is bounded. If $u \in C(\overline{\Omega})$ is harmonic in Ω, then u attains both its maximum and minimum values on $\partial\Omega$. In particular,*

$$\inf_{\partial\Omega} u \le u(x) \le \sup_{\partial\Omega} u \qquad \text{for any } x \in \Omega.$$

Proof. Let u be as in (1). Then $v := -u$ is subharmonic in the mean value sense in Ω and attains its maximum value at x_0. Indeed,

$$v(x_0) = -u(x_0) = -\inf_{\Omega} u = \sup_{\Omega}(-u) = \sup_{\Omega} v.$$

By the strong maximum principle for subharmonic functions (Theorem 2.6), v must be constant, and so is u.

To see (2), we notice that a harmonic function is both subharmonic and superharmonic. The conclusion follows from Theorem 2.6 and (1).

Next we prove (3). The facts that $u \in C(\overline{\Omega})$ and Ω is bounded ensure that u attains its maximum value at a point $x_0 \in \overline{\Omega}$:

$$\sup_{\Omega} u = \max_{\overline{\Omega}} u = u(x_0).$$

Now we have two possibilities. If $x_0 \in \partial\Omega$ then, by definition of x_0,

$$u(x_0) \le \sup_{\partial\Omega} u \le \max_{\overline{\Omega}} u = u(x_0)$$

and $u(x_0) = \sup_{\partial\Omega} u$, as desired. If $x_0 \in \Omega$, then by the strong maximum principle we proved in (2), u must be constant in Ω. Thus, $u(x) = u(x_0)$ for all $x \in \Omega$. But since u is continuous in $\overline{\Omega}$, we get $u \equiv u(x_0)$ in $\overline{\Omega}$, and thus the maximum of u in $\overline{\Omega}$ is the maximum of u on $\partial\Omega$.

As in (1), (4) follows from (3) by considering $v = -u$ and using that $\sup(-u) = -\inf u$.

Finally, as we did in (2), the statement in (5) is deduced from (3) and (4). □

Remark 2.8. It is easy to see in Corollary 2.7 (3), (4) and (5) that the assumption that u is continuous *up to the boundary* $\partial\Omega$ can be dropped. Let Ω be a bounded domain and let $u : \Omega \to \mathbb{R}$. For every $x_0 \in \partial\Omega$ we define

$$\overline{u}(x_0) = \limsup_{\Omega \ni x \to x_0} u(x) \qquad \text{and} \qquad \underline{u}(x_0) = \liminf_{\Omega \ni x \to x_0} u(x).$$

Then we let

$$\sup_{\partial\Omega} u = \sup_{\partial\Omega} \overline{u} \qquad \text{and} \qquad \inf_{\partial\Omega} u = \inf_{\partial\Omega} \overline{u}.$$

We have the following statements.

(*i*) If $u \in C(\Omega)$ is subharmonic in the mean value sense in Ω then $\sup_{\Omega} u \le \sup_{\partial\Omega} u$.

(*ii*) If $u \in C(\Omega)$ be superharmonic in the mean value sense in Ω then $\inf_{\Omega} u \ge \inf_{\partial\Omega} u$.

(*iii*) If $u \in C(\Omega)$ is harmonic in Ω then, for any $x \in \Omega$, $\inf_{\partial\Omega} u \le u(x) \le \sup_{\partial\Omega} u$.

Certainly, the weak maximum principle does not directly imply the strong maximum principle. The strong maximum principle holds even for unbounded domains.

The picture of the weak maximum principle, that requires Ω to be bounded, is the following. A subharmonic function u in a bounded domain that is continuous up to the boundary cannot *go up too much* in the interior, the threshold is given by the maximum of its boundary values. By flipping this picture upside down, a superharmonic function cannot *go down too much* in the interior with respect to its boundary values. A harmonic function continuous up to the boundary is always trapped between its maximum and minimum boundary values. These basic observations are at the center of the proof of the Harnack inequality for solutions to elliptic equations.

Next we precise some consequences of the weak maximum principle.

Corollary 2.9 (Direct consequences of the weak maximum principle)**.** *Let Ω be a bounded domain.*

(1) *Let $u \in C(\overline{\Omega})$ be subharmonic in the mean value sense in Ω. If $u \le 0$ on the boundary $\partial\Omega$ then $u \le 0$ everywhere in $\overline{\Omega}$.*

(2) *Let* $u \in C(\overline{\Omega})$ *be superharmonic in* Ω *in the mean value sense in* Ω. *If* $u \geq 0$ *on the boundary* $\partial\Omega$ *then* $u \geq 0$ *everywhere in* $\overline{\Omega}$.

(3) *Let* $u \in C(\overline{\Omega})$ *be harmonic in* Ω. *If* $u = 0$ *on the boundary* $\partial\Omega$ *then* $u = 0$ *in* $\overline{\Omega}$.

(4) (Ordering and comparison principle) *Suppose that* $u \in C(\overline{\Omega})$ *is subharmonic in the mean value sense in* Ω, $v \in C(\overline{\Omega})$ *is harmonic in* Ω *and* $w \in C(\overline{\Omega})$ *is superharmonic in the mean value sense in* Ω. *If, in addition,* $u \leq v \leq w$ *on the boundary* $\partial\Omega$, *then* $u \leq v \leq w$ *everywhere in* $\overline{\Omega}$. *In particular, if* $u, v \in C^2(\Omega) \cap C(\overline{\Omega})$ *satisfy*

$$\begin{cases} \Delta v \leq \Delta u & \text{in } \Omega \\ u \leq v & \text{on } \partial\Omega \end{cases}$$

then $u \leq v$ *everywhere in* $\overline{\Omega}$.

(5) (Uniqueness) *Let* $u, v \in C^2(\Omega) \cap C(\overline{\Omega})$ *such that*

$$\begin{cases} \Delta u = \Delta v & \text{in } \Omega \\ u = v & \text{on } \partial\Omega. \end{cases}$$

Then $u = v$ *in* $\overline{\Omega}$. *In particular, the Dirichlet problem*

$$\begin{cases} \Delta u = f & \text{in } \Omega \\ u = g & \text{on } \partial\Omega \end{cases} \tag{2.14}$$

for $f \in C(\Omega)$ *and* $g \in C(\partial\Omega)$ *has at most one classical solution* $u \in C^2(\Omega) \cap C(\overline{\Omega})$.

Proof. The statements in (1) and (2) follow from Corollary 2.7(3) and Corollary 2.7(4), respectively. Then (3) is derived from (1) and (2). For (4), let $h := u - v \in C(\overline{\Omega})$. Then h is subharmonic in the mean value sense and $h \leq 0$ on $\partial\Omega$. By part (1), $h \leq 0$ in $\overline{\Omega}$, so $u \leq v$ in $\overline{\Omega}$. The other conclusions follow similarly. Finally (5) is immediate from (4). □

Let Ω be a bounded domain and take $u, v, w \in C(\overline{\Omega})$. Corollary 2.9(4) gives that if u is subharmonic in Ω, v is harmonic in Ω, w is superharmonic in Ω, and u, v and w coincide at the boundary $\partial\Omega$, then u is always *below* v and w is always *above* v in Ω. This finally justifies the names *subharmonic* and *superharmonic.*

Another way of expressing Corollary 2.9(4) for harmonic functions is the following. Two harmonic functions u and v whose boundary values are ordered *cannot cross* in the interior. Moreover, by the strong maximum principle, they cannot even *touch* in the interior unless they are both equal to the same constant.

It is important to remember that the weak maximum principle and the uniqueness result for (2.14) require the domain Ω to be bounded as the following examples show.

Example 2.10. Consider the boundary value problem

$$\begin{cases} \Delta u = 0 & \text{in } \Omega \\ u = 0 & \text{on } \partial\Omega. \end{cases}$$

As we already noticed in Corollary 2.9(3), if Ω is bounded then $u = 0$ is the unique classical solution. Nontrivial examples arise when Ω is unbounded.

(*i*) In the case $\Omega = \mathbb{R}^n \setminus B_1$, the exterior of the unit ball, we find the nontrivial solutions

$$u(x) = \begin{cases} -\log|x| & \text{if } n = 2 \\ \dfrac{1}{|x|^{n-2}} - 1 & \text{if } n \geq 3. \end{cases}$$

Observe that u is bounded in Ω if $n \geq 3$ and unbounded if $n = 2$.

(*ii*) In the case $\Omega = \mathbb{R}^n_+ = \{x \in \mathbb{R}^n : x_n > 0\}$, the upper half space, we find the unbounded solution $u(x) = x_n$.

The proof of the comparison principle in Corollary 2.9(4) depends only on the assumption that the continuous functions involved satisfy the (sub/super)-mean value property. This observation motivates yet another notion of subharmonic function based solely on the comparison principle.

We say that $u \in C(\Omega)$ is **subharmonic in the comparison sense** in Ω if for every ball $B \subset\subset \Omega$ and every $v \in C(\overline{B})$ such that

$$\begin{cases} \Delta v = 0 & \text{in } B \\ u \leq v & \text{on } \partial B \end{cases}$$

we have

$$u \leq v \qquad \text{in } B.$$

Similarly, $u \in C(\Omega)$ is **superharmonic in the comparison sense** in Ω if $-u$ is subharmonic in the comparison sense in Ω.

We claim that this new comparison notion and the mean value notion for subharmonic functions are equivalent. Indeed, any subharmonic function in the mean value sense satisfies the comparison principle with respect to any harmonic function by Corollary 2.9(4). For the converse, assume that u is subharmonic in the comparison sense in Ω. Fix any ball $B = B_R(x) \subset\subset \Omega$ and let v be the harmonic function in B such that $v = u$ on ∂B. Notice that such a v can be constructed using the Poisson formula for balls, see Theorem 2.2. Then $u \leq v$ in B. Therefore, using that v satisfies the mean value property,

$$u(x) \leq v(x) = \frac{1}{|\partial B_R(x)|} \int_{\partial B_R(x)} v\, dS = \frac{1}{|\partial B_R(x)|} \int_{\partial B_R(x)} u\, dS$$

and u is subharmonic in the mean value sense in Ω.

As a consequence, if $u \in C(\Omega)$ is **harmonic in the comparison sense** in Ω, namely, both sub- and super- harmonic in the comparison sense in Ω, then u is harmonic in the classical sense.

The comparison principle notion we presented here is at the basis of the concept of *viscosity solution* for nondivergence form and fully nonlinear equations, see Chapter 8.

2.3.2 The interior derivative estimates and their consequences

The idea behind the interior derivative estimates is the mean value property. A harmonic function is smooth and the usual derivative commutes with the Laplacian. By differentiating the equation $\Delta u = 0$, we see that $\partial_{x_i} u$ is also a harmonic function and as such it satisfies the mean value property. But then we can relate the integral of $\partial_{x_i} u$ back with u with the aid of the divergence theorem. Then the process is repeated to estimate all the derivatives of u of any order by u again.

Theorem 2.11 (Interior derivative estimates for harmonic functions). *Let $u \in C(\Omega)$ be harmonic in Ω. Let $B_R(x_0) \subset\subset \Omega$. Then, for any multi-index $\gamma \in \mathbb{N}_0^n$,*

$$|D^\gamma u(x_0)| \leq \frac{C_{n,\gamma}}{R^{n+|\gamma|}} \|u\|_{L^1(B_R(x_0))}. \tag{2.15}$$

In particular,

$$|D^\gamma u(x_0)| \leq \frac{\omega_n^{1-1/p} C_{n,\gamma}}{R^{n/p+|\gamma|}} \|u\|_{L^p(B_R(x_0))} \qquad \text{for } 1 \leq p \leq \infty. \tag{2.16}$$

Here

$$C_{n,0} = \frac{1}{\omega_n} \qquad \text{and} \qquad C_{n,\gamma} = \frac{(2^{n+1} n |\gamma|)^{|\gamma|}}{\omega_n}, \ |\gamma| \geq 1.$$

In particular, when $p = \infty$, by replacing u by $u - \inf_{B_R(x_0)} u$ we get

$$|D^\gamma u(x_0)| \leq \frac{\omega_n C_{n,\gamma}}{R^{|\gamma|}} \operatorname*{osc}_{B_R(x_0)} u \tag{2.17}$$

where, for any set D, we have defined $\operatorname{osc}_D u = \sup_D u - \inf_D u$.

Remark 2.12. One the most important features of estimates (2.15), (2.16) and (2.17) is that the constant $C_{n,\gamma}$ is independent not only of u, but also of the size and position of the ball $B_R(x_0)$ and the domain Ω. In other words, the estimate is translation invariant. This will be remarkably useful when proving Schauder estimates.

Proof of Theorem 2.11. We first observe that if $1 < p < \infty$, by Hölder's inequality and (2.7),

$$\|u\|_{L^1(B_R(x_0))} \leq |B_R(x_0)|^{1-1/p} \left(\int_{B_R(x_0)} |u|^p \, dz \right)^{1/p} = \omega_n^{1-1/p} R^{n-n/p} \|u\|_{L^p(B_R(x_0))}.$$

In the case $p = \infty$ we have $\|u\|_{L^1(B_R(x_0))} \leq \omega_n R^n \|u\|_{L^\infty(B_R(x_0))}$. Then (2.15) implies (2.16).

The proof of (2.15) is done by induction in $|\gamma|$. Let us begin with the case $|\gamma| = 0$, that is, $\gamma = (0, \dots, 0)$. By the mean value property and (2.7),

$$|u(x_0)| = \left| \frac{1}{|B_R(x_0)|} \int_{B_R(x_0)} u \, dz \right| \leq \frac{1}{\omega_n R^n} \|u\|_{L^1(B_R(x_0))} = \frac{C_{n,0}}{R^n} \|u\|_{L^1(B_R(x_0))}$$

as desired. Suppose now that $|\gamma| = 1$, that is $\gamma = e_i$, for some $i = 1, \dots, n$. Since $u \in C^\infty(\Omega)$,

$$\partial_{x_i}(\Delta u) = \Delta(\partial_{x_i} u) = 0 \qquad \text{in } \Omega.$$

Hence $\partial_{x_i} u$ is harmonic in Ω. By the mean value property, the divergence theorem and (2.7),

$$\begin{aligned} |\partial_{x_i} u(x_0)| &= \frac{1}{|B_{R/2}(x_0)|} \left| \int_{B_{R/2}(x_0)} \partial_{x_i} u \, dx \right| \\ &= \frac{2^n}{\omega_n R^n} \left| \int_{\partial B_{R/2}(x_0)} u \nu_i \, dS \right| \\ &\leq \frac{2^n}{\omega_n R^n} \|u\|_{L^\infty(\partial B_{R/2}(x_0))} \frac{n \omega_n R^{n-1}}{2^{n-1}} \\ &= \frac{2n}{R} \|u\|_{L^\infty(\partial B_{R/2}(x_0))}. \end{aligned} \tag{2.18}$$

Notice now that if $x \in \partial B_{R/2}(x_0)$, then $B_{R/2}(x) \subset B_R(x_0) \subset\subset \Omega$. Hence, by the mean value property, for any $x \in \partial B_{R/2}(x_0)$,

$$\begin{aligned} |u(x)| &= \frac{1}{|B_{R/2}(x)|} \left| \int_{B_{R/2}(x)} u \, dz \right| \\ &\leq \frac{2^n}{\omega_n R^n} \int_{B_R(x_0)} |u| \, dz = \frac{2^n}{\omega_n R^n} \|u\|_{L^1(B_R(x_0))}. \end{aligned} \tag{2.19}$$

Thus, by pasting together (2.18) and (2.19),

$$|\partial_{x_i} u(x_0)| \leq \frac{2^{n+1} n}{\omega_n R^{n+1}} \|u\|_{L^1(B_R(x_0))},$$

which is (2.15) for $|\gamma| = 1$. The rest of the proof is left as an exercise. □

Remark 2.13. We could have proved Theorem 2.11 only in the case when $x_0 = 0$ and $R = 1$, and then obtained the general case by translation and rescaling. Indeed, suppose that for any harmonic function v in B_1, we have

$$|\nabla v(0)| \leq C_n \|v\|_{L^1(B_1)}.$$

Let u be harmonic in Ω and take $B_R(x_0) \subset\subset \Omega$. Define

$$v(x) := u(x_0 - Rx) \qquad \text{for } x \in B_1.$$

Notice that v is well defined in B_1 because $x \in B_1$ if and only if $x_0 - Rx \in B_R(x_0)$. Then

$$\Delta v(x) = R^2 (\Delta u)(x_0 - Rx) = 0 \qquad \text{for all } x \in B_1.$$

Also, $\nabla v(x) = -R(\nabla u)(x_0 - Rx)$, so that $\nabla v(0) = -R \nabla u(x_0)$. Moreover,

$$\|v\|_{L^1(B_1)} = \int_{B_1} |u(x_0 - Rx)| \, dx = R^{-n} \int_{B_R(x_0)} |u(z)| \, dz = R^{-n} \|u\|_{L^1(B_R(x_0))}.$$

Therefore, by applying the estimate for v,

$$|\nabla u(x_0)| = \frac{1}{R}|\nabla v(0)| \le \frac{C_n}{R}\|v\|_{L^1(B_1)} = \frac{C_n}{R^{n+1}}\|u\|_{L^1(B_R(x_0))}.$$

The moral is that, in general, any theorem in PDE has a scaled version depending on the scaling of the equation. We will use this philosophy in a repeated way not only to shorten the statements but also to understand the correct regularity of solutions.

By looking at the method of proof of Theorem 2.11, we can show the following.

Theorem 2.14 (Pointwise gradient estimate for nonnegative harmonic functions and Liouville theorem). *Let $u \in C(\Omega)$ be a nonnegative harmonic function in Ω. Then, for every $B_R(x_0) \subset\subset \Omega$,*

$$|\nabla u(x_0)| \le \frac{n\sqrt{n}}{R} u(x_0).$$

In particular, the Liouville theorem *holds: if $v \in C(\mathbb{R}^n)$ satisfies*

$$\begin{cases} \Delta v = 0 \textit{ in } \mathbb{R}^n \textit{ and} \\ v \textit{ is bounded above or below in } \mathbb{R}^n \end{cases}$$

then v is a constant.

Proof. Let us begin with the gradient estimate. We apply the mean value property to the harmonic function $\partial_{x_i} u$ and the divergence theorem as in (2.18) together with the fact that $u \ge 0$, to get

$$\begin{aligned} |\partial_{x_i} u(x_0)| &\le \frac{1}{\omega_n R^n}\left|\int_{\partial B_R(x_0)} u\nu_i \, dS\right| \\ &\le \frac{n\omega_n R^{n-1}}{\omega_n R^n} \frac{1}{|\partial B_R(x_0)|}\int_{\partial B_R(x_0)} u \, dS = \frac{n}{R} u(x_0) \end{aligned}$$

where in the last identity we used the mean value property for u.

For the Liouville theorem, it is enough to do the proof for v bounded below, for if v is bounded above then it follows by considering $-v$. Also, by adding a constant we can assume that $v \ge 0$. Hence we can apply the gradient estimate for $v \ge 0$ in any ball $B_R(x_0) \subset \mathbb{R}^n$. By taking $R \to \infty$ in such an estimate we see that $\nabla v(x_0) = 0$, for any $x_0 \in \mathbb{R}^n$, and so v is constant. □

Corollary 2.15 (Consequences of the interior derivative estimates).

(1) (Interior L^∞ estimates for derivatives). *Let $u \in C(\Omega)$ be harmonic in a domain Ω with nonempty boundary. Then, for any subdomain $\Omega' \subset\subset \Omega$ and any multi-index $\gamma \in \mathbb{N}_0^n$,*

$$\sup_{\Omega'} |D^\gamma u| \le \frac{c_{n,\gamma}}{\operatorname{dist}(\Omega', \partial\Omega)^{|\gamma|}} \operatorname*{osc}_{\Omega} u$$

where $c_{n,0} = 1$ and $c_{n,\gamma} = (2^{n+1} n|\gamma|)^{|\gamma|}$, for $|\gamma| \ge 1$. In other words, "derivatives of bounded *harmonic functions are bounded in the interior."*

(2) (Real analyticity). *If $u \in C(\Omega)$ is harmonic in Ω then u is real analytic in Ω, namely, given any $x_0 \in \Omega$ there exists a ball $B_R(x_0) \subset \Omega$ such that*

$$u(x) = \sum_{n=0}^{\infty} \sum_{|\gamma|=n} \frac{D^\gamma u(x_0)}{\gamma!}(x - x_0)^\gamma$$

for every $x \in B_R(x_0)$ and the series is absolutely convergent. Here, for $\gamma = (\gamma_1, \ldots, \gamma_n) \in \mathbb{N}_0^n$, we defined $\gamma! = \gamma_1! \cdots \gamma_n!$ and $x^\gamma = x_1^{\gamma_1} \cdots x_n^{\gamma_n}$.

(3) (Polynomial Liouville theorem). *Let $u \in C(\mathbb{R}^n)$ be harmonic in $\mathbb{R}^n$. If u grows like a polynomial of degree k, namely, if there are a constant $M > 0$ and an integer $k \geq 0$ such that*

$$|u(x)| \leq M(1 + |x|^k) \qquad \textit{for all } x \in \mathbb{R}^n$$

then u is a polynomial of degree at most k.

Proof. For (1), let us call $d = \operatorname{dist}(\Omega', \partial\Omega)$. Then $d > 0$ because $\Omega' \subset\subset \Omega$. Let z be any point in Ω'. Define

$$v(x) := u(x + z) \qquad \text{for } x \in B_d.$$

Observe that if $z \in \Omega'$ and $x \in B_d$ then $x+z \in \Omega$, so v is well defined in B_d. Moreover, $\operatorname{osc}_{B_d} v \leq \operatorname{osc}_\Omega u$ and also $D^\gamma v(0) = D^\gamma u(z)$. Now,

$$\Delta v(x) = (\Delta u)(x + z) = 0 \qquad \text{for any } x \in B_d.$$

So we can apply the interior derivative estimate (2.17) to v to get

$$|D^\gamma u(z)| = |D^\gamma v(0)| \leq \frac{\omega_n C_{n,\gamma}}{d^{|\gamma|}} \operatorname*{osc}_{B_d} v \leq \frac{c_{n,\gamma}}{d^{|\gamma|}} \operatorname*{osc}_{\Omega} u.$$

The conclusion follows by taking the supremum in $z \in \Omega'$ in the inequality above.

The idea for (2) is to use the expression of $c_{n,\gamma}$ in the derivative bounds found in part (1) to show that the remainder in the Taylor series expansion of u decays if R is sufficiently small. The proof uses Stirling's formula to estimate $c_{n,\gamma}$ and is left as an exercise.

For the Polynomial Liouville theorem, part (3), notice that we need to show that $D^\gamma u(x) = 0$, for all $|\gamma| = k + 1$ and any $x \in \mathbb{R}^n$. To that end, let us fix $\gamma \in \mathbb{N}_0^n$ such that $|\gamma| = k + 1$ and $x \in \mathbb{R}^n$. Then, for any $R > |x|$, by (2.16) and the assumption,

$$|D^\gamma u(x)| \leq \frac{\omega_n C_{n,\gamma}}{R^{k+1}} \|u\|_{L^\infty(B_R(x))} \leq \frac{\omega_n C_{n,\gamma}}{R^{k+1}} M(1 + (2R)^k).$$

We take $R \to \infty$ above to get $D^\gamma u(x) = 0$. As x was arbitrary, $D^\gamma u \equiv 0$. □

2.3.3 The Harnack inequality

The Harnack inequality is one of the most important estimates for solutions to elliptic PDEs. As we will later see, it plays a crucial role in the regularity theory.

The Harnack inequality for nonnegative harmonic functions establishes that, because of the mean value property, their value at a point, say, the origin, controls the rest of the values in the interior. To quickly see why, suppose that $u \geq 0$ is a harmonic function in B_1. Fix any $x \in B_1$ and let $d_x = \text{dist}(x, \partial B_1)$. Then $B_{d_x}(x) \subset B_1$ and, by the mean value property,

$$u(x) = \frac{1}{|B_{d_x}(x)|} \int_{B_{d_x}(x)} u\,dz \leq \frac{1}{d_x^n} \frac{1}{|B_1|} \int_{B_1} u\,dz = \frac{1}{d_x^n} u(0).$$

By connecting x to any other point in B_1 with a chain of balls, we get the general statement that says that the values of any nonnegative harmonic function in a domain are all comparable in the interior.

Theorem 2.16 (Harnack inequality). *Let Ω be a domain in $\mathbb{R}^n$ with nonempty boundary. For any subdomain $\Omega' \subset\subset \Omega$ there exists a constant $C_H > 0$ depending only on n, Ω' and Ω such that for any $u \in C(\Omega)$ satisfying*

$$\begin{cases} \Delta u = 0 & \text{in } \Omega \\ u \geq 0 & \text{in } \Omega \end{cases}$$

we have

$$\sup_{\Omega'} u \leq C_H \inf_{\Omega'} u. \tag{2.20}$$

Remark 2.17. Some useful remarks to understand the Harnack inequality are in order.

(1) If we want an inequality of the form (2.20) to be true then we must require for u not to change sign in Ω. Indeed, if $u(x_0) < 0$ at a point $x_0 \in \Omega$ then its infimum near x_0 is strictly negative, so the Harnack inequality says that u is striclty negative everywhere around x_0. The same argument works if u is strictly positive at a point x_0, giving $u > 0$ everywhere in Ω. In our statement we assume $u \geq 0$. If $u \geq 0$ and $u(x_0) = 0$ for some $x_0 \in \Omega$, then (2.20) gives that $u \leq 0$ everywhere near x_0, so $u \equiv 0$ in Ω.

(2) The Harnack inequality (2.20) says that all the values of u are comparable in the interior. Indeed, for any $x_1, x_2 \in \Omega'$,

$$u(x_1) \leq \sup_{\Omega'} u \leq C_H \inf_{\Omega'} u \leq C_H u(x_2).$$

In particular, if $u(x_0) = \mu > 0$ for some $x_0 \in \Omega'$ then $u \geq \mu > 0$ everywhere in Ω'.

(3) The Harnack inequality holds for nonnegative *harmonic* functions, namely, for *solutions* to $\Delta u = 0$ such that $u \geq 0$. It is not valid in the form (2.20) for

nonnegative *subharmonic* functions. For example, $u(x) = \frac{|x|^2}{2n}$ and $u(x) = |x|$ are nonnegative and subharmonic in B_2, but $\sup_{B_1} u > 0$ while $\inf_{B_1} u = 0$. The Harnack inequality for nonnegative solutions to $-\Delta u = f$ must take into account the right hand side f, see, for instance, Remarks 9.8 and 10.11.

(4) The constant C_H in Theorem 2.16 is independent of the particular u chosen from the class of nonnegative harmonic functions in Ω. In general, $C_H \to \infty$ as $\Omega' \to \Omega$ because we are not fixing the values of u on the boundary $\partial\Omega$.

(5) We will later see that the Harnack inequality is also valid for nonnegative solutions to divergence (Moser theorem) and nondivergence (Krylov–Safonov theorem) form equations. Notice that the Harnack inequality gives a universal control of the oscillation of u in the interior:

$$1 \le \frac{\sup_{\Omega'} u}{\inf_{\Omega'} u} \le C_H.$$

This idea combined with the equation plus the fact that C_H is independent of the particular u will imply interior C^α regularity.

Proof of Theorem 2.16. We first prove the inequality when Ω' is a small ball inside of Ω. Then the general case will follow by a covering argument.

Step 1. Let $\Omega' = B_R(x) \subset B_{4R}(x) \subset \Omega$. Let us prove that for any x_1, x_2 in $B_R(x)$ we have

$$u(x_1) \le 3^n u(x_2). \tag{2.21}$$

This is enough for our purposes because, by fixing x_2 and taking the supremum in x_1,

$$\sup_{x_1 \in B_R(x)} u(x_1) \le 3^n u(x_2).$$

Now we take the infimum in x_2 above and get the conclusion. To prove (2.21), notice that $B_R(x_1) \subset B_{2R}(x) \subset B_{3R}(x_2)$. Then we can use the mean value property and the nonnegativity of u in the following way:

$$\begin{aligned} u(x_1) &= \frac{1}{|B_R(x_1)|} \int_{B_R(x_1)} u\, dz \\ &\le \frac{1}{|B_R(x_1)|} \int_{B_{3R}(x_2)} u\, dz \\ &= \frac{|B_{3R}(x_2)|}{|B_R(x_1)|} \frac{1}{|B_{3R}(x_2)|} \int_{B_{3R}(x_2)} u\, dz \\ &= 3^n u(x_2). \end{aligned} \tag{2.22}$$

Step 2. Let $\Omega' \subset\subset \Omega$ arbitrary. As u is continuous in Ω, we can find $x_1, x_2 \in \overline{\Omega'}$ such that

$$u(x_1) = \sup_{\Omega'} u \qquad \text{and} \qquad u(x_2) = \inf_{\Omega'} u.$$

Let γ be a closed connected arc in $\overline{\Omega'}$ joining x_1 and x_2. Let $R > 0$ such that $4R < \operatorname{dist}(\gamma, \partial\Omega)$. Since γ is compact, it can be covered by N balls $B^1, \dots, B^N$ of radius R, with $x_1 \in B^1$ and $x_2 \in B^N$. Also, all these balls B^j are well inside Ω, that is, $4B^j \subset \Omega$, $j = 1, \dots, N$. Then we can use Step 1 repeatedly N-times to get

$$\begin{aligned}
\sup_{\Omega'} u = u(x_1) &\le \sup_{B^1} u \le 3^n \inf_{B^1} u \\
&\le 3^n \inf_{B^1 \cap B^2} u \le 3^n \sup_{B^2} u \\
&\le 3^{2n} \inf_{B^2} u \le \dots \le 3^{Nn} \inf_{B_N} u \\
&\le 3^{Nn} u(x_2) = 3^{Nn} \inf_{\Omega'} u.
\end{aligned}$$

□

Remark 2.18. We separated the proof above into two steps in order to clarify the different roles of the equation and the domain Ω'.

(1) Step 1 uses the equation in an essential way (through the mean value property), giving a constant C_H depending only on n. The proof for subharmonic or superharmonic functions fails in the chain of inequalities (2.22). From (2.22) we see that to connect the values of u at x_1 and at x_2 we are passing through an integral quantity that resembles the divergence structure of the Laplacian. This is also the basic idea in Moser's proof of the Harnack inequality for divergence form elliptic equations. If $u \ge 0$ is a subsolution then we have the *local boundedness* estimate

$$\sup u \le C\|u\|_{L^p}$$

while if $u \ge 0$ is a supersolution then we have the *weak Harnack inequality*

$$\|u\|_{L^p} \le C \inf u.$$

Hence if u is a nonnegative *solution*, it will satisfy the Harnack inequality. In this case the integral quantity that connects the values of u is an L^p norm, which in some sense replaces the stronger mean value property.

(2) Step 2 says that the Harnack inequality for subdomains $\Omega' \subset\subset \Omega$ follows at once from that for interior balls. This step does not require to use the equation. Hence, later on, we will omit the general statement and we will present it only for interior balls or cubes. It is also clear that the constant C_H depends on n, Ω' and Ω and that N will increase without bound as $\Omega' \to \Omega$, making $C_H \to \infty$.

We can explicitly see the behavior of the Harnack constant near the boundary for harmonic functions in balls. In fact, using the Poisson formula from Theorem 2.2 and the mean value property, it is not difficult to prove the following result.

Theorem 2.19 (Harnack inequality in balls with explicit constant). *Let u be a nonnegative harmonic function in B_R. Let $0 < r < R$. Then, for any $x \in B_r$,*

$$u(0)\left(\frac{R}{R+r}\right)^{n-2}\frac{R-r}{R+r} \le u(x) \le u(0)\left(\frac{R}{R-r}\right)^{n-2}\frac{R+r}{R-r}. \tag{2.23}$$

In particular,

$$\sup_{B_r} u \leq C_H \inf_{B_r} u \tag{2.24}$$

with

$$C_H = \left(\frac{1 + r/R}{1 - r/R}\right)^n.$$

Moreover

$$C_H \sim (R - r)^{-n} \qquad \text{as } r \to R.$$

The constants in (2.23) are invariant under scaling in the sense that they do not change for different choices of R when $r = \lambda R$, for some $0 < \lambda < 1$. In this case (2.23) reads

$$u(0)\left(\frac{1}{1+\lambda}\right)^{n-2}\frac{1-\lambda}{1+\lambda} \leq u(x) \leq u(0)\left(\frac{1}{1-\lambda}\right)^{n-2}\frac{1+\lambda}{1-\lambda}.$$

In addition, the constants do not depend on the particular point x_0.

This explicit Harnack inequality in balls gives a different proof of the Liouville theorem that does not involve derivative estimates for u. Indeed, if

$$\Delta u = 0 \text{ in } \mathbb{R}^n \qquad \text{and} \qquad u \geq 0 \text{ in } \mathbb{R}^n$$

then, by taking $R \to \infty$ in (2.23), we get $u(x) = u(0)$ for any $x \in \mathbb{R}^n$, so u is constant.

NOTES ON CHAPTER 2

The material in this chapter may be found, perhaps with a different approach and emphasis, in any good book on basic PDEs. We decided to present all the results building up from the mean value property, which is at the core of the divergence structure of the Laplacian. Moreover, we analyze and compare notions of (sub/super)harmonic functions in the classical, mean value, distributional, weak and comparison senses.

Analytic functions and their relation with harmonic functions in dimension 2, including the derivation of the Poisson kernel for the disk, can be seen in Rudin [56]. The derivation of the Poisson formula for the ball in $\mathbb{R}^n$ is in, for example, Evans [32] or Gilbarg–Trudinger [43]. For the theory of harmonic polynomials and spherical harmonics see Stein–Weiss [66]. The proof of Theorem 2.5 is due to Caffarelli [5]. The proofs we skipped in Theorem 2.11 and Corollary 2.15 can be found in Evans [32]. Theorem 2.19 was originally proved in 1887 for dimension $n = 2$ by Carl Gustav Axel von Harnack, see [47] for a proof and a historical account.

We mentioned that the weak maximum principle does not directly imply the strong maximum principle. However, the Hopf lemma together with the weak maximum principle give the strong maximum principle for uniformly elliptic equations, see Gilbarg–Trudinger [43].

CHAPTER 3

The Schauder estimates for the Laplacian

In this chapter we consider regularity estimates for solutions u to the Poisson equation

$$-\Delta u = f \qquad \text{in } \Omega \subset \mathbb{R}^n$$

when f is Hölder continuous. The estimate we want to show says that if f is C^α then u is $C^{2,\alpha}$. In some sense, the inverse of the Laplacian *integrates twice*. This is the *perfect result* because the Laplacian of a $C^{2,\alpha}$ function certainly is C^α. On the other hand, it is surprising that only from the fact that a very special combination of derivatives $u_{xx} + u_{yy}$ is Hölder continuous, it follows that *all* second-order derivatives u_{xx}, u_{yy} and u_{xy} are Hölder continuous as well.

The first step of our analysis is to write down u in terms of f:

$$u = (-\Delta)^{-1} f.$$

This is achieved by using the *method of semigroups*, which is a quite general approach and will be very useful in other situations, for example, to study the fractional Laplacian. We will have to prove that if u is given as above and f is Hölder continuous, then u is C^2 and it is our desired solution.

In this chapter we will prove the $C^{2,\alpha}$ estimate in two ways: first, by using the representation formula for second derivatives of u and, second, by the method of maximum principle. The latter is a powerful, robust technique devised by Luis A. Caffarelli to prove Schauder estimates for second-order fully nonlinear elliptic PDEs.

3.1 REVIEW OF FOURIER TRANSFORM

We want to begin our analysis with the Poisson equation

$$-\Delta u = f \qquad \text{in } \mathbb{R}^n.$$

Our aim is to write u in terms of f by computing the inverse operator $(-\Delta)^{-1}$. The advantage of working in the whole space $\mathbb{R}^n$ is that we can use the Fourier transform. In this section we review its main properties.

DOI: 10.1201/9781032679853-3

Let $u \in L^1(\mathbb{R}^n)$. The **Fourier transform** of u is defined for $\xi \in \mathbb{R}^n$ as

$$\mathcal{F}u(\xi) = \widehat{u}(\xi) = \frac{1}{(2\pi)^{n/2}} \int_{\mathbb{R}^n} u(x) e^{-i\xi\cdot x}\, dx.$$

The **inverse Fourier transform** of u is defined for $x \in \mathbb{R}^n$ as

$$\mathcal{F}^{-1}u(x) = \check{u}(x) = \frac{1}{(2\pi)^{n/2}} \int_{\mathbb{R}^n} u(\xi) e^{ix\cdot\xi}\, d\xi.$$

Observe that $\widehat{u}(\xi) = \check{u}(-\xi)$, so everything we say for $\widehat{u}$ is equally valid for $\check{u}$. Clearly,

$$\widehat{u}(0) = \frac{1}{(2\pi)^{n/2}} \int_{\mathbb{R}^n} u\, dx$$

and

$$\|\widehat{u}\|_{L^\infty(\mathbb{R}^n)} \leq (2\pi)^{-n/2} \|u\|_{L^1(\mathbb{R}^n)}.$$

Moreover, $\widehat{u}$ is continuous and the Riemann–Lebesgue theorem establishes that

$$\widehat{u}(\xi) \to 0 \qquad \text{as } |\xi| \to \infty.$$

Plancherel's identity says that if $u \in L^1(\mathbb{R}^n) \cap L^2(\mathbb{R}^n)$ then $\widehat{u} \in L^2(\mathbb{R}^n)$ and

$$\|\widehat{u}\|_{L^2(\mathbb{R}^n)} = \|u\|_{L^2(\mathbb{R}^n)}.$$

With this result we can extend the definition of the Fourier transform to $L^2(\mathbb{R}^n)$. Indeed, for $u \in L^2(\mathbb{R}^n)$ we take a sequence $\{u_k\}_{k=1}^\infty \subset L^1(\mathbb{R}^n) \cap L^2(\mathbb{R}^n)$ such that $u_k \to u$ in $L^2(\mathbb{R}^n)$, as $k \to \infty$. By Plancherel's identity, $\widehat{u_k}$ is a Cauchy sequence in $L^2(\mathbb{R}^n)$, so $\widehat{u_k}$ converges to some function in $L^2(\mathbb{R}^n)$. We call such limit $\widehat{u}$, and this is our definition of the Fourier transform of u. It is easy to see that the definition does not depend on the choice of the particular sequence $\{u_k\}_{k=1}^\infty$. Notice that this does not say that $\widehat{u}$ is given by the integral definition we introduced at the beginning. Instead, it can be proved that if $u \in L^2(\mathbb{R}^n)$ then

$$\widehat{u}(\xi) = \frac{1}{(2\pi)^{n/2}} \lim_{R\to\infty} \int_{B_R} u(x) e^{-i\xi\cdot x}\, dx \qquad \text{in } L^2(\mathbb{R}^n). \tag{3.1}$$

Theorem 3.1 (Properties of the Fourier transform). *Let $u, v \in L^2(\mathbb{R}^n)$. The following properties hold.*

(*i*) Parseval's identity.

$$\int_{\mathbb{R}^n} u\bar{v}\, dx = \int_{\mathbb{R}^n} \widehat{u}\overline{\widehat{v}}\, d\xi.$$

where $\bar{v}$ denotes the complex conjugate of v.

(*ii*) Inversion.

$$\mathcal{F}^{-1}(\mathcal{F}u)(x) = u(x) \qquad \textit{for a.e. } x \in \mathbb{R}^n.$$

In particular, if $\widehat{u} \in L^1(\mathbb{R}^n)$, then u is continuous and

$$u(x) = \frac{1}{(2\pi)^{n/2}} \int_{\mathbb{R}^n} \widehat{u}(\xi) e^{ix\cdot\xi}\, d\xi.$$

(*iii*) Convolutions.

$$\widehat{u * v} = (2\pi)^{n/2}\widehat{u}\widehat{v}.$$

(*iv*) Derivatives. *If* $D^\gamma u \in L^2(\mathbb{R}^n)$, *for some multi-index* $\gamma \in \mathbb{N}_0^n$, *then*

$$\widehat{D^\gamma u}(\xi) = (i\xi)^\gamma \widehat{u}(\xi).$$

(*v*) Translations. *If we fix* $\tau \in \mathbb{R}^n$ *and define* $u_\tau(x) := u(x+\tau)$, *for* $x \in \mathbb{R}^n$, *then*

$$\widehat{u_\tau}(\xi) = e^{i\tau\cdot\xi}\widehat{u}(\xi).$$

(*vi*) Dilations. *If we fix* $\delta > 0$ *and define* $u_\delta(x) = u(\delta x)$, *for* $x \in \mathbb{R}^n$, *then*

$$\widehat{u_\delta}(\xi) = \frac{1}{\delta^n}\widehat{u}\Big(\frac{\xi}{\delta}\Big).$$

Properties (*ii*), (*iii*) and (*iv*) above make the Fourier transform a powerful tool. Indeed, on the Fourier side, derivatives correspond to multiplication by homogeneous polynomials and a convolution becomes a product. We can perform algebraic manipulations on the Fourier side and then come back to the space variables using (*ii*).

The most important class of test functions for the Fourier transform is the **Schwartz class** of rapidly decreasing smooth functions, which is defined as

$$\mathcal{S} = \Big\{\varphi \in C^\infty(\mathbb{R}^n) : p_{\beta,\gamma}(\varphi) := \sup_{x\in\mathbb{R}^n} |x^\beta D^\gamma \varphi(x)| < \infty, \text{ for every } \beta,\gamma \in \mathbb{N}_0^n\Big\}.$$

We have $C_c^\infty(\mathbb{R}^n) \subset \mathcal{S} \subset L^p(\mathbb{R}^n)$, for every $1 \le p \le \infty$. In particular, $\mathcal{S}$ is dense in $L^p(\mathbb{R}^n)$, for $1 \le p < \infty$, and the Fourier transform is well defined in $\mathcal{S}$. The canonical example of a function in $\mathcal{S}$ that does not have compact support is

$$\varphi(x) = e^{-|x|^2/2} \tag{3.2}$$

for which

$$\widehat{\varphi}(\xi) = e^{-|\xi|^2/2}.$$

The collection $\{p_{\beta,\gamma}\}_{\beta,\gamma\in\mathbb{N}_0^n}$ is a countable family of seminorms on $\mathcal{S}$ that can be used to define a topology on $\mathcal{S}$. We say that $\{\varphi_k\}_{k=1}^\infty$ converges to 0 in $\mathcal{S}$ if $p_{\beta,\gamma}(\varphi_k) \to 0$ as $k \to \infty$, for every β, γ. With this topology $\mathcal{S}$ is a Fréchet space (complete and metrizable). It can be shown that the Fourier transform $\mathcal{F} : \mathcal{S} \to \mathcal{S}$ is continuous and the inversion formula in (*ii*) above holds for $u \in \mathcal{S}$ for every $x \in \mathbb{R}^n$.

The set of bounded linear functionals on $\mathcal{S}$, denoted by $\mathcal{S}'$, is called the space of **tempered distributions**. A linear map $T : \mathcal{S} \to \mathbb{C}$ is in $\mathcal{S}'$ if

$$\lim_{k\to\infty} T(\varphi_k) = 0 \text{ whenever } \lim_{k\to\infty} \varphi_k = 0 \text{ in } \mathcal{S}.$$

For a tempered distribution T we define its Fourier transform $\widehat{T}$ as a new tempered distribution acting on $\varphi \in \mathcal{S}$ via $\widehat{T}(\varphi) = T(\widehat{\varphi})$. With this definition, the Fourier transform is a bounded linear bijection from $\mathcal{S}'$ to $\mathcal{S}'$ whose inverse is also bounded.

Locally integrable functions f with polynomial growth at infinity define tempered distributions via integration $T_f(\varphi) \equiv \langle f, \varphi\rangle \equiv f(\varphi) := \int f\varphi\, dx$. Thus the definition of Fourier transform extends to any such function.

A **Fourier multiplier** operator M is an operator defined by

$$\widehat{Mf}(\xi) = m(\xi)\widehat{f}(\xi)$$

where m is a measurable function on $\mathbb{R}^n$. A Fourier multiplier operator M is bounded in $L^2(\mathbb{R}^n)$ if and only if $m \in L^\infty(\mathbb{R}^n)$ and, in this case,

$$\|M\|_{L^2\to L^2} = \|m\|_{L^\infty(\mathbb{R}^n)}.$$

3.2 THE POISSON EQUATION: IDEAS OF THE METHOD

For a given function $f : \mathbb{R}^n \to \mathbb{R}$ we want to find a function $u \in C^2(\mathbb{R}^n)$ such that

$$-\Delta u = f \qquad \text{in } \mathbb{R}^n.$$

From the very beginning we notice some issues. If u is a solution then $u + C$ is also a solution for any constant C. More generally, if v is a harmonic function on $\mathbb{R}^n$ then $u + v$ is a solution. In general, uniqueness of classical solutions does not hold.

The idea is to find a way of writing u in terms of f by using some (maybe formal) computations. Once a formula is obtained we will need to check it. As we have the equation on $\mathbb{R}^n$, we can use the Fourier transform. Hence, let us suppose that $f \in C_c^\infty(\mathbb{R}^n)$. If we apply the Fourier transform to the equation, by Theorem 3.1(*iv*), we get

$$|\xi|^2\widehat{u}(\xi) = \widehat{f}(\xi).$$

Therefore

$$\widehat{u}(\xi) = |\xi|^{-2}\widehat{f}(\xi).$$

Of course this *pointwise* identity has an issue at $\xi = 0$. We could try to overcome this problem by asking some regularity for $\widehat{f}$, for example, by requiring that $\widehat{f}$ goes to zero near the origin faster than $|\xi|^{-2}$. But let us forget for the moment about this (at the end of the day, the origin has measure zero) and let us continue with the computation. By taking inverse Fourier transform on both sides,

$$u(x) = \mathcal{F}^{-1}\big(|\xi|^{-2}\widehat{f}(\xi)\big)(x).$$

There is another issue with this *integral* identity here as well. The Fourier multiplier $m(\xi) = |\xi|^{-2}$ is not a bounded function, so the corresponding Fourier multiplier operator, that we are going to denote by $M = (-\Delta)^{-1}$, is not bounded on $L^2(\mathbb{R}^n)$. Therefore it is not clear how to take the inverse Fourier transform on the right hand side. Nevertheless, there is just a singularity at the origin that is locally integrable if $n > 2$ and we know that we can give sense to the Fourier transform of an L^1_{loc} function that has polynomial growth at infinity. But still, let us forget for the moment about this problem and think that we can actually invert the Fourier transform of the

product. As products become convolutions under Fourier transform (see Theorem 3.1(iii)), we obtain

$$u(x) = (K * f)(x) = \int_{\mathbb{R}^n} K(x-z)f(z)\,dz.$$

Here $K(x)$ is the *fundamental solution* of the Laplacian and satisfies, at least formally by now, the Fourier transform identity

$$\widehat{K}(\xi) = \frac{|\xi|^{-2}}{(2\pi)^{n/2}} = \widehat{(-\Delta)^{-1}}.$$

These formal computations show that we should be able to write u in terms of f via a convolution with a kernel K.

Now we are ready to outline the program to follow.

Step 1. Compute $K(x)$.

Step 2. Define $u = K * f$ for the largest possible class of functions f.

Step 3. Show that $u \in C^2$ and $-\Delta u = f$.

Step 1 seems to be the most obscure. One possibility is to use the machinery of distributions and the Fourier transform. Another way would be to use the symmetries of the Laplacian. We will take another, more general path: the **method of semigroups**. To that end we need to first understand the heat equation.

After these steps are completed, we will be able to say that, given appropriate functions f, we can indeed find C^2 solutions u to $-\Delta u = f$. The second part of this chapter will focus on proving the Schauder estimates.

3.3 THE CLASSICAL HEAT SEMIGROUP

Given $f \in \mathcal{S}$, we want to find a function $v = v(x,t)$, for $x \in \mathbb{R}^n$ and $t \geq 0$, that solves the **heat equation** on the whole space $\mathbb{R}^n$, that is,

$$\begin{cases} \partial_t v = \Delta v & \text{for } x \in \mathbb{R}^n,\ t > 0 \\ v(x,0) = f(x) & \text{for } x \in \mathbb{R}^n. \end{cases}$$

We denote by $e^{t\Delta}$ the solution operator:

$$f(x) \longmapsto v(x,t) = e^{t\Delta} f(x).$$

This notation is quite convenient because, at least formally,

$$\partial_t(e^{t\Delta} f(x)) = \Delta(e^{t\Delta} f(x))$$

and

$$e^{t\Delta} f(x)\big|_{t=0} = f(x).$$

The family of operators $\{e^{t\Delta}\}_{t\geq 0}$ is called the **classical heat diffusion semigroup (generated by the Laplacian Δ)**. To find $e^{t\Delta}f(x)$, we apply the Fourier transform to the heat equation problem in the variable x for each fixed t to get

$$\begin{cases} \partial_t \widehat{v} = -|\xi|^2 \widehat{v} & \text{for } \xi \in \mathbb{R}^n,\ t>0 \\ \widehat{v}(\xi,0) = \widehat{f}(\xi) & \text{for } \xi \in \mathbb{R}^n. \end{cases}$$

Therefore

$$\widehat{v}(\xi,t) = \widehat{e^{t\Delta} f}(\xi) = e^{-t|\xi|^2}\widehat{f}(\xi). \tag{3.3}$$

Now the Fourier multiplier $e^{-t|\xi|^2}$ belongs to the Schwartz class $\mathcal{S}$ for any $t>0$. We can then use the convolution property of the Fourier transform, see Theorem 3.1(iii), to obtain

$$v(x,t) \equiv e^{t\Delta}f(x) = G_t * f(x) = \int_{\mathbb{R}^n} G_t(x-z)f(z)\,dz$$

for every $x\in\mathbb{R}^n$, $t>0$. Here $G_t(x)$ is the function that satisfies

$$\widehat{G_t}(\xi) = \frac{e^{-t|\xi|^2}}{(2\pi)^{n/2}}$$

for each fixed $t>0$, and is called the **Gauss–Weierstrass heat kernel**.

In general, a one parameter family $\{T_t\}_{t\geq 0}$, of bounded linear operators from a Banach space X into X is a **semigroup of bounded linear operators on X** if $T_0 = \mathrm{Id}$ (the identity operator) and it satisfies the so-called **semigroup property**: for every $t_1, t_2 \geq 0$ we have $T_{t_1+t_2} = T_{t_1}\circ T_{t_2}$. Notice that $\{e^{t\Delta}\}_{t\geq 0}$ is indeed a semigroup of bounded linear operators on $L^2(\mathbb{R}^n)$. Indeed, $e^{t\Delta} : L^2(\mathbb{R}^n) \to L^2(\mathbb{R}^n)$ is continuous for every $t\geq 0$, $e^{0\Delta}f = f$ and $e^{(t_1+t_2)\Delta}f = e^{t_1\Delta}(e^{t_2\Delta}f)$, for every $f\in L^2(\mathbb{R}^n)$, as can be readily see from the Fourier transform definition (3.3).

By using the canonical example (3.2) and the dilation property of the Fourier transform in Theorem 3.1(vi), it is easy to see that

$$G_t(x) = \frac{1}{(4\pi t)^{n/2}} e^{-|x|^2/(4t)} \qquad \text{for } x\in\mathbb{R}^n,\ t>0.$$

Therefore,

$$v(x,t) \equiv e^{t\Delta}f(x) = \frac{1}{(4\pi t)^{n/2}} \int_{\mathbb{R}^n} e^{-|x-z|^2/(4t)} f(z)\,dz. \tag{3.4}$$

Observe that

$$G_t(x) = \frac{1}{t^{n/2}} G_1\left(\frac{x}{t^{1/2}}\right),$$

so $f \longmapsto e^{t\Delta}f = G_t * f$ defines an approximation of the identity. Moreover,

$$e^{t\Delta}1(x) = \int_{\mathbb{R}^n} G_t(x-z)\,dz \equiv 1 \qquad \text{for any } x\in\mathbb{R}^n,\ t>0. \tag{3.5}$$

In particular, (3.5) means that if we start with initial temperature 1 in the whole space, then the temperature remains 1 for all times $t>0$. We summarize next the properties of the semigroup $\{e^{t\Delta}\}_{t\geq 0}$.

Theorem 3.2 (Properties of the heat semigroup). *Let $e^{t\Delta}f(x)$ be defined as in (3.4), for $x \in \mathbb{R}^n$ and $t > 0$.*

(1) *If $f \in C(\mathbb{R}^n) \cap L^\infty(\mathbb{R}^n)$ then $e^{t\Delta}f(x) \in C^\infty(\mathbb{R}^n \times (0,\infty))$ is a classical solution to the heat equation and $e^{t\Delta}f(x) \to f(x)$ for every $x \in \mathbb{R}^n$, as $t \to 0$.*

(2) *If $f \in L^p(\mathbb{R}^n)$, $1 \leq p < \infty$, then $e^{t\Delta}f(x) \in C^\infty(\mathbb{R}^n \times (0,\infty))$ is a classical solution to the heat equation and $e^{t\Delta}f(x) \to f(x)$ in $L^p(\mathbb{R}^n)$ and a.e. in $\mathbb{R}^n$, as $t \to 0$.*

(3) *If f is bounded then $|e^{t\Delta}f(x)| \leq \|f\|_{L^\infty(\mathbb{R}^n)}$ for every $x \in \mathbb{R}^n$, $t > 0$.*

3.4 THE FUNDAMENTAL SOLUTION OF THE LAPLACIAN

We are now in position to apply the method of semigroups to find the inverse of the Laplacian. We start by considering the following numerical formula. Let $\lambda > 0$. It is clear that

$$\lambda^{-1} = \int_0^\infty e^{-t\lambda}\,dt. \tag{3.6}$$

Let us choose $\lambda = |\xi|^2$ (the Fourier multiplier of the Laplacian) and multiply by the Fourier transform of a function $f \in C_c^\infty(\mathbb{R}^n)$. Then, following Section 3.2,

$$\widehat{u}(\xi) = |\xi|^{-2}\widehat{f}(\xi) = \int_0^\infty e^{-t|\xi|^2}\widehat{f}(\xi)\,dt.$$

Next we multiply by $e^{ix\cdot\xi}$ and integrate with respect to $d\xi$:

$$\int_{\mathbb{R}^n} \widehat{u}(\xi)e^{ix\cdot\xi}\,d\xi = \int_{\mathbb{R}^n}\int_0^\infty e^{-t|\xi|^2}\widehat{f}(\xi)e^{ix\cdot\xi}\,dt\,d\xi. \tag{3.7}$$

The double integral above is absolutely convergent if $n \geq 3$ without any additional assumptions on $f \in C_c^\infty(\mathbb{R}^n)$ at 0. Indeed, on one hand,

$$\int_{|\xi|\geq 1}\int_0^\infty e^{-t|\xi|^2}|\widehat{f}(\xi)|\,dt\,d\xi = \int_{|\xi|\geq 1}|\xi|^{-2}|\widehat{f}(\xi)|\,d\xi \leq \int_{|\xi|\geq 1}|\widehat{f}(\xi)|\,d\xi < \infty \tag{3.8}$$

because $\widehat{f} \in \mathcal{S}$. On the other hand, by using polar coordinates,

$$\int_{|\xi|<1}\int_0^\infty e^{-t|\xi|^2}|\widehat{f}(\xi)|\,dt\,d\xi \leq \|\widehat{f}\|_{L^\infty(B_1)}\int_{|\xi|<1}|\xi|^{-2}\,d\xi = C_{n,f}\int_0^1 r^{-2}r^{n-1}\,dr \tag{3.9}$$

which is finite as soon as $n \geq 3$. Observe next that if the dimension is $n = 2$, then (3.8) is still valid. To fix the second part of the computation (3.9), we need to compensate the singularity $|\xi|^{-2}$, which is not locally integrable in $\mathbb{R}^2$, by using $\widehat{f}$. To that end, let us assume that

$$\widehat{f}(0) = \frac{1}{2\pi}\int_{\mathbb{R}^2} f\,dx = 0. \tag{3.10}$$

Then, by the mean value theorem, for every $|\xi| < 1$ we get

$$|\widehat{f}(\xi)| = |\widehat{f}(\xi) - \widehat{f}(0)| \leq \|\nabla\widehat{f}\|_{L^\infty(B_1)}|\xi|.$$

Therefore, if $n = 2$ and f satisfies (3.10),

$$\int_{|\xi|<1}\int_0^\infty e^{-t|\xi|^2}|\widehat{f}(\xi)|\,dt\,d\xi \leq \|\nabla \widehat{f}\|_{L^\infty(B_1)}\int_{|\xi|<1}|\xi|^{-1}\,d\xi = C_f\int_0^1 r^{-1}r\,dr < \infty.$$

Hence we can apply Fubini's Theorem in (3.7) and use the heat semigroup introduced in Section 3.3, see (3.3), to get

$$u(x) = (-\Delta)^{-1}f(x) = \int_0^\infty e^{t\Delta}f(x)\,dt.$$

This is the semigroup formula for the inverse of the Laplacian and it is of significant importance. What we have just done here is to justify the fact that we can replace λ by the positive operator $-\Delta$ in the numerical formula (3.6).

Theorem 3.3 (Inverse of the Laplacian – Fundamental solution). *Let $f \in L_c^\infty(\mathbb{R}^n)$. Suppose in addition that, when $n = 2$, f has zero mean as in* (3.10). *Define*

$$u(x) = (-\Delta)^{-1}f(x) = \int_0^\infty e^{t\Delta}f(x)\,dt$$

for $x \in \mathbb{R}^n$. Then

$$u(x) = K * f(x) = \int_{\mathbb{R}^n} K(x-z)f(z)\,dz \tag{3.11}$$

where $K \in L^1_{\mathrm{loc}}(\mathbb{R}^n)$ is the fundamental solution of the Laplacian *given by*

$$K(x) = \begin{cases} \dfrac{\Gamma(n/2-1)}{4\pi^{n/2}}\cdot\dfrac{1}{|x|^{n-2}} & \text{if } n \geq 3 \\ -\dfrac{1}{2\pi}\cdot\log|x| & \text{if } n = 2. \end{cases} \tag{3.12}$$

Furthermore,

$$\Delta K(x) = 0 \qquad \text{for every } x \in \mathbb{R}^n,\ x \neq 0.$$

Moreover, if $f \in C_c^2(\mathbb{R}^n)$ and u is defined as in (3.11), *then $u \in C^2(\mathbb{R}^n)$ and*

$$-\Delta u(x) = f(x) \qquad \text{for every } x \in \mathbb{R}^n. \tag{3.13}$$

In addition, if $f \in C_c^\infty(\mathbb{R}^n)$ then $u \in C^\infty(\mathbb{R}^n)$ and $D^\beta u = K * D^\beta f$, for all $\beta \in \mathbb{N}_0^n$.

Proof. Let us begin with the case of dimension $n \geq 3$. We first see that u is well defined. Since f is bounded, by Theorem 3.2(3),

$$\int_0^1 |e^{t\Delta}f(x)|\,dt \leq \|f\|_{L^\infty(\mathbb{R}^n)}$$

for every $x \in \mathbb{R}^n$. Observe next that, as $f \in L^1(\mathbb{R}^n)$,

$$|e^{t\Delta}f(x)| \leq \frac{1}{(4\pi t)^{n/2}}\int_{\mathbb{R}^n} e^{-|x-z|^2/(4t)}|f(z)|\,dz \leq \frac{C_n}{t^{n/2}}\|f\|_{L^1(\mathbb{R}^n)}$$

for any $t > 0$ and every $x \in \mathbb{R}^n$. Then,

$$\int_1^\infty |e^{t\Delta} f(x)|\, dt \leq C \int_1^\infty \frac{1}{t^{n/2}}\, dt < \infty$$

because $n \geq 3$. Hence the integral defining $u(x)$ is absolutely convergent. Now, by writing down the formula for the heat kernel (3.4) and interchanging the integrals,

$$\begin{aligned} u(x) &= \int_0^\infty e^{t\Delta} f(x)\, dt \\ &= \int_0^\infty \frac{1}{(4\pi t)^{n/2}} \int_{\mathbb{R}^n} e^{-|x-z|^2/(4t)} f(z)\, dz\, dt \\ &= \int_{\mathbb{R}^n} \left[\frac{1}{4\pi^{n/2}} \int_0^\infty \frac{e^{-|x-z|^2/(4t)}}{(4t)^{n/2-1}} \frac{dt}{t} \right] f(z)\, dz \\ &= \int_{\mathbb{R}^n} K(x-z) f(z)\, dz. \end{aligned} \tag{3.14}$$

For the computation of the kernel, we perform the change of variables $r = |x|^2/(4t)$, which gives $dt/t = -dr/r$, to get

$$\begin{aligned} K(x) &= \frac{1}{4\pi^{n/2}} \int_0^\infty \frac{e^{-|x|^2/(4t)}}{(4t)^{n/2-1}} \frac{dt}{t} \\ &= \frac{1}{4\pi^{n/2}} \cdot \frac{1}{|x|^{n-2}} \int_0^\infty e^{-r} r^{n/2-1} \frac{dr}{r} \\ &= \frac{\Gamma(n/2-1)}{4\pi^{n/2}} \cdot \frac{1}{|x|^{n-2}}. \end{aligned}$$

The interchange of integrals we did in (3.14) is justified because of the initial computations in this proof, or because now we know that the kernel $K(x)$ is a nonnegative locally integrable function and $f \in L_c^\infty(\mathbb{R}^n)$.

Let us next consider the case of dimension $n = 2$. We cannot repeat the same computation because if $n = 2$ then the integral in the variable t in brackets appearing in (3.14) is divergent at infinity. Instead, we use the fact that f has zero mean. By the change of variables $s = 1/(4t)$, for which $dt/t = -ds/s$,

$$\begin{aligned} u(x) &= \int_0^\infty e^{t\Delta} f(x)\, dt \\ &= \frac{1}{4\pi} \int_0^\infty \int_{\mathbb{R}^2} e^{-|z|^2/(4t)} f(x-z)\, dz\, \frac{dt}{t} \\ &= \frac{1}{4\pi} \int_0^\infty \int_{\mathbb{R}^2} e^{-s|z|^2} f(x-z)\, dz\, \frac{ds}{s} \\ &= \frac{1}{4\pi} \int_0^\infty \int_{\mathbb{R}^2} (e^{-s|z|^2} - \psi(s)) f(x-z)\, dz\, \frac{ds}{s} \end{aligned} \tag{3.15}$$

where $\psi(s)$ is *any* function of $s > 0$. We choose ψ in a convenient way so we can interchange the integrals. Let $\psi(s) = \chi_{(0,1)}(s)$. To check that the double integral in the last line of (3.15) is absolutely convergent, we write

$$\int_0^\infty \int_{\mathbb{R}^2} |e^{-s|z|^2} - \chi_{(0,1)}(s)||f(x-z)|\, dz\, \frac{ds}{s} = \int_{\mathbb{R}^2} |f(x-z)| \int_0^\infty |e^{-s|z|^2} - \chi_{(0,1)}(s)|\, \frac{ds}{s}.$$

We now compute by changing variables and integrating by parts:

$$\begin{aligned}\int_0^\infty |e^{-s|z|^2} - \chi_{(0,1)}(s)| \frac{ds}{s} &= \int_0^1 (1 - e^{-s|z|^2}) \frac{ds}{s} + \int_1^\infty e^{-s|z|^2} \frac{ds}{s} \\ &= \int_0^{|z|^2} (1 - e^{-r}) \frac{dr}{r} + \int_{|z|^2}^\infty e^{-r} \frac{dr}{r} \\ &= \int_0^{|z|^2} (1 - e^{-r}) \frac{d}{dr}(\log r) \, dr + \int_{|z|^2}^\infty e^{-r} \frac{d}{dr}(\log r) \, dr \\ &= (1 - e^{-|z|^2}) \log(|z|^2) - \int_0^{|z|^2} e^{-r} \log r \, dr \\ &\quad - e^{-|z|^2} \log(|z|^2) + \int_{|z|^2}^\infty e^{-r} \log r \, dr \\ &\leq C|\log|z|| + \int_0^\infty e^{-r} |\log r| \, dr \leq C + C|\log|z||.\end{aligned} \tag{3.16}$$

Since $C + C|\log|z|| \in L^1_{\text{loc}}(\mathbb{R}^2)$ and $f \in L^\infty_c(\mathbb{R}^2)$, we get that the last double integral in (3.15) is absolutely convergent. Then, by Fubini's theorem,

$$u(x) = \frac{1}{4\pi} \int_{\mathbb{R}^2} f(x - z) F(z) \, dz$$

where, by following (3.16)

$$\begin{aligned}F(z) &= \int_0^1 (e^{-s|z|^2} - 1) \frac{ds}{s} + \int_1^\infty e^{-s|z|^2} \frac{ds}{s} \\ &= (e^{-|z|^2} - 1) \log(|z|^2) + \int_0^{|z|^2} e^{-r} \log r \, dr \\ &\quad - e^{-|z|^2} \log(|z|^2) + \int_{|z|^2}^\infty e^{-r} \log r \, dr \\ &= -\log(|z|^2) + \int_0^\infty e^{-r} \log r \, dr \\ &= -2 \log|z| - \gamma.\end{aligned}$$

The constant $\gamma \approx 0.577215$ is the Euler–Mascheroni constant. Therefore, by recalling that f has zero mean, we finally get

$$u(x) = \frac{1}{4\pi} \int_{\mathbb{R}^2} f(x - z) F(z) \, dz = -\frac{1}{2\pi} \int_{\mathbb{R}^2} f(x - z) \log|z| \, dz.$$

The fact that $K(x)$ is a harmonic function outside of the origin is a simple verification.

Suppose next that $f \in C^2_c(\mathbb{R}^n)$ and define $u(x) = K * f(x)$. For $h \neq 0$, and $i = 1, \ldots, n$ consider the incremental quotient

$$\frac{u(x + he_i) - u(x)}{h} = \int_{\mathbb{R}^n} K(z) \frac{f(x - z + he_i) - f(x - z)}{h} \, dz.$$

Notice that ∇f is uniformly continuous on $\mathbb{R}^n$. Then, for any $\varepsilon > 0$ and every h sufficiently small, by the mean value theorem,

$$\left|\frac{f(x-z+he_i)-f(x-z)}{h} - \partial_{x_i} f(x-z)\right| = |\partial_{x_i} f(\eta) - \partial_{x_i} f(x-z)| < \varepsilon$$

where η is an intermediate point between $x-z+he_i$ and $x-z$. This says that

$$\frac{f(x-z+he_i)-f(x-z)}{h} \to \partial_{x_i} f(x-z) \qquad \text{uniformly in } \mathbb{R}^n \text{ as } h \to 0.$$

Therefore we can take the limit as $h \to 0$ inside the integral and get

$$\partial_{x_i} u(x) = \int_{\mathbb{R}^n} K(z)\partial_{x_i} f(x-z)\,dz.$$

In a similar way,

$$D^2 u(x) = \int_{\mathbb{R}^n} K(z) D^2 f(x-z)\,dz.$$

As the integral above is a continuous function of x, we get $u \in C^2(\mathbb{R}^n)$. Observe that these arguments imply that if $f \in C_c^\infty(\mathbb{R}^n)$, then $u \in C^\infty(\mathbb{R}^n)$ and $D^\beta u = K * D^\beta f$, for any multi-index $\beta \in \mathbb{N}_0^n$.

In order to prove (3.13), we use integration by parts twice to write

$$\begin{aligned}
-\Delta u(x) &= -\int_{\mathbb{R}^n} K(z)\Delta_x f(x-z)\,dz = -\int_{\mathbb{R}^n} K(z)\Delta_z f(x-z)\,dz \\
&= -\lim_{\varepsilon\to 0}\int_{|z|>\varepsilon} K(z)\Delta_z f(x-z)\,dz \\
&= \lim_{\varepsilon\to 0}\left[\int_{|z|>\varepsilon} \nabla K(z)\nabla f(x-z)\,dz - \int_{|z|=\varepsilon} K(z)\partial_\nu f(x-z)\,dS\right] \\
&= \lim_{\varepsilon\to 0}\left[-\int_{|z|>\varepsilon} \Delta K(z) f(x-z)\,dz\right. \\
&\qquad \left. + \int_{|z|=\varepsilon} \partial_\nu K(z) f(x-z)\,dS - \int_{|z|=\varepsilon} K(z)\partial_\nu f(x-z)\,dS\right] \\
&= \lim_{\varepsilon\to 0}\left[\int_{|z|=\varepsilon} \partial_\nu K(z) f(x-z)\,dS - \int_{|z|=\varepsilon} K(z)\partial_\nu f(x-z)\,dS\right].
\end{aligned}$$

Here ∂_ν denotes the exterior normal derivative on the boundary of $\mathbb{R}^n \setminus B_\varepsilon$, or, which is the same, the interior normal derivative on the boundary of B_ε. We claim that

$$\lim_{\varepsilon\to 0}\int_{|z|=\varepsilon} \partial_\nu K(z) f(x-z)\,dS = f(x)$$

and

$$\lim_{\varepsilon\to 0}\int_{|z|=\varepsilon} K(z)\partial_\nu f(x-z)\,dS = 0$$

from which the identity $-\Delta u(x) = f(x)$ follows. We only consider the case $n \geq 3$ because for $n = 2$ analogous arguments work. On one hand, by using the value of the measure of the sphere of radius $R = \varepsilon$, see (2.7),

$$\begin{aligned}\int_{|z|=\varepsilon} \partial_\nu K(z) f(x-z)\, dS &= -\int_{|z|=\varepsilon} \nabla K(z) \cdot \frac{z}{\varepsilon} f(x-z)\, dS \\ &= \frac{\Gamma(n/2-1)}{4\pi^{n/2}}(n-2) \int_{|z|=\varepsilon} \frac{z}{|z|^n} \cdot \frac{z}{\varepsilon} f(x-z)\, dS \\ &= \frac{\Gamma(n/2)}{2\pi^{n/2}\varepsilon^{n-1}} \int_{|z|=\varepsilon} f(x-z)\, dS \\ &= \frac{1}{|\partial B_\varepsilon|} \int_{\partial B_\varepsilon} f(x-z)\, dS \to f(x)\end{aligned}$$

as $\varepsilon \to 0$, since f is continuous at x. On the other hand,

$$\begin{aligned}\int_{|z|=\varepsilon} |K(z)||\partial_\nu f(x-z)|\, dS &\leq C_n \|\nabla f\|_{L^\infty(\mathbb{R}^n)} \int_{|z|=\varepsilon} \frac{1}{|z|^{n-2}}\, dS \\ &\leq C_n \|\nabla f\|_{L^\infty(\mathbb{R}^n)} \frac{\varepsilon^{n-1}}{\varepsilon^{n-2}} = C\varepsilon \to 0\end{aligned}$$

as $\varepsilon \to 0$. This completes the proof. □

Let us stress the importance of the proof of Theorem 3.3 and the method of semigroups. There are at least two classical ways of obtaining the fundamental solution of the Laplacian. One is by noticing that the Laplacian is invariant under rotations, which suggests to look for radially symmetric solutions. The special symmetries of the Laplacian come into play. With this approach it is not clear how to choose the constants in (3.12). The other one is by taking the inverse Fourier transform of the locally integrable Fourier multiplier $|\xi|^{-2}$. This requires a nontrivial use of the symmetry properties of the Fourier transform as well as a justification of the computation using distributions because $|\xi|^{-2}$ is just a locally integrable function if $n \geq 3$. The computation for $n = 2$ seems to be even more complex because the local integrability of the multiplier is lost. However, with the semigroup approach presented here, the use of explicit symmetries of the Laplacian or the Fourier transform is much avoided. Instead, we need to know the heat kernel. On top of that, the constants are explicitly computed. Unlike the previous two approaches, the semigroup method can be applied in situations where the Fourier transform is not available (variable coefficients, bounded domains, manifolds, etc.)

3.5 SOLVABILITY OF THE POISSON EQUATION

From now on we will only focus in the case of dimension $n \geq 3$. We will find minimal conditions on f such that $-\Delta u = f$ can be solved classically.

Using the divergence theorem, we have seen in Theorem 3.3 that if $u = K * f$, where K is as in (3.12), and $f \in C_c^2(\mathbb{R}^n)$ then $-\Delta u = f$ in $\mathbb{R}^n$. If f were not C^2, one might be tempted to compute

$$\Delta u(x) = \int_{\mathbb{R}^n} \Delta K(x-z) f(z)\, dz$$

which is obviously incorrect. First, $\Delta K(x-z)=0$ for every $z\neq x$, which would give $\Delta u=0$. Secondly,

$$|D^2K(x)|\leq \frac{C_n}{|x|^n}$$

which is not locally integrable, so we cannot differentiate under the integral sign. We will see how the Hölder continuity of f comes naturally into play to be able to prove that $-\Delta u = f$.

The fundamental solution of the Laplacian $K(x)$ obtained in (3.12) is a locally integrable function. Therefore, the expression

$$u(x)=\int_{\mathbb{R}^n} K(x-z)f(z)\,dz$$

has no sense for f just bounded. Something must be assumed about the integrability or decay of f at infinity. The most simple situation is to suppose that f has compact support as we have done in Theorem 3.3. As this is a book on regularity theory, we will not care about the most general statements, but about the ones that manifest the ideas. Our first result establishes that for f bounded with compact support we can always compute $\nabla(K*f)=(\nabla K)*f$.

Lemma 3.4. *Let f be a bounded function supported on a bounded domain Ω. Define*

$$u(x)=(-\Delta)^{-1}f(x)=\int_\Omega K(x-z)f(z)\,dz$$

for $x\in\mathbb{R}^n$. Then $u\in C^1(\mathbb{R}^n)$ and

$$\nabla u(x)=\int_\Omega \nabla K(x-z)f(z)\,dz. \tag{3.17}$$

Moreover, there is a constant $C>0$ depending only on n such that

$$\|u\|_{L^\infty(\mathbb{R}^n)}+\|\nabla u\|_{L^\infty(\mathbb{R}^n)}\leq C(1+|\Omega|)\|f\|_{L^\infty(\Omega)}.$$

Proof. Let us first verify the estimates for u and ∇u. By passing to polar coordinates,

$$\begin{aligned}|u(x)| &\leq C\Big[\int_{|x-z|<1}\frac{|f(z)|}{|x-z|^{n-2}}\,dz+\int_{|x-z|\geq 1}\frac{|f(z)|}{|x-z|^{n-2}}\,dz\Big]\\ &\leq C\|f\|_{L^\infty(\Omega)}\int_0^1\frac{r^{n-1}}{r^{n-2}}\,dz + C\int_\Omega |f(z)|\,dz\\ &\leq C(1+|\Omega|)\|f\|_{L^\infty(\Omega)}\end{aligned}$$

for any $x\in\mathbb{R}^n$, where C depends only on n. Suppose that (3.17) is true. Since

$$|\nabla K(x-z)|\leq \frac{C}{|x-z|^{n-1}}$$

for all $x\neq z$, we can repeat the estimate we just did and see that $\|\nabla u\|_{L^\infty(\mathbb{R}^n)}\leq C(1+|\Omega|)\|f\|_{L^\infty(\Omega)}$.

Let v be the right hand side of (3.17):

$$v(x) := \int_\Omega \nabla K(x-z) f(z)\, dz.$$

We are only left to show that u is $C^1(\mathbb{R}^n)$ and that $\nabla u = v$. To that end we need to cut away the singularity at $z = x$ in the definition of u in a smooth way to be able to introduce the derivates inside the integral. For this we pick $\eta \in C^1(\mathbb{R})$ such that

$$0 \le \eta \le 1, \qquad 0 \le \eta' \le 2, \qquad \eta = 0 \text{ for } t \le 1, \qquad \eta = 1 \text{ for } t \ge 2.$$

For $\varepsilon > 0$ we let

$$u_\varepsilon(x) = \int_\Omega \eta(|x-z|/\varepsilon) K(x-z) f(z)\, dz.$$

Notice that $\eta(|x-z|/\varepsilon) = 0$ whenever $|x-z| < \varepsilon$. It is easy to check that u_ε is a smooth function and we can compute its derivatives by taking the derivatives inside the integral sign. The first observation is that u_ε converges uniformly to u on $\mathbb{R}^n$. Indeed, by using polar coordinates,

$$\begin{aligned}
|u_\varepsilon(x) - u(x)| &\le C\|f\|_{L^\infty(\Omega)} \int_\Omega (\eta(|x-z|/\varepsilon) - 1) K(x-z)\, dz \\
&\le C\|f\|_{L^\infty(\Omega)} \int_{|x-z|<2\varepsilon} \frac{1}{|x-z|^{n-2}}\, dz \\
&= C\|f\|_{L^\infty(\Omega)} \int_0^{2\varepsilon} r\, dr = C_n \|f\|_{L^\infty(\Omega)} \varepsilon^2
\end{aligned}$$

uniformly in $x \in \mathbb{R}^n$. The second observation is that ∇u_ε converges to v uniformly on $\mathbb{R}^n$. For any $x \in \mathbb{R}^n$,

$$\begin{aligned}
|\nabla u_\varepsilon(x) - v(x)| &= \left| \int_\Omega \nabla\Big[(\eta(|x-z|/\varepsilon) - 1) K(x-z)\Big] f(z)\, dz \right| \\
&\le \|f\|_{L^\infty(\Omega)} \int_{|x-z|<2\varepsilon} \left(\tfrac{2}{\varepsilon}|K(x-z)| + |\nabla K(x-z)|\right) dz \\
&\le C\|f\|_{L^\infty(\Omega)} \int_{|x-z|<2\varepsilon} \left(\frac{1}{\varepsilon |x-z|^{n-2}} + \frac{1}{|x-z|^{n-1}} \right) dz \\
&= C_n \|f\|_{L^\infty(\Omega)} \varepsilon.
\end{aligned}$$

In conclusion, u_ε and ∇u_ε converge uniformly on $\mathbb{R}^n$ to u and v, respectively. Then u is C^1 and $\nabla u = v$, namely, (3.17) holds. □

Let f be a bounded function supported on a bounded domain Ω. Lemma 3.4 says that the computation

$$\nabla u(x) = \nabla(-\Delta)^{-1} f(x) = \int_\Omega \nabla K(x-z) f(z)\, dz$$

is correct. As we pointed out before, the identity

$$D^2 u(x) = D^2(-\Delta)^{-1} f(x) = \int_\Omega D^2 K(x-z) f(z)\, dz$$

is incorrect due to the singularity $|D^2K(x)| \sim |x|^{-n}$ near $x = 0$. One way to compensate this is to ask for some regularity of f at x, in order to be able to write

$$D^2u(x) = \int_\Omega D^2K(x-z)(f(z) - f(x))\,dz + f(x)\int_\Omega D^2K(x-z)\,dz. \tag{3.18}$$

The first term above is well defined if f is, for instance, Hölder continuous of order $0 < \alpha \leq 1$ at $x \in \Omega$:

$$\int_\Omega |D^2K(x-z)||f(z) - f(x)|\,dz \leq C_n[f]_{C^\alpha(x)} \int_\Omega \frac{|x-z|^\alpha}{|x-z|^n}\,dz < \infty.$$

In fact, this computation only requires f to have a modulus of continuity $\omega(r)$ that compensates r^{-n} in an integrable way at $r = 0$, namely, f should be at least Dini continuous:

$$\int_0^1 \frac{\omega(r)}{r}\,dr < \infty.$$

The new problem now comes from the second term in (3.18), which is not finite. But, if we could use the divergence theorem, a cancelation would appear:

$$\begin{aligned}\int_\Omega \partial_{x_i x_j} K(x-z)\,dz &= \int_\Omega \partial_{z_i z_j} K(x-z)\,dz \\ &= \int_{\partial\Omega} \partial_{z_i} K(x-z)\nu_j\,dS = -\int_{\partial\Omega} \partial_{x_i} K(x-z)\nu_j\,dS.\end{aligned}$$

The last integral is well defined. All these considerations lead to the following result.

Theorem 3.5 (Solvabiliy of the Poisson problem in a bounded domain and representation of second derivatives). *Let f be a bounded function supported on a bounded domain Ω. Suppose that $f \in C^\alpha_{\mathrm{loc}}(\Omega)$ for some $0 < \alpha < 1$. Let*

$$u(x) = (-\Delta)^{-1} f(x) = \int_\Omega K(x-z)f(z)\,dz.$$

Then $u \in C^2(\Omega)$ and, for every $x \in \Omega$,

$$\partial_{ij}u(x) = \int_{\Omega_0} \partial_{ij}K(x-z)(f(z) - f(x))\,dz - f(x)\int_{\partial\Omega_0} \partial_i K(x-z)\nu_j(z)\,dS_z$$

where Ω_0 is a any bounded domain such that $\Omega \subseteq \Omega_0$ and for which the divergence theorem holds. In particular,

$$-\Delta u(x) = f(x) \qquad \text{for every } x \in \Omega.$$

Proof. As we have already noticed, the function

$$w(x) := \int_{\Omega_0} \partial_{ij}K(x-z)(f(z) - f(x))\,dz - f(x)\int_{\partial\Omega_0} \partial_i K(x-z)\nu_j(z)\,dS_z$$

is well defined for every $x \in \Omega$ because $f \in C^{\alpha}_{\rm loc}(\Omega) \cap L^{\infty}(\mathbb{R}^n)$ and $\operatorname{dist}(x, \partial\Omega_0) > 0$. As in the proof of Lemma 3.4 we consider the function $\eta \in C^1(\mathbb{R})$ that will help us cut away the singularities. Let

$$v(x) := \partial_i u(x) = \int_\Omega \partial_i K(x-z) f(z)\, dz$$

for $x \in \mathbb{R}^n$ and for some $i = 1, \dots, n$. Then v is well defined because of Lemma 3.4. As in the proof of that lemma, we define, for every $\varepsilon > 0$,

$$v_\varepsilon(x) := \int_\Omega \eta(|x-z|/\varepsilon)\partial_i K(x-z) f(z)\, dz.$$

Then, $v_\varepsilon \in C^1(\Omega)$ and converges to $v = \partial_i u$ uniformly in compact subsets of Ω. For any $x \in \Omega$ such that $0 < \varepsilon < \operatorname{dist}(x, \partial\Omega)/2$,

$$\begin{aligned}
\partial_j v_\varepsilon(x) &= \int_{\Omega_0} \partial_j(\eta(|x-z|/\varepsilon)\partial_i K(x-z)) f(z)\, dz \\
&= \int_{\Omega_0} \partial_j(\eta(|x-z|/\varepsilon)\partial_i K(x-z))(f(z)-f(x))\, dz \\
&\quad - f(x)\int_{\Omega_0} \partial_{z_j}(\eta(|x-z|/\varepsilon)\partial_i K(x-z))\, dz \\
&= \int_{\Omega_0} \partial_j(\eta(|x-z|/\varepsilon)\partial_i K(x-z))(f(z)-f(x))\, dz \\
&\quad - f(x)\int_{\partial\Omega_0} \partial_i K(x-z)\nu_j(z)\, dz
\end{aligned}$$

where in the last line we used the divergence theorem and the fact that $\eta(|x-z|/\varepsilon) = 1$ because $x \in \Omega$, $z \in \partial\Omega_0$ and $\varepsilon < \operatorname{dist}(x, \partial\Omega)/2$, so that $|x-z|/\varepsilon > 2$. With this,

$$\begin{aligned}
|\partial_j v_\varepsilon(x) - w(x)| &= \left| \int_\Omega \partial_j\Big[(\eta(|x-z|/\varepsilon) - 1)\partial_i K(x-z)\Big](f(z)-f(x))\, dz \right| \\
&\le \int_{|x-z|<2\varepsilon} \Big(\tfrac{2}{\varepsilon}|\nabla K(x-z)| + |D^2 K(x-z)|\Big)|f(z)-f(x)|\, dz \\
&\le C[f]_{C^\alpha(x)} \int_{|x-z|<2\varepsilon} \left(\frac{1}{\varepsilon|x-z|^{n-1}} + \frac{1}{|x-z|^n}\right)|x-z|^\alpha\, dz \\
&= C_{n,\alpha}[f]_{C^\alpha(x)}\varepsilon^\alpha
\end{aligned}$$

provided $2\varepsilon < \operatorname{dist}(x, \partial\Omega)$. Therefore $\partial_j v_\varepsilon$ converges to w uniformly on compact subsets of Ω. As v_ε converges to $\partial_i u$ uniformly on compact subsets of Ω, we conclude that $u \in C^2(\Omega)$ and that $\partial_{ij} u = w$.

For fixed $x \in \Omega$, let us take $\Omega_0 = B_R(x)$ for some $R > 0$ sufficiently large. Then, using that $s\Gamma(s) = \Gamma(s+1)$, for $s > 0$,

$$\begin{aligned}
-\Delta u(x) &= -\int_{B_R(x)} \Delta K(x-z)(f(z)-f(x))\,dz \\
&\quad + f(x)\sum_{i=1}^{n}\int_{\partial B_R(x)} \partial_i K(x-z)\nu_i(z)\,dS_z \\
&= f(x)\frac{\Gamma(n/2-1)}{4\pi^{n/2}}\sum_{i=1}^{n}\int_{\partial B_R(x)} \partial_i\left(|x-z|^{2-n}\right)\frac{(z_i-x_i)}{R}\,dS_z \\
&= f(x)\frac{\Gamma(n/2-1)}{4\pi^{n/2}R^{n+1}}n(n-2)\int_{\partial B_R(x)} (z_1-x_1)^2\,dS_z \\
&= f(x)\frac{\Gamma(n/2+1)}{\pi^{n/2}}\int_{\partial B_1} z_1^2\,dS_z.
\end{aligned}$$

The conclusion follows by noticing that, by using polar coordinates,

$$\int_{B_1} (x_1)^2\,dx = \frac{1}{n+2}\int_{\partial B_1} z_1^2\,dS_z = \frac{\pi^{n/2}}{2\Gamma(n/2+2)}. \tag{3.19}$$

□

We have completed the 3 steps of our program delineated at the end of Section 3.2, constructing a solution to $-\Delta u = f$ for $f \in C^\alpha$. From Theorem 3.5 some simple consequences can now be drawn.

Corollary 3.6. *Let f be a function supported on a bounded domain Ω and $0 < \alpha < 1$.*

(1) (Existence and uniqueness for the Dirichlet problem). *Suppose that for any given $g \in C(\partial\Omega)$, there is a solution $v \in C(\overline{\Omega})$ to*

$$\begin{cases} \Delta v = 0 & \text{in } \Omega \\ v = g & \text{on } \partial\Omega. \end{cases} \tag{3.20}$$

In other words, suppose that Ω is a regular domain *for the Laplacian. Then, if $f \in L^\infty(\Omega) \cap C^\alpha_{\mathrm{loc}}(\Omega)$, the Dirichlet problem*

$$\begin{cases} -\Delta u = f & \text{in } \Omega \\ u = g & \text{on } \partial\Omega \end{cases} \tag{3.21}$$

has a unique classical solution $u \in C^2(\Omega) \cap C(\overline{\Omega})$, for any $g \in C(\partial\Omega)$.

(2) (Representation of general solutions). *Suppose that $u \in C^2(\Omega)$ and $f \in L^\infty(\Omega) \cap C^\alpha_{\mathrm{loc}}(\Omega)$ satisfy*

$$-\Delta u = f \qquad \text{in } \Omega.$$

Then, for every $x \in \Omega$,

$$u(x) = v(x) + (-\Delta)^{-1} f(x)$$

where v is harmonic in Ω.

(3) (Representation of global solutions). *Let $f \in C_c^\alpha(\Omega)$ and $u \in L^\infty(\mathbb{R}^n)$ such that*

$$-\Delta u = f \qquad \text{in } \mathbb{R}^n.$$

Then, for every $x \in \mathbb{R}^n$,

$$u(x) = (-\Delta)^{-1} f(x) + C$$

for some constant $C \in \mathbb{R}$.

Proof. For (1), let $\widetilde{u} = (-\Delta)^{-1} f$. Then $\widetilde{u} \in C^1(\mathbb{R}^n) \cap C^2(\Omega)$ and $-\Delta \widetilde{u} = f$ in Ω. Fix any $g \in C(\partial\Omega)$. Let $v \in C(\overline{\Omega})$ be the unique solution to

$$\begin{cases} \Delta v = 0 & \text{in } \Omega \\ v = g - \widetilde{u} & \text{on } \partial\Omega \end{cases}$$

which exists by assumption because $g - \widetilde{u} \in C(\partial\Omega)$. It follows that $u = v + \widetilde{u}$ is the unique classical solution to (3.21).

Let u be as in (2). Then $v = u - (-\Delta)^{-1} f$ is harmonic in Ω.

To prove (3), we have seen that if $\widetilde{u}(x) = (-\Delta)^{-1} f(x)$ then $\widetilde{u}$ is bounded on $\mathbb{R}^n$ (see Lemma 3.4) and $-\Delta \widetilde{u} = f$ on $\mathbb{R}^n$. Hence $v = u - \widetilde{u}$ is bounded and harmonic. By the Liouville theorem (see Theorem 2.14), v is constant and $u = \widetilde{u} + C$. □

Remark 3.7 (Regular domains for the Laplacian). The question of which domains are regular for the Laplacian has been widely studied. We will not focus on giving an answer to it in full generality. Theorem 2.2 says that we can always solve the Dirichlet problem (3.20) when Ω is a ball, and provides an explicit integral formula for v in terms of g. On the other hand, we say that a bounded domain Ω satisfies an **exterior cone condition** if every point of its boundary is the vertex of a circular cone opening outside of Ω or, more precisely, if for every $x \in \partial\Omega$ there exists a finite circular cone K with vertex at x such that $\overline{K} \cap \overline{\Omega} = \{x\}$. It can be proved that if Ω is bounded and satisfies an exterior cone condition at every $x \in \partial\Omega$, then for every $g \in C(\partial\Omega)$ we can ensure the existence and uniqueness of the solution $v \in C(\overline{\Omega})$ to (3.20). The solution is constructed by using **Perron's method**: v is the largest subharmonic function that is below g at the boundary:

$$v(x) = \sup \{u(x) : \Delta u \geq 0 \text{ in } \Omega, u \leq g \text{ on } \partial\Omega\}.$$

The exterior cone condition allows to prove that $v = g$ on $\partial\Omega$. Examples of domains satisfying the exterior cone condition are Lipschitz domains.

Conclusions. Let Ω be a bounded domain and let $f \in L^\infty(\Omega) \cap C^\alpha_{\text{loc}}(\Omega)$, for some $0 < \alpha \leq 1$.

(A) There always exists $u \in C^2(\Omega)$ such that

$$-\Delta u = f \qquad \text{in } \Omega.$$

Observe that here we are not imposing any regularity assumptions on $\partial\Omega$.

(B) If, in addition, we are given $g \in C(\partial\Omega)$ and Ω is a regular domain for the Laplacian (for example, Ω is a Lipschitz domain) then there exists a unique solution $u \in C^2(\Omega) \cap C(\overline{\Omega})$ to

$$\begin{cases} -\Delta u = f & \text{in } \Omega \\ u = g & \text{on } \partial\Omega. \end{cases}$$

3.6 SCHAUDER ESTIMATES BY REPRESENTATION FORMULAS

In this section we use the representation formula for D^2u found in Theorem 3.5 to show that if $f \in C^\alpha$ and $u = (-\Delta)^{-1}f$, then $u \in C^{2,\alpha}$ locally. In general, a C^2 solution v to $-\Delta v = f$ is not necessarily represented as a convolution with the fundamental solution, but differs from u only by a harmonic function, see Corollary 3.6(2). By using interior derivative estimates for harmonic functions, we then deduce the Schauder estimates for v.

Theorem 3.8 (Schauder estimates by representation). *Suppose that $f \in C^{0,\alpha}(B_2)$ for some $0 < \alpha < 1$. Let*

$$u(x) = (-\Delta)^{-1}f(x) = \int_{B_2} K(x-z)f(z)\,dz \qquad x \in B_2.$$

Then $u \in C^{2,\alpha}(B_1)$ and there exists a constant $C > 0$ depending only on n and α such that

$$\|D^2u\|_{L^\infty(B_1)} + [D^2u]_{C^\alpha(B_1)} \le C\|f\|_{C^{0,\alpha}(B_2)}.$$

Proof. For any $x \in B_1$, by Theorem 3.5, we have the representation

$$\partial_{ij}u(x) = \int_{B_2} \partial_{ij}K(x-z)(f(z)-f(x))\,dz - f(x)\int_{\partial B_2} \partial_i K(x-z)\nu_j(z)\,dS_z.$$

Therefore,

$$\begin{aligned} |D^2u(x)| &\le C[f]_{C^\alpha(x)} \int_{B_2} \frac{1}{|x-z|^{n-\alpha}}\,dz + C|f(x)| \int_{\partial B_2} \frac{1}{|x-z|^{n-1}}\,dS_z \\ &\le C_{n,\alpha}[f]_{C^\alpha(x)} + C_n|f(x)|. \end{aligned}$$

Let $x_1, x_2 \in B_1$, define $\rho = |x_1 - x_2|$ and the intermediate point $\bar{x} = \frac{1}{2}(x_1 + x_2)$. We begin by writing

$$\begin{aligned}
\partial_{ij}u(x_1) - \partial_{ij}u(x_2) &= \int_{B_\rho(\bar{x})} \partial_{ij}K(x_1 - z)(f(z) - f(x_1))\,dz \\
&\quad - f(x_1)\int_{\partial B_2} \partial_i K(x_1 - z)\nu_j(z)\,dS_z \\
&\quad + \int_{B_2\setminus B_\rho(\bar{x})} \partial_{ij}K(x_1 - z)(f(z) - f(x_1))\,dz \\
&\quad - \int_{B_\rho(\bar{x})} \partial_{ij}K(x_2 - z)(f(z) - f(x_2))\,dz \\
&\quad + f(x_2)\int_{\partial B_2} \partial_i K(x_2 - z)\nu_j(z)\,dS_z \\
&\quad - \int_{B_2\setminus B_\rho(\bar{x})} \partial_{ij}K(x_2 - z)(f(z) - f(x_2))\,dz \\
&= I_1 - I_2 + I_3 - I_4 + I_5 - I_6.
\end{aligned}$$

The term I_1 involves an integral over $B_\rho(\bar{x})$, where z is close to x_1. Therefore, it is estimated by using the Hölder continuity of f, the size estimate for D^2K and the fact that $B_\rho(\bar{x}) \subset B_{3\rho/2}(x_1)$:

$$|I_1| \le C[f]_{C^\alpha(x_1)} \int_{B_{3\rho/2}(x_1)} \frac{1}{|x_1 - z|^{n-\alpha}}\,dz = C_{n,\alpha}[f]_{C^\alpha(x_1)}|x_1 - x_2|^\alpha.$$

Similarly, I_4 is estimated using that $f \in C^\alpha(x_2)$ and $B_\rho(\bar{x}) \subset B_{3\rho/2}(x_2)$:

$$|I_4| \le C_{n,\alpha}[f]_{C^\alpha(x_2)}|x_1 - x_2|^\alpha.$$

Next, we consider the difference of boundary terms $I_5 - I_2$. We add and subtract to it the cross-term

$$f(x_2)\int_{\partial B_2} \partial_i K(x_1 - y)\nu_j(z)\,dS_z$$

and end up with

$$\begin{aligned}
I_5 - I_2 &= (f(x_2) - f(x_1))\int_{\partial B_2} \partial_i K(x_1 - z)\nu_j(z)\,dS_z \\
&\quad + f(x_2)\int_{\partial B_2} (\partial_i K(x_2 - z) - \partial_i K(x_1 - z))\nu_j(z)\,dS_z \\
&= J_1 + J_2.
\end{aligned}$$

In J_1 we use the Hölder continuity of f, the size estimate for ∇K and the fact that $|x_1 - z| \ge 1$ to bound

$$|J_1| \le C[f]_{C^\alpha(x_1)} \int_{\partial B_2} \frac{|x_1 - x_2|^\alpha}{|x_1 - z|^{n-1}}\,dS_z \le C_n[f]_{C^\alpha(x_1)}|x_1 - x_2|^\alpha.$$

For J_2, we need to use some regularity of ∇K. By the mean value theorem, for some intermediate point $\hat{x}$ between x_1 and x_2,

$$|\partial_i K(x_2 - z) - \partial_i K(x_1 - z)| \le |\nabla \partial_i K(\hat{x} - z)||x_1 - x_2| \le C_{n,\alpha}|x_1 - x_2|^\alpha$$

where in the second inequality we used the size estimate of D^2K and that $|\hat{x}-z| \geq 1$ for $z \in \partial B_2$, and $|x_1 - x_2|/2 < 1$. We conclude that

$$|J_2| \leq C_{n,\alpha}|f(x_2)||x_1 - x_2|^\alpha.$$

Thus,

$$|I_5 - I_2| \leq C_{n,\alpha}([f]_{C^\alpha(x_1)} + |f(x_2)|)|x_1 - x_2|^\alpha.$$

We are left to bound the difference $I_3 - I_6$ of the integrals over $B_2 \setminus B_\rho(\bar{x})$, where z is far from x_1 and x_2. As we did for the difference of the boundary integrals, we add and subtract the corresponding cross-term to obtain

$$\begin{aligned} I_3 - I_6 &= (f(x_2) - f(x_1)) \int_{B_2 \setminus B_\rho(\bar{x})} \partial_{ij}K(x_2 - z)\, dz \\ &\quad + \int_{B_2 \setminus B_\rho(\bar{x})} (\partial_{ij}K(x_1 - z) - \partial_{ij}K(x_2 - z))(f(z) - f(x_1))\, dz \\ &= J_3 + J_4. \end{aligned}$$

Observe that, by the divergence theorem,

$$\begin{aligned} \int_{B_2 \setminus B_\rho(\bar{x})} \partial_{ij}K(x_2 - z)\, dz &= \int_{\partial(B_2 \setminus B_\rho(\bar{x}))} \partial_i K(x_2 - z)\nu_j(z)\, dS \\ &= \int_{\partial B_2} \partial_i K(x_2 - z)\nu_j(z)\, dS_z + \int_{\partial B_\rho(\bar{x})} \partial_i K(x_2 - z)\nu_j(z)\, dS \\ &\leq C + \frac{C}{\rho^{n-1}}|\partial B_\rho(\bar{x})| \leq C_n. \end{aligned}$$

where we used the size bound for ∇K and the estimates $|x_2 - z| \geq 1$ for the first integral and $|x_2 - z| \geq \rho/2$ for the second one, respectively. With this,

$$|J_3| \leq C_n[f]_{C^\alpha(x_1)}|x_1 - x_2|^\alpha.$$

We estimate the integrand of J_4 by using the mean value theorem with some intermediate point $\hat{x}$ between x_1 and x_2, the estimate $D^3K(x) \leq \frac{C_n}{|x|^{n+1}}$, $x \neq 0$, and the fact that if $z \in B_2 \setminus B_\rho(\bar{x})$ then $|x_1 - z| \leq \frac{3}{2}|\bar{x} - z| \leq 3|\hat{x} - z|$:

$$\begin{aligned} &|(\partial_{ij}K(x_1 - z) - \partial_{ij}K(x_2 - z))(f(z) - f(x_1))| \\ &\qquad \leq |\nabla \partial_{ij}K(\hat{x} - z)||x_1 - x_2||f(z) - f(x_1)| \\ &\qquad \leq C[f]_{C^\alpha(x_1)}|x_1 - x_2|\frac{|x_1 - z|^\alpha}{|\hat{x} - z|^{n+1}} \\ &\qquad \leq C_{n,\alpha}[f]_{C^\alpha(x_1)}\frac{|x_1 - x_2|}{|\bar{x} - z|^{n+1-\alpha}}. \end{aligned}$$

Then

$$|J_4| \leq C[f]_{C^\alpha(x_1)} \int_{|\bar{x}-z|\geq\rho} \frac{|x_1 - x_2|}{|\bar{x} - z|^{n+1-\alpha}}\, dz = C_{n,\alpha}[f]_{C^\alpha(x_1)}|x_1 - x_2|^\alpha.$$

Thus

$$|I_3 - I_6| \leq C_{n,\alpha}[f]_{C^\alpha(x_1)}|x_1 - x_2|^\alpha.$$

We conclude by pasting together the estimates for I_1, I_4, $I_5 - I_2$ and $I_3 - I_6$ above. □

Corollary 3.9 (Interior Schauder estimates). *If $u \in C^2(B_2)$ and $f \in C^\alpha(B_2)$ satisfy the Poisson equation $-\Delta u = f$ in B_2 then $u \in C^{2,\alpha}(B_1)$ and there exists a constant $C > 0$ depending only on n and α such that*

$$\|u\|_{C^{2,\alpha}(B_1)} \le C\big(\|u\|_{L^\infty(B_2)} + \|f\|_{C^{0,\alpha}(B_2)}\big).$$

Proof. By Corollary 3.6(2), we can write $u = v + (-\Delta)^{-1}f$, where v is harmonic in B_2 and f is extended to be zero outside of B_2. By Lemma 3.4 and Theorem 3.8,

$$\|(-\Delta)^{-1}f\|_{C^{2,\alpha}(B_1)} \le C_{n,\alpha}\|f\|_{C^{0,\alpha}(B_2)}.$$

By interior derivative estimates for harmonic functions (see Corollary 2.15(1))

$$\|v\|_{C^{2,\alpha}(B_1)} \le C_{n,\alpha}\|v\|_{L^\infty(B_2)} \le C_{n,\alpha}\big(\|u\|_{L^\infty(B_2)} + \|f\|_{L^\infty(B_2)}\big).$$

Thus, the estimate for u follows. □

Remark 3.10 (Interior Schauder estimates in bounded domains). The passage from B_2 to B_1 in Theorem 3.8 and Corollary 3.9 is not restrictive in the sense that we can pass from any ball $B(x)$ to any smaller ball $B'(x) \subset\subset B(x)$, and the constants will depend on the distance between $B'(x)$ and $\partial B(x)$, see also Theorem 3.12. More generally, by using an argument as in the proof of Corollary 2.15(1), these results can also be stated with B_2 and B_1 replaced by a bounded domain Ω and a subdomain $\Omega' \subset\subset \Omega$, respectively. The constant in the $C^{2,\alpha}$ estimate will depend on the distance between Ω' and $\partial\Omega$.

Remark 3.11 (Global Schauder estimates). If $f \in C^\alpha(\mathbb{R}^n)$ has compact support in a bounded domain Ω and $u = (-\Delta)^{-1}f$, then the representation formula for $D^2u(x)$ of Theorem 3.5 holds for any $x \in \mathbb{R}^n$. Indeed, the arguments in the proof work for any $x \in \mathbb{R}^n$ because they only depend on the fact that f is supported in Ω. See also Theorem 4.25 in Chapter 4 for a more general result. Then, in this case, the proof of Theorem 3.8 can be done for any $x_1, x_2 \in \mathbb{R}^n$, and we obtain that $u \in C^{2,\alpha}(\mathbb{R}^n)$.

3.7 SCHAUDER ESTIMATES BY THE METHOD OF MAXIMUM PRINCIPLE

Our aim is to prove the Schauder estimate for the Poisson equation using Caffarelli's method of maximum principle. The technique does not rely on explicit representation formulas for D^2u but on the solvability of the Dirichlet problem, the comparison principle and the regularity of harmonic functions.

Theorem 3.12 (Interior Schauder estimates). *Let $f \in C^{0,\alpha}(B_1)$, for $0 < \alpha < 1$, and let $u \in C^2(B_1)$, solve*

$$-\Delta u = f \qquad \text{in } B_1.$$

Then $u \in C^{2,\alpha}(B_{1/2})$ and there exists a constant $C > 0$ depending only on n and α such that

$$\|u\|_{C^{2,\alpha}(B_{1/2})} \le C\big(\|u\|_{L^\infty(B_1)} + \|f\|_{C^{0,\alpha}(B_1)}\big).$$

The proof is based on the following characterization of the $C^{k,\alpha}$ spaces due to Campanato.

Theorem 3.13 (Campanato's theorem). *Let u be a measurable function defined on a Lipschitz domain Ω. Let $k \geq 0$ and $0 < \alpha < 1$. Then $u \in C^{k,\alpha}(\overline{\Omega})$ if and only if the following holds. There exists a constant $D > 0$ such that for every $z \in \overline{\Omega}$ there exists a polynomial $P_z(x)$ of degree k such that*

$$|u(x) - P_z(x)| \leq D|x - z|^{k+\alpha}$$

holds for every $x \in B_1(z) \cap \Omega$. In this case, the norm $\|u\|_{C^{k,\alpha}(\Omega)}$ is equivalent to

$$\sup\left\{|P_z| + \frac{|u(x) - P_z(x)|}{|x - z|^{k+\alpha}} : x \in B_1(z) \cap \Omega, x \neq z, z \in \overline{\Omega}\right\}$$

where $|P_z|$ denotes the sum of the modulus of the coefficients of the polynomial P_z.

We are going to show that $u \in C^{2,\alpha}(0)$, namely, that there exists a second-order polynomial $P(x)$ with bounded coefficients such that

$$|u(x) - P(x)| \leq D|x|^{2+\alpha} \qquad \text{for all } |x| < 1.$$

The reader will see that this estimate can also be obtained around any point $z \in \overline{B_{1/2}}$ instead of just the origin. Thus, Theorem 3.13 will immediately give Theorem 3.12. Observe that, in view of Theorem 3.13, if we write $P(x) = \frac{1}{2}x^T Ax + B \cdot x + C$, then $C = u(0)$, $B = \nabla u(0)$ and $A = D^2u(0)$. Thus, if the coefficients of P are bounded, then u, ∇u and D^2u will also be bounded. Moreover, the constant D above is comparable to $[D^2u]_{C^\alpha(0)}$.

Let us explain the strategy. To begin with, notice that the difficulty with the estimate $|u(x) - P(x)| \leq D|x|^{2+\alpha}$ is that it involves uncountably many inequalities, one for each x. Suppose we were able to find P. Then the estimate is equivalent to showing that there is a scale $0 < \rho < 1$ such that for every $k = 0, 1, 2, \ldots,$

$$|u(x) - P(x)| \leq D\rho^{k(2+\alpha)} \qquad \text{for all } |x| < \rho^k.$$

Indeed, if $|x| < 1$, there exists $k \geq 0$ such that $\rho^{k+1} \leq |x| < \rho^k$, and then

$$|u(x) - P(x)| \leq D\rho^{k(2+\alpha)} = \frac{D}{\rho^{2+\alpha}}\rho^{(k+1)(2+\alpha)} \leq \tilde{D}|x|^{2+\alpha}$$

where $\tilde{D}$ now depends on D, α and ρ. The key idea is to show the desired inequality when $k = 1$ and then reduce the general case for $k > 1$ to this one by scaling.

Since we are assuming that $u \in C^2$, we already know that u has a second-order Taylor polynomial $P(x)$ at the origin. But we still have the problem of estimating $P(x)$ and $u - P$. However, we will avoid a direct computation of $P(x)$ and, instead, construct it "from scratch." This is important because for other elliptic equations, in general, u will only be a weak solution (Sobolev, continuous) and we will not know a

priori that u has a second-order Taylor polynomial expansion. Thus, we are going to show that for any $k = 0, 1, 2, \dots$, there is a quadratic polynomial $P_k(x)$ such that

$$|u(x) - P_k(x)| \leq D\rho^{k(2+\alpha)} \qquad \text{for all } |x| < \rho^k.$$

If we can prove that the sequence $\{P_k\}_{k=1}^{\infty}$ converges to a limiting quadratic polynomial P_∞ in a sufficiently fast way, say, as

$$|P_k(x) - P_\infty(x)| \leq C_0\rho^{k(2+\alpha)} \qquad \text{for all } |x| < \rho^k$$

for some $C_0 > 0$, then we would be done. Indeed, for any $|x| < \rho^k$, we get

$$|u(x) - P_\infty(x)| \leq |u(x) - P_k(x)| + |P_k(x) - P_\infty(x)| \leq D\rho^{k(2+\alpha)} + C_0\rho^{k(2+\alpha)}.$$

The next question is, who are the polynomials P_k? Notice that P_k should be very close to u in the small ball B_{ρ^k}. Since f is continuous, at small scales f looks like a constant. If such a constant is very small (something that we will be able to assume by scaling), then u should be very close to being a harmonic function: $\Delta u \approx 0$. Hence we will approximate u by a harmonic function h. Next, since harmonic functions are smooth and their derivatives are controlled by their oscillation, h is going to be close to its second-order Taylor polynomial in a uniform way in B_{ρ^k}.

In this method we measure "closeness" by using the maximum principle. As we mentioned before, the idea is robust enough to be adaptable to other settings, for instance, for equations in nondivergence form and fully nonlinear equations, where the ABP estimate is available. The same ideas work for variational, divergence form equations, but replacing the maximum principle by the energy estimate or Caccioppoli inequality. For these results, see Sections 7.1 and 11.1.

Let us begin with a precise statement of the maximum principle.

Theorem 3.14 (Maximum principle). *Let $u \in C^2(B_R)$ be a solution to $-\Delta u = f$ in B_R. Then*

$$\|u\|_{L^\infty(B_R)} \leq \|u\|_{L^\infty(\partial B_R)} + \frac{R^2}{2n}\|f\|_{L^\infty(B_R)}.$$

Proof. By scaling, it is enough to prove the estimate for $R = 1$. Indeed, suppose the result is true for a solution to $-\Delta v = g$ in B_1. Given u, we define $v(x) = u(Rx)$, for every $x \in B_1$. Then $-\Delta v(x) = R^2(-\Delta u)(Rx) = R^2 f(Rx) =: g(x)$, for every $x \in B_1$. Therefore

$$\begin{aligned} \|u\|_{L^\infty(B_R)} = \|v\|_{L^\infty(B_1)} &\leq \|v\|_{L^\infty(\partial B_1)} + \frac{1}{2n}\|g\|_{L^\infty(B_1)} \\ &= \|u\|_{L^\infty(\partial B_R)} + \frac{R^2}{2n}\|f\|_{L^\infty(B_R)}. \end{aligned}$$

Hence we let $R = 1$. The idea is to use the weak maximum principle for sub- and superharmonic functions, see Corollary 2.9. To that end we need to add or subtract a barrier function to u in order to make it sub/superharmonic in B_1. Let us define

$$w(x) = u(x) - \sup_{\partial B_1} u - \frac{1 - |x|^2}{2n} \sup_{B_1} f.$$

Then

$$\Delta w = -f + \sup_{B_1} f \geq 0 \qquad \text{in } B_1$$

and $w \leq 0$ on ∂B_1, that is, w is subharmonic in B_1 and $w \leq 0$ on ∂B_1. By the weak maximum principle for subharmonic functions, $w \leq 0$ everywhere in B_1, so that,

$$u(x) \leq \sup_{\partial B_1} u + \frac{1-|x|^2}{2n} \sup_{B_1} f \leq \sup_{\partial B_1} u + \frac{1}{2n} \sup_{B_1} f \qquad \text{for every } x \in B_1.$$

On the other hand, $-u$ satisfies $-\Delta(-u) = -f$ in B_1 so, by the estimate above,

$$-u(x) \leq \sup_{\partial B_1}(-u) + \frac{1}{2n} \sup_{B_1}(-f) = -\inf_{\partial B_1} u - \frac{1}{2n} \inf_{B_1} f.$$

Collecting the estimates we obtain

$$\inf_{\partial B_1} u + \frac{1}{2n} \inf_{B_1} f \leq u(x) \leq \sup_{\partial B_1} u + \frac{1}{2n} \sup_{B_1} f,$$

for every $x \in B_1$. Observe that, as $u \leq |u|$ and $f \leq |f|$,

$$\sup_{\partial B_1} u + \frac{1}{2n} \sup_{B_1} f \leq \sup_{\partial B_1} |u| + \frac{1}{2n} \sup_{B_1} |f| = \|u\|_{L^\infty(\partial B_1)} + \frac{1}{2n} \|f\|_{L^\infty(B_1)}$$

and, as $u \geq -|u|$ and $f \geq -|f|$,

$$\begin{aligned} \inf_{\partial B_1} u + \frac{1}{2n} \inf_{B_1} f &\geq \inf_{\partial B_1}(-|u|) + \frac{1}{2n} \inf_{B_1}(-|f|) \\ &= -\sup_{\partial B_1} |u| - \frac{1}{2n} \sup_{B_1} |f| \\ &= -\Big(\|u\|_{L^\infty(\partial B_1)} + \frac{1}{2n} \|f\|_{L^\infty(B_1)} \Big). \end{aligned}$$

In conclusion, for every $x \in B_1$,

$$-\Big(\|u\|_{L^\infty(\partial B_1)} + \frac{1}{2n} \|f\|_{L^\infty(B_1)} \Big) \leq u(x) \leq \|u\|_{L^\infty(\partial B_1)} + \frac{1}{2n} \|f\|_{L^\infty(B_1)}.$$

□

We are ready to begin with the proof of the interior Schauder estimates by the method of maximum principle. The following is the crucial step: an initial approximation of u by a quadratic polynomial coming from a harmonic function h that is close to u.

Lemma 3.15 (Key lemma). *For any $0 < \alpha < 1$ there exist $0 < \varepsilon_0, \rho < 1$ such that for any solution $u \in C^2(B_1)$ to $-\Delta u = f$ in B_1 such that $|u| \leq 1$ in B_1 and $\|f\|_{L^\infty(B_1)} \leq \varepsilon_0$, there is a quadratic polynomial*

$$P(x) = \frac{1}{2} x^T A x + B \cdot x + C$$

such that $\Delta P = 0$,

$$|u(x) - P(x)| < \rho^{2+\alpha} \qquad \textit{for every } |x| < \rho$$

and

$$|A| + |B| + |C| \leq c_0$$

where $c_0 > 0$ *is a constant depending only on* n.

Remark 3.16. By scaling, the key lemma can be applied to any C^2 solution u to $-\Delta u = f$ in B_1, giving an estimate of the coefficients of P in terms of $\|u\|_{L^\infty(B_1)}$ and $\|f\|_{L^\infty(B_1)}$. Indeed, fix any $0 < \alpha < 1$. Define

$$v(x) = \frac{u(x)}{\|u\|_{L^\infty(B_1)} + \|f\|_{L^\infty(B_1)}/\varepsilon_0}.$$

Then $\Delta v = g$ in B_1, where

$$g(x) = \frac{f(x)}{\|u\|_{L^\infty(B_1)} + \|f\|_{L^\infty(B_1)}/\varepsilon_0}.$$

It is clear that $|v| \leq 1$ in B_1 and that $\|g\|_{L^\infty(B_1)} \leq \varepsilon_0$. Then, by the key lemma, there is a harmonic polynomial $P(x) = \frac{1}{2}x^T Ax + B \cdot x + C$ such that

$$|v(x) - P(x)| < \rho^{2+\alpha} \qquad \text{for every } |x| < \rho.$$

Multiplying this identity by $\|u\|_{L^\infty(B_1)} + \|f\|_{L^\infty(B_1)}/\varepsilon_0$ we see that, for every $|x| < \rho$,

$$|u(x) - \tilde{P}(x)| < \rho^{2+\alpha}\Big(\|u\|_{L^\infty(B_1)} + \|f\|_{L^\infty(B_1)}/\varepsilon_0\Big)$$

where now $\tilde{P}(x) = \frac{1}{2}x^T \tilde{A}x + \tilde{B} \cdot x + \tilde{C}$ is a harmonic polynomial such that

$$|\tilde{A}| + |\tilde{B}| + |\tilde{C}| \leq c_0\Big(\|u\|_{L^\infty(B_1)} + \|f\|_{L^\infty(B_1)}/\varepsilon_0\Big).$$

It is important to notice that c_0 and ε_0 depend only on n and α.

Proof of Lemma 3.15. We divide the proof into 3 steps.

Step 1. Approximation by a harmonic function. Let h be the solution to

$$\begin{cases} \Delta h = 0 & \text{in } B_{1/2} \\ h = u & \text{on } \partial B_{1/2}. \end{cases} \tag{3.22}$$

Since $u \in C(\partial B_{1/2})$, we know that such a classical solution $h \in C^\infty(B_{1/2}) \cap C(\overline{B_{1/2}})$ exists and is unique, see Theorem 2.2. To see how close u and h are, we look at the equation for the difference $u - h$:

$$\begin{cases} -\Delta(u - h) = f & \text{in } B_{1/2} \\ u - h = 0 & \text{on } \partial B_{1/2}. \end{cases}$$

By the maximum principle, Theorem 3.14,

$$|u(x) - h(x)| \leq \frac{1}{8n}\|f\|_{L^\infty(B_{1/2})} \qquad \text{for every } |x| < 1/2.$$

The right hand side f controls how close u and h are in L^∞. Hence, the smaller the right hand side, the closer u to being a harmonic function.

Step 2. Estimates for h and $h - P$. First we estimate the size of h. By the weak maximum principle for harmonic functions (or by Theorem 3.14) and the hypothesis $|u| \leq 1$ in B_1, the solution h to (3.22) satisfies

$$\|h\|_{L^\infty(B_{1/2})} \leq \|u\|_{L^\infty(\partial B_{1/2})} \leq 1.$$

Next we use this to estimate the size of the derivatives of h. Since h is harmonic in $B_{1/2}$, by the interior derivative estimates (see Corollary 2.15(1)), there is a constant $C_n > 0$ such that, for every $x \in B_{1/4}$,

$$|h(x)| + |\nabla h(x)| + |D^2h(x)| + |D^3h(x)| \leq C_n\|h\|_{L^\infty(B_{1/2})} \leq C_n. \tag{3.23}$$

It is important to notice that, once the geometry is fixed and we are working in the normalized situation, that is, $|u| \leq 1$, the constant C_n above depends only on dimension. We are ready to choose the polynomial P. We take $P(x) = \frac{1}{2}x^TAx + B \cdot x + C$, where

$$A = D^2h(0) \qquad B = \nabla h(0) \qquad C = h(0).$$

By (3.23), $|A| + |B| + |C| \leq C_n$. We let $c_0 = C_n$ for the statement. Moreover, $P(x)$ is a harmonic polynomial in $B_{1/2}$, see Example 2.1. We finally estimate the distance between h and P. By Taylor's theorem and (3.23),

$$|h(x) - P(x)| \leq \frac{1}{6}|x|^3\|D^3h\|_{L^\infty(B_{1/4})} \leq \frac{c_0}{6}|x|^3 \qquad \text{for any } x \in B_{1/4}.$$

Step 3. Estimate for $u - P$ and conclusion. From steps 1 and 2, for any $x \in B_{1/4}$,

$$\begin{aligned} |u(x) - P(x)| &\leq |u(x) - h(x)| + |h(x) - P(x)| \\ &\leq \frac{1}{8n}\|f\|_{L^\infty(B_1)} + \frac{c_0}{6}|x|^3. \end{aligned}$$

Let $0 < \rho < 1/4$ small enough depending only on n and α such that

$$\frac{c_0}{6}|x|^3 < \frac{\rho^{2+\alpha}}{2} \qquad \text{for every } |x| < \rho.$$

Once ρ is fixed, we choose $\varepsilon_0 > 0$ small enough depending only on n and α such that

$$\frac{1}{8n}\|f\|_{L^\infty(B_1)} \leq \frac{1}{8n}\varepsilon_0 < \frac{\rho^{2+\alpha}}{2}.$$

Thus

$$|u(x) - P(x)| < \rho^{2+\alpha} \qquad \text{for every } |x| < \rho.$$

□

Remark 3.17. The key lemma is not valid in general for $\alpha = 1$. The proof shows that if $\alpha = 1$, then

$$|u(x) - P(x)| < \rho^{3-\varepsilon}$$

for any $\varepsilon > 0$. This reflects the fact that solutions to $-\Delta u = f$ with $f \in \mathrm{Lip}$ are not locally $C^{2,1}$ in general if $n \geq 2$, see the next example.

Example 3.18 ($-\Delta v \in L^\infty$ does not imply $v \in C^{1,1}$)**.** Let

$$v(x_1, x_2) = |x_1||x_2| \log(|x_1| + |x_2|) \qquad \text{for } (x_1, x_2) \in \mathbb{R}^2.$$

Let $x_1, x_2 > 0$. We have

$$\begin{aligned} v_{11}(x_1, x_2) &= \partial_{x_1}\left[x_2 \log(x_1 + x_2) + \frac{x_1 x_2}{x_1 + x_2}\right] \\ &= \frac{2x_2}{x_1 + x_2} - \frac{x_1 x_2}{(x_1 + x_2)^2} \end{aligned}$$

and

$$v_{22}(x_1, x_2) = \frac{2x_1}{x_1 + x_2} - \frac{x_1 x_2}{(x_1 + x_2)^2}$$

so that

$$-\Delta v = -2 + \frac{2x_1 x_2}{(x_1 + x_2)^2} =: g(x_1, x_2).$$

Then $\|g\|_{L^\infty(\overline{\mathbb{R}_+ \times \mathbb{R}_+})} \leq 3$ and the right hand side is bounded. However,

$$v_{21}(x_1, x_2) = \log(x_1 + x_2) + 1 - \frac{x_1 x_2}{(x_1 + x_2)^2}$$

which is unbounded at 0 and at infinity.

The key lemma shows that there is a fixed scale ρ for which we can approximate u by a quadratic polynomial P. To show that u is $C^{2,\alpha}$ at the origin, we need to show that u is geometrically close to a quadratic polynomial *at every scale.* This is achieved by iterating the key lemma using a scaling. As it will be evident from the proof, it is not enough to have f only bounded or just continuous, since we need to control the flatness of f at every scale in a uniform way.

Theorem 3.19. *Let $u \in C^2(B_1)$ be a bounded solution to $-\Delta u = f$ in B_1. Suppose that $f \in C^\alpha(0)$ for some $0 < \alpha < 1$, that is,*

$$[f]_{C^\alpha(0)} = \sup_{|x|<1} \frac{|f(x) - f(0)|}{|x|^\alpha} < \infty.$$

Then $u \in C^{2,\alpha}(0)$, namely, there is a quadratic polynomial $P_\infty(x) = \frac{1}{2}x^T Ax + B\cdot x + C$ such that

$$|u(x) - P_\infty(x)| < D|x|^{2+\alpha} \qquad \textit{for all } |x| < 1$$

with

$$|A| + |B| + |C| + |D| \leq c_1([f]_{C^\alpha(0)} + |f(0)| + \|u\|_{L^\infty(B_1)})$$

where $c_1 > 0$ is a constant depending only on n and α.

Proof. We split the proof into three steps.

Step 1. Two reductions. First, we can assume that $f(0) = 0$. Indeed, if not, take

$$v(x) = u(x) + f(0)\frac{|x|^2}{2n},$$

so

$$-\Delta v = f - f(0) = g$$

and $g(0) = 0$. Clearly, the estimates for v translate to u. Second, we can suppose that $\|u\|_{L^\infty(B_1)} \leq 1$ and $[f]_{C^\alpha(0)} \leq \varepsilon_0$, where $0 < \varepsilon_0 < 1$ is as in the key lemma. Let

$$v(x) = \varepsilon_0 \frac{u(x)}{\|u\|_{L^\infty(B_1)} + [f]_{C^\alpha(0)}}.$$

Then $\|v\|_{L^\infty(B_1)} \leq 1$ and

$$-\Delta v = \varepsilon_0 \frac{f}{\|u\|_{L^\infty(B_1)} + [f]_{C^\alpha(0)}} = g$$

with $[g]_{C^\alpha(0)} \leq \varepsilon_0$. Again, the estimates on v pass to u. Therefore, we assume that

$$f(0) = 0 \qquad \|u\|_{L^\infty(B_1)} \leq 1 \qquad [f]_{C^\alpha(0)} \leq \varepsilon_0. \tag{3.24}$$

Step 2. A claim and the conclusion. We claim that for every $k \geq 0$ there exists a harmonic polynomial

$$P_k(x) = \frac{1}{2}x^T A_k x + B_k \cdot x + C_k$$

such that

$$|u(x) - P_k(x)| < \rho^{k(2+\alpha)} \qquad \text{for any } |x| < \rho^k$$

and

$$\rho^{2k}|A_k - A_{k+1}| + \rho^k|B_k - B_{k+1}| + |C_k - C_{k+1}| \leq c_0\rho^{k(2+\alpha)} \tag{3.25}$$

where $c_0 > 0$ is a constant depending only on n.

From this claim the conclusion follows. Indeed, let k and m be positive integers. We can assume that $m = k + j$ for some positive integer j. By (3.25),

$$\begin{aligned}
|A_k - A_m| = |A_k - A_{k+j}| &\leq \sum_{\ell=0}^{j-1} |A_{k+\ell} - A_{k+\ell+1}| \\
&\leq c_0 \sum_{\ell=0}^{j-1} \rho^{\alpha(k+\ell)} = c_0\rho^{\alpha k} \sum_{\ell=0}^{j-1} \rho^{\alpha\ell} \\
&= c_0\rho^{\alpha k}\frac{1-\rho^{\alpha j}}{1-\rho^\alpha} = \frac{c_0}{1-\rho^\alpha}|(\rho^\alpha)^k - (\rho^\alpha)^m|.
\end{aligned}$$

As $k, m \to \infty$, the right hand side above tends to 0 and A_k is a Cauchy sequence. A similar reasoning shows that B_k and C_k are also Cauchy sequences. Thus there are $A_\infty, B_\infty, C_\infty$ such that

$$A_k, B_k, C_k \to A_\infty, B_\infty, C_\infty \qquad \text{as } k \to \infty.$$

Moreover, by taking $j \to \infty$ in the estimate for $|A_k - A_m|$ above, and similarly for B_k and C_k,

$$\rho^{2k}|A_k - A_\infty| + \rho^k|B_k - B_\infty| + |C_k - C_\infty| \le c_{\rho,\alpha,n}\rho^{k(2+\alpha)}. \tag{3.26}$$

Define

$$P_\infty(x) = \frac{1}{2}x^T A_\infty x + B_\infty \cdot x + C_\infty.$$

From (3.26) we see that if $|x| < \rho^k$ then $|P_k(x) - P_\infty(x)| \le c_{\rho,\alpha,n}\rho^{k(2+\alpha)}$. Then, for any $x \in B_1$ with $\rho^{k+1} \le |x| < \rho^k$,

$$\begin{aligned} |u(x) - P_\infty(x)| &\le |u(x) - P_k(x)| + |P_k(x) - P_\infty(x)| \\ &\le \rho^{k(2+\alpha)} + c_{\rho,\alpha,n}\rho^{k(2+\alpha)} \\ &= c_1\rho^{(k+1)(2+\alpha)} \le c_1|x|^{2+\alpha} \end{aligned}$$

with $c_1 > 0$ depending on n, α and ρ, and ρ depends only on α.

Step 3. Proof of claim. The proof is done by induction on k using the scaling of the equation.

If $k = 0$ then the claim is clearly true with $P_0 = 0$ and $P_1 = P$ as in the key lemma, Lemma 3.15, because of (3.24).

For the induction step, assume that the claim is true for some $k \ge 0$. For $x \in B_1$, we define

$$\tilde{u}(x) = \frac{(u - P_k)(\rho^k x)}{\rho^{k(2+\alpha)}}.$$

Observe that if $x \in B_1$ then $\rho^k x \in B_{\rho^k}$. This means that we are *zooming out* $u - P_k$ from near the origin to B_1. We have

$$-\Delta\tilde{u} = \frac{\rho^{2k}(-\Delta u + \Delta P_k)(\rho^k x)}{\rho^{k(2+\alpha)}} = \frac{f(\rho^k x)}{\rho^{k\alpha}} = \tilde{f}(x) \qquad \text{in } B_1.$$

We want to apply the key lemma to $\tilde{u}$, so we need to check the hypotheses. By the induction hypothesis and (3.24), for any $x \in B_1$, as $|\rho^k x| < \rho^k$,

$$|\tilde{u}(x)| = \frac{|u(\rho^k x) - P_k(\rho^k x)|}{\rho^{k(2+\alpha)}} \le 1$$

and

$$|\tilde{f}(x)| = \frac{|f(\rho^k x)|}{\rho^{k\alpha}} = \frac{|f(\rho^k x) - f(0)|}{|\rho^k x|^\alpha}|x|^\alpha \le [f]_{C^\alpha(0)}|x| < \varepsilon_0.$$

Therefore, by the key lemma, there is a quadratic harmonic polynomial P with bounded coefficients such that

$$|\tilde{u}(x) - P(x)| < \rho^{2+\alpha} \qquad \text{for } |x| < \rho.$$

By writing down the definition of $\tilde{u}$,

$$|u(\rho^k x) - P_k(\rho^k x) - \rho^{k(2+\alpha)}P(x)| < \rho^{(k+1)(2+\alpha)} \qquad \text{for } |x| < \rho.$$

If we call $z = \rho^k x$ and let

$$P_{k+1}(z) := P_k(z) + \rho^{k(2+\alpha)}P(z/\rho^k)$$

we finally see that

$$|u(z) - P_{k+1}(z)| < \rho^{(k+1)(2+\alpha)} \qquad \text{for } |z| < \rho^{k+1}.$$

This is the $(k+1)$-step of the induction. To finish we notice that, for any $x \in B_1$,

$$\begin{aligned} P_{k+1}(x) - P_k(x) &= \rho^{k(2+\alpha)}\left[\frac{1}{2}\left(\frac{x}{\rho^k}\right)^T A\left(\frac{x}{\rho^k}\right) + B\cdot\frac{x}{\rho^k} + C\right] \\ &= \frac{1}{2}x^T(\rho^{k\alpha}A)x + (\rho^{k(1+\alpha)}B)\cdot x + \rho^{k(2+\alpha)}C. \end{aligned}$$

By evaluating this identity at $x = 0$ we obtain $|C_k - C_{k+1}| < c_0\rho^{k(2+\alpha)}$. Similarly, by considering the gradient and the Hessian of $P_k(x) - P_{k+1}(x)$ and evaluating at $x = 0$ we get $|B_k - B_{k+1}| < c_0\rho^{k(1+\alpha)}$ and $|A_k - A_{k+1}| < c_0\rho^{k\alpha}$, respectively. □

Remark 3.20 (Schauder estimates up to the boundary)**.** The method of maximum principle is quite robust. It allows us to prove boundary Schauder estimates as well. We consider here the case of a flat boundary and sketch the proof. Let $B_r^+ = B_r \cap \{x_n > 0\}$, $B_r^- = B_r \cap \{x_n < 0\}$ and $T_r = B_r \cap \{x_n = 0\}$, for $r > 0$. Suppose that $u = u(x) = u(x', x_n) \in C^2(B_1^+)$ solves

$$\begin{cases} -\Delta u = f & \text{in } B_1^+ \\ u = 0 & \text{on } T_1 \end{cases} \tag{3.27}$$

with $|u| \leq 1$ in B_1^+ and $\|f\|_{L^\infty(B_1^+)} \leq \varepsilon_0$. Consider the solution h to

$$\begin{cases} \Delta h = 0 & \text{in } B_{1/2}^+ \\ h = u & \text{on } \partial B_{1/2}^+. \end{cases}$$

The maximum principle in Theorem 3.14 can be extended to hold in any bounded domain. Then, for every $x \in B_{1/2}^+$,

$$|u(x) - h(x)| \leq C_n\|f\|_{L^\infty(B_{1/2}^+)} \leq C_n\varepsilon_0.$$

Since $h = 0$ on $T_{1/2}$, its odd reflection to $B^-_{1/2}$ defined as

$$\tilde{h}(x) = \begin{cases} h(x', x_n) & \text{if } x_n \geq 0 \\ -h(x', -x_n) & \text{if } x_n < 0 \end{cases}$$

is continuous in $\overline{B_{1/2}}$. Moreover, it is easy to check that $\tilde{h}$ is harmonic in the weak sense in $B_{1/2}$ (see Section 2.2), so that

$$\begin{cases} \Delta\tilde{h} = 0 & \text{in } B_{1/2} \\ \tilde{h} = \tilde{u} & \text{on } \partial B_{1/2} \end{cases}$$

where $\tilde{u} \in C(B_1)$ denotes the odd reflection of u to B^-_1. By the weak maximum principle, $|\tilde{h}| \leq 1$ in $B_{1/2}$ and, by interior derivative estimates for harmonic functions, $|D^k\tilde{h}(0)| \leq C_0$, for $k = 1, 2, 3$, with $C_0 > 0$ depending only on n. Hence, if P is the second-order Taylor polynomial of $\tilde{h}$ at the origin, then

$$|\tilde{u} - P| \leq |\tilde{u} - \tilde{h}| + |\tilde{h} - P| < C_n\varepsilon_0 + C_0|x|^3 \qquad \text{in } B_{1/4}.$$

By choosing first $0 < \rho < 1/4$ and then $\varepsilon_0 > 0$, and restricting back to $B^+_{1/2}$ we get

$$|u - h| < \rho^{2+\alpha} \qquad \text{in } B^+_\rho.$$

This is the key lemma up to a flat boundary. The rest of the proof follows in the same way as before by replacing B_ρ by B^+_ρ. Therefore, we obtain that if u is a bounded solution to (3.27) and $f \in C^\alpha(0)$ then $u \in C^{2,\alpha}(0)$.

NOTES ON CHAPTER 3

Details about the Fourier transform can be found in Duoandikoetxea [30], Evans [32], Folland [37] and Stein–Weiss [66]. Proofs of the properties of the heat semigroup and approximations of the identity can be seen in Duoandikoetxea [30], Evans [32], Stein [63] and Wheeden–Zygmund [74].

The presentation of the method of semigroups and the derivation of the fundamental solution follows Stinga [68].

For the Euler–Mascheroni constant, see [51].

To find the fundamental solution of the Laplacian by looking at radially symmetric functions, see Evans [32]. For a Fourier transform derivation using distribution theory, see Folland [38].

Perron's method for harmonic functions can be found in Gilbarg–Trudinger [43] and Han–Lin [44].

The Schauder estimates by integral representation formulas proof follows the lines of Gilbarg–Trudinger [43].

The method of maximum principle is due to Caffarelli [3,4], and our presentation is based on his works. See the papers by Campanato [19,20] for the proof of Theorem 3.13.

CHAPTER 4

The Calderón–Zygmund estimates for the Laplacian

In this chapter we consider regularity estimates for solutions u to the Poisson equation

$$-\Delta u = f \qquad \text{in } \Omega \subset \mathbb{R}^n$$

when f is in L^p, $1 < p < \infty$. We want to show the Calderón–Zygmund estimate, which says that D^2u is in L^p. This is equivalent to establishing that the second-order Riesz transforms

$$R_{ij}f = \partial_{ij}(-\Delta)^{-1}f \qquad i, j = 1, \ldots, n$$

are bounded operators on L^p. Moreover, we will find a pointwise formula for $D^2u(x)$ for almost every x. Parallel to the case of the Schauder estimate, the inverse of the Laplacian *integrates twice* at the scale of L^p spaces. However, proving a pointwise formula for $D^2u(x)$ will be much more involved compared to what we did in Theorem 3.5. When $p = 1$ and $p = \infty$, weak-type and BMO estimates for D^2u will be obtained, respectively. These L^p Sobolev estimates were first obtained by Alberto Pedro Calderón and Antoni Zygmund in 1952. The proof is based on a powerful geometric construction called the Calderón–Zygmund decomposition.

Throughout this chapter, Ω denotes a bounded domain and, for simplicity, we let $n \geq 3$. The reader should recall the definition and properties of the fundamental solution for the Laplacian $K(x) = c_n|x|^{2-n}$ given in Theorem 3.3.

4.1 SOLVABILITY WITH L^p RIGHT HAND SIDE

Let $f \in \mathcal{D}'(\Omega)$. We say that $u \in \mathcal{D}'(\mathbb{R}^n)$ is a **distributional solution** to the Poisson equation $-\Delta u = f$ in Ω if

$$u(-\Delta\varphi) = f(\varphi)$$

for every $\varphi \in C_c^\infty(\Omega)$.

Theorem 4.1 (Distributional solvability in $L^p(\Omega)$). *Let $f \in L^p(\mathbb{R}^n)$, $1 \leq p \leq \infty$, with support contained in Ω. Then*

$$u(x) = (-\Delta)^{-1}f(x) = \int_\Omega K(x - z)f(z)\,dz$$

DOI: 10.1201/9781032679853-4

is in $L^p_{\text{loc}}(\mathbb{R}^n)$ *and there exists a constant* $C > 0$ *depending only on* n *such that, for any compact set* $A \subset \mathbb{R}^n$,

$$\|u\|_{L^p(A)} \le C|\Omega|^{2/(np')}|A|^{2/(np)}\|f\|_{L^p(\Omega)} \tag{4.1}$$

where $1/p + 1/p' = 1$. *If* $1 \le p < \infty$ *then* $-\Delta u = f$ *in* Ω *in the sense of distributions. In particular, for* $1 \le p < \infty$,

$$-\Delta u \in L^p(\Omega) \qquad \text{and} \qquad -\Delta u(x) = f(x) \text{ for a.e. } x \in \Omega.$$

Proof. Let A be a compact set. Fix any $x \in \mathbb{R}^n$ and let $R > 0$ such that $|B_R(x)| = |A|$. We first notice that

$$\begin{aligned}\int_A K(x-z)\,dz &= c\int_A \frac{1}{|x-z|^{n-2}}\,dz \\ &\le c\int_{B_R(x)} \frac{1}{|x-z|^{n-2}}\,dz \\ &= C_n \int_0^R \frac{r^{n-1}}{r^{n-2}}\,dr = C_n R^2 = C_n|A|^{2/n}\end{aligned}$$

so that

$$\sup_{x\in\mathbb{R}^n}\int_A K(x-z)\,dz \le C_n|A|^{2/n}.$$

If $p = \infty$ then, for any $x \in \mathbb{R}^n$,

$$\|u\|_{L^\infty(\mathbb{R}^n)} \le \|f\|_{L^\infty(\Omega)} \sup_{x\in\mathbb{R}^n}\int_\Omega K(x-z)\,dz \le C_n|\Omega|^{2/n}\|f\|_{L^\infty(\Omega)}.$$

Let $1 \le p < \infty$. By Hölder's inequality,

$$\begin{aligned}|u(x)| &\le \int_\Omega K(x-z)^{1/p}|f(z)|K(x-z)^{1/p'}\,dz \\ &\le \left(\int_\Omega K(x-z)|f(z)|^p\,dz\right)^{1/p}\left(\int_\Omega K(x-z)\,dz\right)^{1/p'} \\ &\le C_{n,p}|\Omega|^{2/(np')}\left(\int_\Omega K(x-z)|f(z)|^p\,dz\right)^{1/p}.\end{aligned}$$

Therefore,

$$\begin{aligned}\|u\|^p_{L^p(A)} &\le C_{n,p}|\Omega|^{2p/(np')}\int_\Omega\left[\int_A K(x-z)\,dx\right]|f(z)|^p\,dz \\ &\le C_{n,p}|\Omega|^{2p/(np')}|A|^{2/n}\|f\|^p_{L^p(\Omega)}.\end{aligned}$$

Thus (4.1) holds for any $1 \le p \le \infty$.

Let $1 \le p < \infty$. Since $u \in L^p(\Omega) \subset \mathcal{D}'(\Omega)$, we can compute the derivatives $\partial_{ij}u$ in the sense of distributions. Recall from Theorem 3.3 that if $\varphi \in C_c^\infty(\Omega)$, then $K * \varphi \in C^\infty(\Omega)$ and

$$-\Delta(K * \varphi)(x) = \varphi(x) \qquad \text{in } \Omega.$$

Let $f_k \in C_c^\infty(\Omega)$ such that $f_k \to f$ in $L^p(\Omega)$. Let $u_k = K * f_k$. Then $u_k \in C^\infty(\mathbb{R}^n)$ and $-\Delta u_k = f_k$. By (4.1),

$$\|u_k - u\|_{L^p(\Omega)} = \|K * (f_k - f)\|_{L^p(\Omega)} \leq C_{n,p,|\Omega|}\|f_k - f\|_{L^p(\Omega)} \to 0$$

as $k \to \infty$, and $u_k \to u$ in $L^p(\Omega)$. Now we compute $-\Delta u$ in the sense of distributions. For any $\varphi \in C_c^\infty(\Omega)$,

$$\begin{aligned} -\Delta u(\varphi) = u(-\Delta\varphi) &= \int_\Omega u(-\Delta\varphi)\,dx \\ &= \lim_{k\to\infty}\int_\Omega u_k(-\Delta\varphi)\,dx \\ &= \lim_{k\to\infty}\int_\Omega (-\Delta u_k)\varphi\,dx \\ &= \lim_{k\to\infty}\int_\Omega f_k\varphi\,dx = \int_\Omega f\varphi\,dx. \end{aligned}$$

In other words, $-\Delta u = f$ in $\mathcal{D}'(\Omega)$. Since $f \in L^p(\Omega)$, we get $-\Delta u \in L^p(\Omega)$ and $-\Delta u = f$ a.e. in Ω. □

Corollary 4.2 (Representation of general solutions). *Let $f \in L^p(\mathbb{R}^n)$, $1 \leq p < \infty$, with support contained in Ω. Let $u \in L^1_{\text{loc}}(\Omega)$ satisfy*

$$-\Delta u = f \qquad \text{in the sense of distributions in } \Omega.$$

Then, for almost every $x \in \Omega$,

$$u(x) = v(x) + (-\Delta)^{-1}f(x)$$

where v is harmonic in Ω. In particular, $u \in L^p_{\text{loc}}(\Omega)$, $-\Delta u \in L^p(\Omega)$ and $-\Delta u(x) = f(x)$ for a.e. $x \in \Omega$.

Proof. By Theorem 4.1, $v = u - (-\Delta)^{-1}f$ is a locally integrable distributional solution to $\Delta v = 0$ in Ω. But harmonic functions in the distributional sense have continuous representatives that are classical harmonic functions, see Section 2.2. Thus v is harmonic in Ω. Theorem 4.1 and the local boundedness of v give that $u \in L^p_{\text{loc}}(\Omega)$ and the rest of the conclusion. □

By interior derivative estimates (see Theorem 2.11 and Corollary 2.15(1)), any harmonic function in Ω is in $W^{k,p}_{\text{loc}}(\Omega)$, for any $k \geq 0$ and $1 \leq p \leq \infty$. Therefore, in view of the representation given by Corollary 4.2, to prove Sobolev estimates for solutions u to $-\Delta u = f$, we only need to focus on the regularity of $(-\Delta)^{-1}f$.

4.2 L^2 ESTIMATE FOR SECOND DERIVATIVES

We know that if $f \in L^p(\mathbb{R}^n)$, $1 \leq p < \infty$, has support contained in Ω then $u = (-\Delta)^{-1}f = K * f$ satisfies $-\Delta u \in L^p(\Omega)$ and $-\Delta u = f$ a.e. in Ω. We want to prove that $\partial_{ij}u$ are not only distributions, but also measurable functions in $L^p(\mathbb{R}^n)$ when $1 < p < \infty$, and that satisfy a weak-type inequality when $p = 1$.

We consider then the **second-order Riesz transform** operator

$$f \longmapsto D^2u = D^2(-\Delta)^{-1}f \equiv R_2 f.$$

We want to prove that $R_2 f$, which is well defined in the sense of distributions for any f in L^p with compact support (see Theorem 4.1), extends as a bounded operator on $L^p(\mathbb{R}^n)$, $1 < p < \infty$, and of weak-type $(1,1)$. The case $p = 2$ is rather simple and is based on integration by parts or the Fourier transform.

Theorem 4.3. *Let $f \in L^2(\mathbb{R}^n)$ with support contained in Ω. Let $u = (-\Delta)^{-1}f$ be as in Theorem 4.1. Then $u \in L^2_{\mathrm{loc}}(\mathbb{R}^n)$,*

$$D^2u \in L^2(\mathbb{R}^n)$$

and

$$\|D^2u\|_{L^2(\mathbb{R}^n)} = \|f\|_{L^2(\Omega)}.$$

In particular, the second-order Riesz transform R_2 extends as a bounded operator on $L^2(\mathbb{R}^n)$ such that, for any $f \in L^2(\mathbb{R}^n)$,

$$\|R_2 f\|_{L^2(\mathbb{R}^n)} = \|f\|_{L^2(\mathbb{R}^n)}.$$

Proof. Suppose first that $f \in C_c^\infty(\mathbb{R}^n)$. Then $u = (-\Delta)^{-1}f = K * f \in C^\infty(\mathbb{R}^n)$ and $-\Delta u = f$ in $\mathbb{R}^n$. Let $R > 0$ such that $\mathrm{supp}(f) \subset B_R$. Now we compute

$$\begin{aligned}
\int_{B_R} |D^2u|^2\,dx &= \int_{B_R} (\partial_{ij}u)(\partial_{ij}u)\,dx \\
&= -\int_{B_R} (\partial_j u)(\partial_{iij}u)\,dx + \int_{\partial B_R} (\partial_j u)(\partial_{ij}u)\nu_i\,dS \\
&= \int_{B_R} (\partial_{jj}u)(\partial_{ii}u)\,dx - \int_{\partial B_R} (\partial_j u)(\partial_{ii}u)\nu_j\,dS + \int_{\partial B_R} (\partial_j u)(\partial_{ji}u)\nu_i\,dS \\
&= \int_{B_R} (\Delta u)^2\,dx - \int_{\partial B_R} (\Delta u)(\partial_\nu u)\,dS + \int_{\partial B_R} \nabla u \nabla(\partial_\nu u)\,dS \\
&= \int_{B_R} f^2\,dx + \int_{\partial B_R} \nabla u \nabla(\partial_\nu u)\,dS.
\end{aligned}$$

We take the limit as $R \to \infty$ to cancel the boundary term. Indeed, observe that if $x \in \partial B_R$, then $x \notin \mathrm{supp}(f)$ and, for any multi-index $\gamma \in \mathbb{N}_0^n$,

$$D^\gamma u(x) = \int_\Omega D^\gamma K(x-z) f(z)\,dz.$$

If R is sufficiently large so that $\Omega \subset B_{R/2}$ then, for any $x \in \partial B_R$ and $z \in \Omega$, we have $R = |x| \le |x-z| + |z| \le |x-z| + R/2$, from which $|x-z| > R/2$. Hence,

$$|D^\gamma u(x)| \le C_{n,\gamma} \int_\Omega \frac{|f(z)|}{|x-z|^{n-2+|\gamma|}}\,dz \le C_{n,\gamma} \|f\|_{L^1(\Omega)} \frac{1}{R^{n-2+|\gamma|}}$$

uniformly in $x \in \partial B_R$. Therefore,

$$\int_{\partial B_R} \nabla u \nabla(\partial_\nu u)\,dS \le C_n R^{n-1} \|\nabla u\|_{L^\infty(\partial B_R)} \|D^2 u\|_{L^\infty(\partial B_R)} \le \frac{C_{n,f}}{R^n} \to 0$$

as $R \to \infty$. Thus

$$\int_{\mathbb{R}^n} |D^2 u|^2 \, dx = \int_\Omega f^2 \, dx.$$

Next, for the general case of $f \in L^2(\mathbb{R}^n)$ with support contained in Ω, we take a sequence $\{f_k\}_{k=1}^\infty \subset C_c^\infty(\Omega)$ such that $f_k \to f$ in $L^2(\Omega)$ and define $u_k = (-\Delta)^{-1} f_k = K * f_k \in C^\infty(\mathbb{R}^n)$. Theorem 4.1 implies that $u \in L^2_{\mathrm{loc}}(\mathbb{R}^n)$ and that for any compact set A we have

$$\|u_k - u\|_{L^2(A)} = \|K * (f_k - f)\|_{L^2(A)} \le C_{n,|\Omega|,|A|} \|f_k - f\|_{L^2(\Omega)} \to 0$$

as $k \to \infty$. In other words, $u_k \to u$ locally in $L^2(\mathbb{R}^n)$. Also, we proved above that

$$\int_{\mathbb{R}^n} |D^2 u_k|^2 \, dx = \int_\Omega f_k^2 \, dx.$$

This gives, in particular, that $D^2 u_k$ is a Cauchy sequence in $L^2(\mathbb{R}^n)$, so it converges in $L^2(\mathbb{R}^n)$ to some v in $L^2(\mathbb{R}^n)$ such that

$$\int_{\mathbb{R}^n} |v|^2 \, dx = \int_\Omega f^2 \, dx.$$

We claim that $v = D^2 u$ in $\mathcal{D}'(\mathbb{R}^n)$ and, therefore, $D^2 u \in L^2(\mathbb{R}^n)$ and $\|D^2 u\|_{L^2(\mathbb{R}^n)} = \|f\|_{L^2(\Omega)}$. Indeed, for any $\varphi \in C_c^\infty(\mathbb{R}^n)$, since $u_k \to u$ locally in $L^2(\mathbb{R}^n)$,

$$\begin{aligned}
(D^2 u)(\varphi) = u(D^2 \varphi) &= \int_{\operatorname{supp}(\varphi)} u D^2 \varphi \, dx \\
&= \lim_{k \to \infty} \int_{\operatorname{supp}(\varphi)} u_k D^2 \varphi \, dx \\
&= \lim_{k \to \infty} \int_{\operatorname{supp}(\varphi)} (D^2 u_k) \varphi \, dx = \int_{\mathbb{R}^n} v \varphi \, dx.
\end{aligned}$$

To show that R_2 (initially defined for $f \in L_c^2(\mathbb{R}^n)$) extends to a bounded operator on $L^2(\mathbb{R}^n)$, we consider $f \in L^2(\mathbb{R}^n)$ and a sequence $\{f_k\}_{k=1}^\infty \subset L^2(\mathbb{R}^n)$ of functions with compact support such that $f_k \to f$ in $L^2(\mathbb{R}^n)$. Then

$$\int_{\mathbb{R}^n} |R_2 f_k - R_2 f_j|^2 \, dx = \int_{\mathbb{R}^n} |f_k - f_j|^2 \, dx \to 0$$

as $k, j \to \infty$. Therefore, there exists $R_2 f \in L^2(\mathbb{R}^n)$ such that $R_2 f_k \to R_2 f$ in $L^2(\mathbb{R}^n)$. Moreover,

$$\int_{\mathbb{R}^n} |R_2 f|^2 \, dx = \lim_{k \to \infty} \int_{\mathbb{R}^n} |R_2 f_k|^2 \, dx = \lim_{k \to \infty} \int_{\mathbb{R}^n} |f_k|^2 \, dx = \int_{\mathbb{R}^n} |f|^2 \, dx.$$

It is easy to see that the definition of $R_2 f$ is independent of the choice of the approximating sequence $\{f_k\}_{k=1}^\infty$ and that $R_2 f = D^2(-\Delta)^{-1} f$ whenever $f \in L_c^2(\mathbb{R}^n)$. □

Remark 4.4. Even if we know that $D^2 u \in L^2(\mathbb{R}^n)$, we still do not have a formula for $D^2 u(x)$ for a.e. $x \in \mathbb{R}^n$. We will address this point in Section 4.6.

Theorem 4.5 (Fourier transform of D^2u). *Let $f \in L^2(\mathbb{R}^n)$ with support contained in Ω. Let $u = (-\Delta)^{-1}f$ be as in Theorem 4.1. Then $D^2u \in L^2(\mathbb{R}^n)$ and*

$$\widehat{\partial_{ij}u}(\xi) = \xi_i\xi_j|\xi|^{-2}\widehat{f}(\xi).$$

In particular, for any $f \in L^2(\mathbb{R}^n)$ the second-order Riesz transforms $R_{ij} = \partial_{ij}(-\Delta)^{-1}$, $i, j = 1, \dots, n$, satisfy

$$\widehat{R_{ij}f}(\xi) = \xi_i\xi_j|\xi|^{-2}\widehat{f}(\xi).$$

Proof. Let $f \in L^2(\mathbb{R}^n)$ with support contained in Ω and $f_k \in C_c^\infty(\mathbb{R}^n)$ such that $f_k \to f$ in $L^2(\mathbb{R}^n)$. Denote $u_k = (-\Delta)^{-1}f_k$ as in the proof of Theorem 4.3. Recall that $D^2u_k \to D^2u$ in $L^2(\mathbb{R}^n)$. For any $\varphi \in C_c^\infty(\mathbb{R}^n)$,

$$\begin{aligned}
\widehat{\partial_{ij}u}(\varphi) = \partial_{ij}u(\widehat{\varphi}) &= \int_{\mathbb{R}^n} \partial_{ij}u\widehat{\varphi}\,dx \\
&= \lim_{k\to\infty} \int_{\mathbb{R}^n} \partial_{ij}u_k\widehat{\varphi}\,dx = \lim_{k\to\infty} \int_{\mathbb{R}^n} \widehat{\partial_{ij}u_k}\varphi\,d\xi \\
&= \lim_{k\to\infty} \int_{\mathbb{R}^n} \xi_i\xi_j|\xi|^{-2}\widehat{f_k}\varphi\,d\xi = \int_{\mathbb{R}^n} \xi_i\xi_j|\xi|^{-2}\widehat{f}\varphi\,d\xi.
\end{aligned}$$

In the last identity we used that $\xi_i\xi_j|\xi|^{-2}$ is bounded on $\mathbb{R}^n$ in order to pass to the limit. Since the identity above is true for any $\varphi \in C_c^\infty(\mathbb{R}^n)$, the conclusion follows. The proof for R_{ij} is similar. □

4.3 THE CALDERÓN–ZYGMUND THEOREM

The situation so far is the following. Given $f \in L^p(\mathbb{R}^n)$, $1 \le p < \infty$, with support contained in Ω, the function $u = (-\Delta)^{-1}f$ is in $L^p_{\text{loc}}(\mathbb{R}^n)$ so, in particular, D^2u is well defined in the sense of distributions, and $-\Delta u = f$ for a.e. $x \in \Omega$. When $p = 2$, the second derivatives D^2u are in fact functions in $L^2(\mathbb{R}^n)$ and $\|D^2u\|_{L^2(\mathbb{R}^n)} = \|f\|_{L^2(\Omega)}$.

In particular, the second-order Riesz transform operator

$$R_2 = D^2(-\Delta)^{-1} : L^2(\mathbb{R}^n) \to L^2(\mathbb{R}^n)$$

is well defined, linear and bounded, and R_2f coincides with D^2u when $f \in L^2_c(\mathbb{R}^n)$. Moreover, for any $x \notin \operatorname{supp}(f)$ it is clear that we can compute

$$R_2f(x) = D^2u(x) = D^2 \int_\Omega K(x-z)f(z)\,dz = \int_\Omega D^2K(x-z)f(z)\,dz$$

because x is away from the singularity the integral has at $z = x$. In addition, since

$$\partial_{ij}K(x) = \frac{\Gamma(n/2)}{2\pi^{n/2}}\left[n\frac{x_ix_j}{|x|^{n+2}} - \frac{\delta_{ij}}{|x|^n}\right]$$

it follows that the kernel $D^2K(x-z)$ satisfies the symmetry property $D^2K(x-z) = D^2K(z-x)$, the size estimate

$$|D^2K(x-z)| \le \frac{C_n}{|x-z|^n}$$

and the smoothness estimate

$$|\nabla D^2 K(x-z)| \leq \frac{C_n}{|x-z|^{n+1}}$$

for $x \neq z$. The Calderón–Zygmund theorem says that an operator of this form is bounded on $L^p(\mathbb{R}^n)$ and of weak-type $(1,1)$.

Definition 4.6. Let $\mathcal{K}(x,z)$ be a measurable function defined for $x, z \in \mathbb{R}^n$, $x \neq z$. We say that $\mathcal{K}$ is a **standard kernel** if there exists a constant $C > 0$ such that

$$|\mathcal{K}(x,z)| \leq \frac{C}{|x-z|^n}$$

and

$$|\nabla_x \mathcal{K}(x,z)| + |\nabla_z \mathcal{K}(x,z)| \leq \frac{C}{|x-z|^{n+1}}$$

for all $x \neq z$.

We just saw that $\mathcal{K}(x,z) = D^2 K(x-z)$ is a standard kernel.

Definition 4.7. A linear operator T is a **Calderón–Zygmund operator** if the following two conditions hold.

(1) T is a bounded operator on $L^2(\mathbb{R}^n)$.

(2) There exists a standard kernel $\mathcal{K}(x,z)$ such that, for every $f \in L_c^\infty(\mathbb{R}^n)$,

$$Tf(x) = \int_{\mathbb{R}^n} \mathcal{K}(x,z) f(z)\, dx \qquad \text{for all } x \notin \operatorname{supp}(f).$$

Clearly, $T = R_2$ is a Calderón–Zygmund operator.

Theorem 4.8 (Calderón–Zygmund theorem)**.** *Let T be a Calderón–Zygmund operator. Then T extends as a weak-type $(1,1)$ operator and as a bounded operator on $L^p(\mathbb{R}^n)$, for $1 < p < \infty$. In other words, there exists a constant $C > 0$ depending only on n and p such that, when $p = 1$,*

$$|\{x \in \mathbb{R}^n : |Tf(x)| > \lambda\}| \leq \frac{C}{\lambda}\|f\|_{L^1(\mathbb{R}^n)} \qquad \textit{for all } f \in L^1(\mathbb{R}^n)$$

and every $\lambda > 0$ and, when $1 < p < \infty$,

$$\|Tf\|_{L^p(\mathbb{R}^n)} \leq C\|f\|_{L^p(\mathbb{R}^n)} \qquad \textit{for all } f \in L^p(\mathbb{R}^n).$$

The Calderón–Zygmund theorem gives that the second-order Riesz transform R_2 extends as a weak-type $(1,1)$ operator and as a bounded operator on $L^p(\mathbb{R}^n)$ for $1 < p < \infty$.

Corollary 4.9 (Calderón–Zygmund estimates). *Let $f \in L^p(\mathbb{R}^n)$, $1 \leq p < \infty$, with support contained in Ω and let*

$$u(x) = (-\Delta)^{-1} f(x) = \int_\Omega K(x-z) f(z)\, dz$$

for $x \in \mathbb{R}^n$. There exists a constant $C > 0$ depending only on n and p such that if $1 < p < \infty$ then $u \in L^p(\Omega)$, $D^2 u \in L^p(\mathbb{R}^n)$, $-\Delta u = f$ a.e. in Ω and

$$\|D^2 u\|_{L^p(\mathbb{R}^n)} \leq C \|f\|_{L^p(\Omega)}.$$

Furthermore, for $p = 2$,

$$\|D^2 u\|_{L^2(\mathbb{R}^n)} = \|f\|_{L^2(\Omega)}.$$

If $p = 1$ then $D^2 u$ satisfies a weak-type $(1,1)$ inequality, namely,

$$|\{x \in \mathbb{R}^n : |D^2 u(x)| > \lambda\}| \leq \frac{C}{\lambda} \|f\|_{L^1(\Omega)} \qquad \text{for all } \lambda > 0.$$

In particular, if $u \in W_0^{2,p}(\Omega)$, $1 < p < \infty$, then

$$\|D^2 u\|_{L^p(\Omega)} \leq C \|\Delta u\|_{L^p(\Omega)}$$

and, when $p = 2$, $\|D^2 u\|_{L^2(\Omega)} = \|\Delta u\|_{L^2(\Omega)}$.

Proof. The case $p = 2$ was established in Theorem 4.3. We already know that the operator $f \in L_c^\infty(\mathbb{R}^n) \longmapsto R_2 f = D^2 u$ is a Calderón–Zygmund operator. Therefore, by the Calderón–Zygmund theorem and an approximation argument by smooth functions as in the proof of Theorem 4.3, we get the global estimates for $D^2 u$. If $u \in W_0^{2,p}(\Omega)$ then there is a sequence $u_k \in C_c^\infty(\Omega)$ such that $u_k \to u$ in $W^{2,p}(\Omega)$. Since $u_k = (-\Delta)^{-1}(-\Delta u_k)$, we have $\|D^2 u_k\|_{L^p(\Omega)} \leq C \|\Delta u_k\|_{L^p(\Omega)}$ and the same estimate holds for u by taking $k \to \infty$. □

Remark 4.10. Corollary 4.9 gives that $D^2 u \in L^p(\mathbb{R}^n)$, $1 < p < \infty$. However, it does not provide an explicit pointwise formula for $D^2 u(x)$ for a.e. $x \in \mathbb{R}^n$. We will address this in Section 4.6.

To prove the Calderón–Zygmund theorem we need two preliminary tools. The first one is a real analysis interpolation lemma, whose proof is omitted.

Lemma 4.11 (Marcinkiewicz interpolation lemma). *Let $1 \leq p_0 < p_1 \leq \infty$. Let T be a sublinear operator, that is, $T(f_1 + f_2) \leq T f_1 + T f_2$, that maps*

$$L^{p_0}(\mathbb{R}^n) + L^{p_1}(\mathbb{R}^n) = \{f = g + h : g \in L^{p_0}(\mathbb{R}^n), h \in L^{p_1}(\mathbb{R}^n)\}$$

into the set of measurable functions on $\mathbb{R}^n$. Assume that T is of weak-type (p_0, p_0) and of weak-type (p_1, p_1), that is, there exists a constant $A > 0$ such that, if $p_i < \infty$,

$$|\{x \in \mathbb{R}^n : |Tf(x)| > \lambda\}| \leq \frac{A}{\lambda^{p_i}} \|f\|_{L^{p_i}(\mathbb{R}^n)}^{p_i}$$

for all $f \in L^{p_i}(\mathbb{R}^n)$, while T is bounded on $L^\infty(\mathbb{R}^n)$ if $p_i = \infty$, $i = 0, 1$. Then $T : L^p(\mathbb{R}^n) \to L^p(\mathbb{R}^n)$ is bounded for $p_0 < p < p_1$ and there exists a constant $C > 0$ depending on A, p_0 and p_1 such that

$$\|Tf\|_{L^p(\mathbb{R}^n)} \leq C \|f\|_{L^p(\mathbb{R}^n)}.$$

By Chebyshev's inequality, if T is a bounded operator on $L^p(\mathbb{R}^n)$ then it is of weak-type (p, p), for $1 \le p \le \infty$.

The second tool is the famous Calderón–Zygmund decomposition. We will give the proof as it will be important for other regularity estimates such as the Calderón–Zygmund estimates for divergence form equations (Chapter 7) and the Harnack inequality for nondivergence form equations (Chapter 9). Furthermore, it is used to prove the John–Nirenberg inequality for functions in BMO that is a key component for obtaining Moser's Harnack inequality for divergence form equations (Chapter 6).

Lemma 4.12 (Calderón–Zygmund decomposition). *Let $f \in L^1(\mathbb{R}^n)$. Given $\lambda > 0$ there exist a constant $C > 0$ depending only on n, disjoint cubes $\{Q_j\}_{j=1}^{\infty}$ and a decomposition*

$$f = g + b = g + \sum_{j=1}^{\infty} b_j$$

such that

(*i*) $|g(x)| \le C\lambda$ *and* $\|g\|_{L^1(\mathbb{R}^n)} \le \|f\|_{L^1(\mathbb{R}^n)}$;

(*ii*) $\operatorname{supp}(b_j) \subset Q_j$;

(*iii*) $\displaystyle\int_{Q_j} b_j\,dx = 0$ *and* $\displaystyle\frac{1}{|Q_j|}\int_{Q_j} |b_j|\,dx \le C\lambda$;

(*iv*) $\displaystyle\sum_{j=1}^{\infty} |Q_j| \le \frac{\|f\|_{L^1(\mathbb{R}^n)}}{\lambda}$;

(*v*) $|f(x)| \le \lambda$ *for almost every* $\displaystyle x \notin \bigcup_{j=1}^{\infty} Q_j$;

(*vi*) $\displaystyle\lambda < \frac{1}{|Q_j|}\int_{Q_j} |f|\,dx \le 2^n\lambda.$

Proof. Since $f \in L^1(\mathbb{R}^n)$, we can divide $\mathbb{R}^n$ in a net of cubes Q, all of them disjoint, such that

$$\frac{1}{|Q|}\int_Q |f|\,dx \le \lambda.$$

For example, we can choose Q sufficiently large and then translate it. Each cube Q in this family is subdivided into 2^n congruent subcubes by just splitting each edge of Q into two equal segments. For each of these subcubes, Q is their "predecessor." In these subcubes Q' we have two possibilities: either

(1) $\displaystyle\frac{1}{|Q'|}\int_{Q'} |f|\,dx \le \lambda$ or

(2) $\displaystyle\frac{1}{|Q'|}\int_{Q'} |f|\,dx > \lambda$

We take the cubes Q' that satisfy (2) and we save them. For the cubes Q' that satisfy (1) we repeat the process. That is, we divide Q' into 2^n subcubes and take those subcubes Q'' that satisfy (2) above, and so on. This process continues and we obtain a sequence $\{Q_j\}_{j=1}^{\infty}$ of disjoint cubes where (2) happens.

Since the cubes Q_j are disjoint and on each of them (2) holds, we get

$$\sum_{j=1}^{\infty} |Q_j| \leq \frac{1}{\lambda} \sum_{j=1}^{\infty} \int_{Q_j} |f| \, dx = \frac{1}{\lambda} \int_{\cup Q_j} |f| \, dx \leq \frac{\|f\|_{L^1(\mathbb{R}^n)}}{\lambda}$$

so (iv) is proved.

Next, if $\tilde{Q}_j$ is the unique predecessor of Q_j then $\tilde{Q}_j$ is not a member of the family $\{Q_j\}_{j=1}^{\infty}$ that satisfies (2). This is because $\tilde{Q}_j$ was subdivided to obtain Q_j, so that $\tilde{Q}_j$ was already satisfying (1). With this observation we obtain (vi):

$$\lambda < \frac{1}{|Q_j|} \int_{Q_j} |f| \, dx \leq \frac{|\tilde{Q}_j|}{|Q_j|} \frac{1}{|\tilde{Q}_j|} \int_{\tilde{Q}_j} |f| \, dx \leq 2^n \lambda.$$

Now we define

$$g(x) = \begin{cases} \dfrac{1}{|Q_j|} \displaystyle\int_{Q_j} f \, dz & \text{for } x \in Q_j \\ f(x) & \text{for } x \notin Q_j \end{cases}$$

and

$$b_j(x) = \left(f(x) - \frac{1}{|Q_j|} \int_{Q_j} f \, dz \right) \chi_{Q_j}(x).$$

Clearly $f = g + b$, (ii) holds, as well as the first equality in (iii). To check the second estimate in (iii), let $\tilde{Q}_j$ be the unique predecessor of Q_j, which satisfies (1). Then

$$\begin{aligned} \frac{1}{|Q_j|} \int_{Q_j} |b_j| \, dx &\leq \frac{2}{|Q_j|} \int_{Q_j} |f| \, dx \\ &\leq \frac{2}{|Q_j|} \int_{\tilde{Q}_j} |f| \, dx \\ &\leq \frac{2}{|Q_j|} |\tilde{Q}_j| \lambda = C_n \lambda. \end{aligned}$$

To verify (i), we have

$$\begin{aligned} \int_{\mathbb{R}^n} |g| \, dx &= \int_{\cup Q_j} |g| \, dx + \int_{\mathbb{R}^n \setminus (\cup Q_j)} |g| \, dx \\ &\leq \sum_{j=1}^{\infty} \int_{Q_j} |f| \, dx + \int_{\mathbb{R}^n \setminus (\cup Q_j)} |f| \, dx = \|f\|_{L^1(\mathbb{R}^n)}. \end{aligned}$$

If $x \in Q_j$, by considering its unique predecessor $\tilde{Q}_j$, which satisfies (1), we get

$$|g(x)| \leq \frac{1}{|Q_j|} \int_{Q_j} |f| \, dx \leq \frac{1}{|Q_j|} \int_{\tilde{Q}_j} |f| \, dx \leq C\lambda.$$

If $x \notin \bigcup_{j=1}^{\infty} Q_j$, then all the cubes Q that contain x satisfy (1):

$$\frac{1}{|Q|}\int_Q |f|\,dx \le \lambda.$$

These cubes form a nested sequence shrinking to x because they were all subdivided in the process as they all fell into case (1). Therefore, by the Lebesgue differentiation theorem,

$$|g(x)| = |f(x)| \le \lambda \qquad \text{for all } x \notin \bigcup_{j=1}^{\infty} Q_j.$$

This completes the proof of (i) and also gives (v). □

We are ready to prove the Calderón–Zygmund theorem. The idea is the following. Since T is bounded in $L^2(\mathbb{R}^n)$, by Chebyshev's inequality, T is of weak-type $(2,2)$:

$$|\{x \in \mathbb{R}^n : |Tf(x)| > \lambda\}| \le \frac{1}{\lambda^2}\int_{|Tf|>\lambda} |Tf|^2\,dx \le \frac{C}{\lambda^2}\|f\|^2_{L^2(\mathbb{R}^n)}$$

for any $\lambda > 0$. Next, using the Calderón–Zygmund decomposition we prove that T is of weak-type $(1,1)$. Therefore, by the Marcinkiewicz interpolation lemma (Lemma 4.11), T is bounded on $L^p(\mathbb{R}^n)$ for $1 < p < 2$. Finally, by using a duality argument, we will see that T is bounded on $L^p(\mathbb{R}^n)$ for $2 < p < \infty$.

Proof of Theorem 4.8. We divide the proof into 3 steps.

Step 1. ***T* is of weak-type** $(1,1)$**.** Let $f \in L^\infty_c(\mathbb{R}^n)$. Fix $\lambda > 0$. We have to show that there exists a constant $C > 0$ depending only on n such that

$$|\{x \in \mathbb{R}^n : |Tf(x)| > \lambda\}| \le \frac{C}{\lambda}\|f\|_{L^1(\mathbb{R}^n)}.$$

Since $f \in L^1(\mathbb{R}^n)$, we can form the Calderón–Zygmund decomposition of f at level λ and write $f = g + b$, where g and b are as in Lemma 4.12. Since T is linear, $Tf = Tg + Tb$ and

$$\begin{aligned}&|\{x \in \mathbb{R}^n : |Tf(x)| > \lambda\}| \\ &\qquad\qquad \le |\{x \in \mathbb{R}^n : |Tg(x)| > \lambda/2\}| + |\{x \in \mathbb{R}^n : |Tb(x)| > \lambda/2\}|.\end{aligned}$$

We first estimate Tg. This only uses the L^2 boundedness of T. In fact, by Chebyshev's inequality, the L^2 boundedness of T and using that $|g(x)|^2 = |g(x)||g(x)| \le C\lambda|g(x)|$ and $\|g\|_{L^1(\mathbb{R}^n)} \le \|f\|_{L^1(\mathbb{R}^n)}$,

$$\begin{aligned}|\{x \in \mathbb{R}^n : |Tg(x)| > \lambda/2\}| &\le \frac{4}{\lambda^2}\int_{\mathbb{R}^n} |Tg(x)|^2\,dx \\ &\le \frac{C}{\lambda^2}\int_{\mathbb{R}^n} |g(x)|^2\,dx \\ &\le \frac{C}{\lambda}\int_{\mathbb{R}^n} |g(x)|\,dx \le \frac{C}{\lambda}\|f\|_{L^1(\mathbb{R}^n)}.\end{aligned}$$

Next, we estimate Tb. Here the size and smoothness conditions on $\mathcal{K}$ will come into play. Recall that $b = \sum_{j=1}^{\infty} b_j$. By construction, each $b_j \in L_c^\infty(Q_j)$, so they are in $L^2(\mathbb{R}^n)$ and, because the cubes Q_j are disjoint,

$$\int_{\mathbb{R}^n} |b|^2\, dx = \sum_{j=1}^{\infty} \int_{Q_j} |b_j|^2\, dx.$$

Hence $b = \sum_{j=1}^{\infty} b_j$ in $L^2(\mathbb{R}^n)$. Since T is bounded on $L^2(\mathbb{R}^n)$,

$$Tb = \sum_{j=1}^{\infty} Tb_j \qquad \text{in } L^2(\mathbb{R}^n).$$

Thus

$$|Tb(x)| \le \sum_{j=1}^{\infty} |Tb_j(x)| \qquad \text{for a.e. } x \in \mathbb{R}^n.$$

Recall that we want to estimate the measure of

$$E_\lambda = \{x \in \mathbb{R}^n : |Tb(x)| > \lambda/2\}.$$

For any given cube Q_j in the Calderón–Zygmund decomposition, we denote by Q_j^* the cube with the same center z_j as Q_j and having side length twice the side length of Q_j. By using part (iv) of the Calderón–Zygmund decomposition in Lemma 4.12,

$$\begin{aligned}
|E_\lambda| &= \left|E_\lambda \cap \left(\bigcup_{j=1}^{\infty} Q_j^*\right)\right| + \left|E_\lambda \cap \left(\bigcup_{j=1}^{\infty} Q_j^*\right)^c\right| \\
&\le \left|\bigcup_{j=1}^{\infty} Q_j^*\right| + \left|\left\{x \in \left(\bigcup_{j=1}^{\infty} Q_j^*\right)^c : |Tb(x)| > \lambda/2\right\}\right| \\
&\le \sum_{j=1}^{\infty} |Q_j^*| + \frac{2}{\lambda}\int_{(\cup Q_\ell^*)^c} |Tb(x)|\, dx \qquad (4.2)\\
&\le 2^n \sum_{j=1}^{\infty} |Q_j| + \frac{2}{\lambda}\sum_{j=1}^{\infty} \int_{(\cup Q_\ell^*)^c} |Tb_j(x)|\, dx \\
&\le \frac{2^n}{\lambda}\|f\|_{L^1(\mathbb{R}^n)} + \frac{2}{\lambda}\sum_{j=1}^{\infty}\int_{(Q_j^*)^c} |Tb_j(x)|\, dx.
\end{aligned}$$

We now have to estimate each one of the integrals in the second term above. Recall that $b_j \in L_c^\infty(Q_j)$. Then, for $x \notin \operatorname{supp}(b_j) \subset Q_j$, since b_j has zero mean over Q_j,

$$Tb_j(x) = \int_{Q_j} \mathcal{K}(x,z) b_j(z)\, dz = \int_{Q_j} (\mathcal{K}(x,z) - \mathcal{K}(x,z_j)) b_j(z)\, dz$$

where z_j is the center of Q_j. Therefore

$$\begin{aligned}
\int_{(Q_j^*)^c} |Tb_j(x)|\, dx &\le \int_{(Q_j^*)^c} \int_{Q_j} |\mathcal{K}(x,z) - \mathcal{K}(x,z_j)||b_j(z)|\, dz\, dx \\
&= \int_{Q_j} |b_j(z)| \left[\int_{(Q_j^*)^c} |\mathcal{K}(x,z) - \mathcal{K}(x,z_j)|\, dx\right] dz.
\end{aligned} \qquad (4.3)$$

Let us estimate the integral in brackets. Notice that if $\xi \in Q_j$ and $x \in (Q_j^*)^c$ then $|\xi - x|$ is comparable to $|z_j - x|$. In particular, if $\ell(Q_j)$ denotes the side length of the cube Q_j then

$$|x - z_j| \leq |x - \xi| + |\xi - z_j| \leq |x - \xi| + \sqrt{n}\ell(Q_j) \leq C_n|x - \xi|.$$

By the mean value theorem and the smoothness condition on $\mathcal{K}$ we then have, for any $z \in Q_j$ and some intermediate point ξ between z and z_j,

$$\begin{aligned}|\mathcal{K}(x, z) - \mathcal{K}(x, z_j)| &= |\nabla_z \mathcal{K}(x, \xi) \cdot (z - z_j)| \\ &\leq \frac{C}{|x - \xi|^{n+1}}|z - z_j| \leq C\frac{\ell(Q_j)}{|x - z_j|^{n+1}}.\end{aligned}$$

Thus,

$$\begin{aligned}\int_{(Q_j^*)^c} |\mathcal{K}(x, z) - \mathcal{K}(x, z_j)|\, dx &\leq C\ell(Q_j) \int_{(Q_j^*)^c} \frac{1}{|x - z_j|^{n+1}}\, dx \\ &\leq C\ell(Q_j) \int_{[B_{\ell(Q_j)}(z_j)]^c} \frac{1}{|x - z_j|^{n+1}}\, dx \\ &= C\ell(Q_j) \int_{\ell(Q_j)}^{\infty} \frac{r^{n-1}}{r^{n+1}}\, dr = C\end{aligned} \tag{4.4}$$

where $C > 0$ is a constant independent of Q_j. Therefore, we can bound in (4.3) as

$$\int_{(Q_j^*)^c} |Tb_j(x)|\, dx \leq C \int_{Q_j} |b_j(z)|\, dz$$

for C independent of j. With this and Lemma 4.12(*iii*)–(*iv*), we continue in (4.2):

$$\begin{aligned}|E_\lambda| &\leq \frac{2^n}{\lambda}\|f\|_{L^1(\mathbb{R}^n)} + \frac{2C}{\lambda}\sum_{j=1}^{\infty}\int_{Q_j} |b_j(z)|\, dz \\ &\leq \frac{2^n}{\lambda}\|f\|_{L^1(\mathbb{R}^n)} + C\sum_{j=1}^{\infty}|Q_j| \\ &\leq \frac{C}{\lambda}\|f\|_{L^1(\mathbb{R}^n)}.\end{aligned}$$

This completes the estimate we needed. Therefore, T is of weak-type $(1, 1)$.

Step 2. T is bounded on $L^p(\mathbb{R}^n)$ for $1 < p < 2$. This is a direct application of the Marcinkiewicz interpolation lemma, see Lemma 4.11.

Step 3. T is bounded on $L^p(\mathbb{R}^n)$ for $2 < p < \infty$. Let T^* denote the adjoint of T in $L^2(\mathbb{R}^n)$. By general functional analysis, T^* is a bounded operator in $L^2(\mathbb{R}^n)$ with the same operator norm as T. We claim that, for any $f \in L_c^\infty(\mathbb{R}^n)$,

$$T^*f(x) = \int_{\mathbb{R}^n} \mathcal{K}^*(x, z)f(z)\, dz \qquad \text{for } x \notin \operatorname{supp}(f)$$

where we define $\mathcal{K}^*(x,z) = \mathcal{K}(z,x)$ for $x \neq z$. Indeed, for any $\varphi \in C_c^\infty(\mathbb{R}^n)$ such that $\operatorname{supp}(\varphi) \cap \operatorname{supp}(f) = \varnothing$,

$$\begin{aligned}\int_{\mathbb{R}^n} T^* f(x)\varphi(x)\,dx &= \int_{\mathbb{R}^n} f(x) T\varphi(x)\,dx \\ &= \int_{\mathbb{R}^n} f(x) \int_{\mathbb{R}^n} \mathcal{K}(x,z)\varphi(z)\,dz\,dx \\ &= \int_{\mathbb{R}^n} \varphi(z) \int_{\mathbb{R}^n} \mathcal{K}(x,z) f(x)\,dx\,dz.\end{aligned}$$

Since $\mathcal{K}$ is a standard kernel, $\mathcal{K}^*$ is also a standard kernel and therefore T^* is a Calderón–Zygmund operator. Steps 1 and 2 apply to give that T^* is of weak-type $(1,1)$ and bounded on $L^p(\mathbb{R}^n)$ for $1 < p < 2$.

Let $2 < p < \infty$. To finally prove that T is bounded on $L^p(\mathbb{R}^n)$ observe that, for any $f \in L_c^\infty(\mathbb{R}^n)$, by the boundedness of T^* in $L^{p'}(\mathbb{R}^n)$, $1 < p' < 2$,

$$\begin{aligned}\|Tf\|_{L^p(\mathbb{R}^n)} &= \sup_{\|g\|_{L^{p'}(\mathbb{R}^n)}=1} \left| \int_{\mathbb{R}^n} Tf(x) g(x)\,dx \right| \\ &= \sup_{\|g\|_{L^{p'}(\mathbb{R}^n)}=1} \left| \int_{\mathbb{R}^n} f(x) T^* g(x)\,dx \right| \\ &\leq \|f\|_{L^p(\mathbb{R}^n)} \sup_{\|g\|_{L^{p'}(\mathbb{R}^n)}=1} \|T^* g\|_{L^{p'}(\mathbb{R}^n)} \\ &\leq C\|f\|_{L^p(\mathbb{R}^n)} \sup_{\|g\|_{L^{p'}(\mathbb{R}^n)}=1} \|g\|_{L^{p'}(\mathbb{R}^n)} = C\|f\|_{L^p(\mathbb{R}^n)}.\end{aligned}$$

□

Remark 4.13. Recall that, for $f \in L^1(\mathbb{R}^n)$,

$$\int_{\mathbb{R}^n} |f(x)|\,dx = \int_{\mathbb{R}^n} \int_0^\infty \chi_{(0,|f(x)|)}(\lambda)\,d\lambda\,dx = \int_0^\infty |\{x \in \mathbb{R}^n : |f(x)| > \lambda\}|\,d\lambda$$

since $\chi_{(0,|f(x)|)}(\lambda) = \chi_{\{x:|f(x)|>\lambda\}}(x)$, and the last integral is finite. If T is an operator of weak-type $(1,1)$, the estimate

$$|\{x \in \mathbb{R}^n : |Tf(x)| > \lambda\}| \leq \frac{C}{\lambda}\|f\|_{L^1(\mathbb{R}^n)}$$

is not enough to ensure that $Tf \in L^1(\mathbb{R}^n)$.

Example 4.14 (Hilbert transform)**.** For a measurable function $f = f(x)$ of $x \in \mathbb{R}$, we define the **Hilbert transform** by

$$Hf(x) = \lim_{\varepsilon\to 0} H_\varepsilon f(x) = \lim_{\varepsilon\to 0} \frac{1}{\pi} \int_{|z|>\varepsilon} \frac{f(x-z)}{z}\,dz = \lim_{\varepsilon\to 0} \frac{1}{\pi} \int_{|x-z|>\varepsilon} \frac{f(z)}{x-z}\,dz$$

whenever the limit exists. Since $K_\varepsilon(z) = z^{-1}\chi_{|z|>\varepsilon} \in L^q(\mathbb{R})$ for $1 < q \le \infty$ then, by Hölder's inequality, $H_\varepsilon f$ is well defined for every $\varepsilon > 0$ and $f \in L^p(\mathbb{R})$, $1 \le p < \infty$. For $f \in L^\infty_c(\mathbb{R})$ we obviously have

$$Hf(x) = \frac{1}{\pi}\int_{\mathbb{R}} \frac{f(z)}{x-z}\,dz \qquad \text{for } x \notin \operatorname{supp}(f).$$

Moreover, the kernel $\mathcal{K}(x,z) = 1/(x-z)$ is a standard kernel. Let us check that H is bounded in $L^2(\mathbb{R})$. Observe that, by (3.1),

$$\begin{aligned}\widehat{K_\varepsilon}(\xi) &= \frac{1}{(2\pi)^{1/2}} \lim_{N\to\infty} \int_{\varepsilon<|z|<N} \frac{e^{-i\xi z}}{z}\,dz \\ &= -\frac{i}{(2\pi)^{1/2}} \lim_{N\to\infty} \int_{\varepsilon<|z|<N} \frac{\sin(\xi z)}{z}\,dz \\ &= -\frac{2i}{(2\pi)^{1/2}} \operatorname{sgn}(\xi) \lim_{N\to\infty} \int_{\varepsilon|\xi|}^{N|\xi|} \frac{\sin(r)}{r}\,dr\end{aligned}$$

and this is uniformly bounded in $\varepsilon > 0$ and $\xi \in \mathbb{R}^n$. Hence, H_ε are bounded operators in $L^2(\mathbb{R})$, with norm independent of ε. Next, if $\varphi \in C^1_c(\mathbb{R})$ then, since $1/z$ is odd,

$$H_\varepsilon\varphi(x) = \frac{1}{\pi}\int_{\varepsilon<|z|<1} \frac{\varphi(x-z)-\varphi(x)}{z}\,dz + \frac{1}{\pi}\int_{|z|>1} \frac{\varphi(x-z)}{z}\,dz.$$

The first integral is supported in a compact set of x and converges uniformly as $\varepsilon \to 0$ because $|\varphi(x-z)-\varphi(x)| \le C|z|$. The second integral is a function in $L^2(\mathbb{R})$ because $\varphi \in C^1_c(\mathbb{R})$. Therefore, $H_\varepsilon\varphi$ converges in $L^2(\mathbb{R})$ as $\varepsilon \to 0$. Now, let $f \in L^2(\mathbb{R})$ and $\varphi_k \in C^1_c(\mathbb{R})$ such that $\varphi_k \to f$ in $L^2(\mathbb{R})$. Then, for any $\varepsilon, \varepsilon' > 0$,

$$\begin{aligned}\|H_\varepsilon f - H_{\varepsilon'} f\|_{L^2(\mathbb{R})} &\le \|H_\varepsilon f - H_\varepsilon \varphi_k\|_{L^2(\mathbb{R})} + \|H_{\varepsilon'}\varphi_k - H_{\varepsilon'} f\|_{L^2(\mathbb{R})} \\ &\quad + \|H_\varepsilon\varphi_k - H_{\varepsilon'}\varphi_k\|_{L^2(\mathbb{R})} \\ &\le C\|f-\varphi_k\|_{L^2(\mathbb{R})} + \|H_\varepsilon\varphi_k - H_{\varepsilon'}\varphi_k\|_{L^2(\mathbb{R})}.\end{aligned}$$

It follows that $H_\varepsilon f$ converges in $L^2(\mathbb{R})$ to Hf and, since H_ε is bounded in $L^2(\mathbb{R})$ with norm independent of ε, we have that H is bounded in $L^2(\mathbb{R})$. We conclude that H is a Calderón–Zygmund operator, thus it extends as a weak-type $(1,1)$ operator and as a bounded operator on $L^p(\mathbb{R})$, $1 < p < \infty$. Now, if $g(x) = \chi_{(0,1)}(x)$ then, for $x \notin (0,1)$,

$$Hg(x) = \frac{1}{\pi}\int_0^1 \frac{1}{x-z}\,dz = -\frac{1}{\pi}\log|x-z|\Big|_{z=0}^{z=1} = \frac{1}{\pi}\log\left|\frac{x}{x-1}\right|$$

while for $x \in (0,1)$ and all ε sufficiently small,

$$\begin{aligned}H_\varepsilon g(x) &= \frac{1}{\pi}\int_{x+\varepsilon}^1 \frac{1}{x-z}\,dz + \int_0^{x-\varepsilon} \frac{1}{x-z}\,dz \\ &= \frac{1}{\pi}\left[\log\left|\frac{\varepsilon}{x-1}\right| + \log\left|\frac{x}{\varepsilon}\right|\right] = \frac{1}{\pi}\log\left|\frac{x}{x-1}\right|.\end{aligned}$$

Hence, Hg is neither in $L^1(\mathbb{R})$ nor in $L^\infty(\mathbb{R})$.

Remark 4.15 (Generalizations of the Calderón–Zygmund theorem)**.** There are many useful generalizations of Theorem 4.8. Here we mention three of them that will be important for our purposes.

(i) The definition of Calderón–Zygmund operator T requires that T is initially bounded on $L^2(\mathbb{R}^n)$, see Definition 4.7(1). This is in fact the natural case for the second-order Riesz transform, where we can check its L^2 boundedness by using integration by parts. Similarly, by means of the Fourier transform, we showed in Example 4.14 that the Hilbert transform H is bounded on $L^2(\mathbb{R})$. However, there are other applications where one only knows that T is initially bounded in $L^q(\mathbb{R}^n)$ for some $1 \leq q \leq \infty$, not necessarily $L^2(\mathbb{R}^n)$. Under this new assumption, the conclusions of Theorem 4.8 continue to hold. Indeed, the only thing that needs to be verified when $q > 1$ is the validity of the weak-type $(1, 1)$ estimate, as the rest follows from interpolation and duality. But the reader can easily verify that the estimates for Tg and Tb given in the proof of Theorem 4.8 follow through under the new assumption that T is bounded in $L^q(\mathbb{R}^n)$, $1 < q < \infty$. For $q = \infty$, observe that in the Calderón–Zygmund decomposition of f at level λ we have $|g(x)| \leq C\lambda$ so that, since T is bounded in $L^\infty(\mathbb{R}^n)$, we get $\{x \in \mathbb{R}^n : |Tf(x)| > D\lambda\} \subset \{x \in \mathbb{R}^n : |Tb(x)| > \lambda\}$ for some constant $D > 0$, and the proof continues as before.

(ii) In addition to assuming (i), one can also replace the size and smoothness estimates of $\mathcal{K}$ in the definition of standard kernel, see Definition 4.6, by weaker conditions, the so-called **Hörmander conditions**:

$$\int_{|x-z|>2|z-\bar{z}|} \left|\mathcal{K}(x,z) - \mathcal{K}(x,\bar{z})\right| dx \leq C$$

and

$$\int_{|x-z|>2|x-\bar{x}|} \left|\mathcal{K}(x,z) - \mathcal{K}(\bar{x},z)\right| dz \leq C$$

for some constant $C > 0$. Indeed, the first Hörmander condition is exactly what one needs in order to have (4.4) and complete the proof of the weak-type $(1, 1)$ estimate. The second one ensures that the weak-type $(1, 1)$ inequality obtained in Step 1 in the proof of Theorem 4.8 also holds for the adjoint of T so as to run Step 3. The Hörmander conditions are important in that there are many examples where the kernel $\mathcal{K}$ is nonsmooth, and thus the smoothness conditions of Definition 4.6 cannot be directly checked.

(iii) Furthermore, one can extend the theory to vector-valued Calderón–Zygmund operators. We consider linear operators T defined over real-valued functions $f(x)$, $x \in \mathbb{R}^n$, such that $Tf(x) \in X$, where X is a Banach space. For technical reasons of measurability, we need to assume that X is separable. An operator T of this form is said to be a **vector-valued Calderón–Zygmund operator** if the following conditions hold.

- T is bounded from $L^q(\mathbb{R}^n)$ to $L^q(\mathbb{R}^n; X)$ for some $1 \leq q \leq \infty$, that is,

$$\int_{\mathbb{R}^n} \|Tf(x)\|_X^q \, dx \leq C \int_{\mathbb{R}^n} |f(x)|^q \, dx \qquad \text{if } q < \infty$$

  or $\sup_{x \in \mathbb{R}^n} \|Tf(x)\|_X \leq C\|f\|_{L^\infty(\mathbb{R}^n)}$ if $q = \infty$.
- There is a kernel $\mathcal{K}(x, z)$ defined for $x \neq z$ with values in X that satisfies the Hörmander conditions with $\|\cdot\|_X$ in place of $|\cdot|$: for some constant $C > 0$,

$$\int_{|x-z|>2|z-\bar{z}|} \left\|\mathcal{K}(x, z) - \mathcal{K}(x, \bar{z})\right\|_X dx \leq C \tag{4.5}$$

  and

$$\int_{|x-z|>2|x-\bar{x}|} \left\|\mathcal{K}(x, z) - \mathcal{K}(\bar{x}, z)\right\|_X dz \leq C \tag{4.6}$$

- For every $f \in L_c^\infty(\mathbb{R}^n)$ and $x \notin \operatorname{supp}(f)$, we can represent

$$Tf(x) = \int_{\mathbb{R}^n} \mathcal{K}(x, z) f(z) \, dz$$

  where $\mathcal{K}(x, z)f(z)$ denotes the product of the scalar $f(z)$ and the vector $\mathcal{K}(x, z) \in X$ for every x and z, and the integral is understood in the sense of Bochner integration of functions with values in Banach spaces.

The reader can verify, by replacing $|\cdot|$ by $\|\cdot\|_X$ in the proofs where necessary, that the Calderón–Zygmund theorem (Theorem 4.8) holds in this vector-valued setting, giving that T extends as a weak-type $(1, 1)$ operator, namely,

$$|\{x \in \mathbb{R}^n : \|Tf(x)\|_X > \lambda\}| \leq \frac{C}{\lambda}\|f\|_{L^1(\mathbb{R}^n)}$$

and as a bounded operator from $L^p(\mathbb{R}^n)$ to $L^p(\mathbb{R}^n; X)$ for $1 < p < \infty$. Furthermore, one can also assume that the functions f are vector-valued (so that the kernel $\mathcal{K}$ will be a bounded linear operator), giving the parallel result.

One may naively think that this vector-valued generalization is just a mere curiosity. But it turns out to be a very powerful technique as it allows to linearize nonlinear operators that otherwise would be extremely difficult to analyze. We will not get into all these important applications here, but consider the example of the Hardy–Littlewood maximal function next.

Example 4.16 (Hardy–Littlewood maximal function)**.** For a locally integrable function f, the **Hardy–Littlewood maximal function** is defined as

$$Mf(x) = \sup_{r>0} \frac{1}{|B_r(x)|} \int_{B_r(x)} |f(z)| \, dz.$$

If one takes supremum over all the cubes Q centered at x (instead of Euclidean balls) then the resulting maximal operator is pointwise equivalent to M. Similarly, one can consider maximal operators where the supremum is taken over all cubes or balls

that contain the point x (not necessarily centered at x), obtaining again pointwise equivalent definitions.

Clearly, M is a sublinear operator on $L^\infty(\mathbb{R}^n)$ with $\|Mf\|_{L^\infty(\mathbb{R}^n)} \leq \|f\|_{L^\infty(\mathbb{R}^n)}$. What is not so clear is that M is in fact a (vector-valued) Calderón–Zygmund operator. Thus, by the vector-valued Calderón–Zygmund theory, see Remark 4.15(iii), we have that M is of weak-type $(1,1)$ and bounded on $L^p(\mathbb{R}^n)$ for $1 < p \leq \infty$.

To prove this, note first that if Φ is a Schwartz class function such that $\frac{1}{|B_1|}\chi_{B_1} \leq \Phi$ then, by denoting $\Phi_r(x) = \frac{1}{r^n}\Phi(\frac{x}{r})$,

$$\begin{aligned} Mf(x) &= \sup_{r>0} \int_{\mathbb{R}^n} \frac{1}{r^n|B_1|}\chi_{B_1}\Big(\frac{x-z}{r}\Big)|f(z)|\,dz \\ &\leq \sup_{r>0} \int_{\mathbb{R}^n} \Phi_r(x-z)|f(z)|\,dz. \end{aligned}$$

Therefore, if we define the maximal operator

$$M_\Phi f(x) = \sup_{r>0} \left| \int_{\mathbb{R}^n} \Phi_r(x-z) f(z)\,dz \right|,$$

then $Mf(x) \leq M_\Phi|f|(x)$. If we show that M_Φ is of weak-type $(1,1)$ and bounded on $L^p(\mathbb{R}^n)$, $1 < p \leq \infty$, then so will be the Hardy–Littlewood maximal function M.

Now, since $(\Phi_r * f)(x)$ is a continuous function of $r > 0$, it is enough to take the supremum in the definition of M_Φ over all positive rationals $r \in \mathbb{Q}_+$. In particular, M_Φ is measurable. Consider then the Banach space $X = \ell^\infty(\mathbb{Q}_+)$ of bounded sequences over $\mathbb{Q}_+$. Clearly,

$$M_\Phi f(x) = \left\| \int_{\mathbb{R}^n} \Phi_r(x-z) f(z)\,dz \right\|_X.$$

Therefore, we can look at M_Φ as the norm in X of a *linear* operator T mapping $f(x)$ into functions $Tf(x)$ having values in X:

$$M_\Phi f(x) = \|Tf(x)\|_X$$

where

$$Tf(x) = \left\{ \int_{\mathbb{R}^n} \Phi_r(x-z) f(z)\,dz \right\}_{r\in\mathbb{Q}_+} = \int_{\mathbb{R}^n} \{\Phi_r(x-z)\}_{r\in\mathbb{Q}_+} f(z)\,dz.$$

Thus, if we define the vector-valued kernel $\mathcal{K}(x,z)$ by

$$(x,z) \longmapsto \mathcal{K}(x,z) = \{\Phi_r(x-z)\}_{r\in\mathbb{Q}_+},$$

then T can be expressed as

$$Tf(x) = \int_{\mathbb{R}^n} \mathcal{K}(x,z) f(z)\,dz.$$

This formula is true for every x and every $f \in L_c^\infty(\mathbb{R}^n)$. We check that T is bounded from $L^\infty(\mathbb{R}^n)$ into $L^\infty(\mathbb{R}^n; X)$:

$$\sup_{x\in\mathbb{R}^n} \|Tf(x)\|_X = \|M_\Phi f\|_{L^\infty(\mathbb{R}^n)} \leq \|\Phi\|_{L^1(\mathbb{R}^n)} \|f\|_{L^\infty(\mathbb{R}^n)}.$$

Next, since Φ is a Schwartz class function,

$$\begin{aligned}\|\mathcal{K}(x,z)\|_X &= \sup_{r\in\mathbb{Q}_+} |\Phi_r(x-z)| \\ &= \frac{1}{|x-z|^n} \sup_{r\in\mathbb{Q}_+} \left| \left(\frac{|x-z|}{r}\right)^n \Phi\left(\frac{x-z}{r}\right)\right| \\ &\le \frac{C}{|x-z|^n}\end{aligned}$$

and, similarly,

$$\begin{aligned}\|\nabla_x \mathcal{K}(x,z)\|_X + \|\nabla_z \mathcal{K}(x,z)\|_X &\le \frac{2}{|x-z|^{n+1}} \sup_{r\in\mathbb{Q}_+} \left| \left(\frac{|x-z|}{r}\right)^{n+1} \nabla\Phi\left(\frac{x-z}{r}\right)\right| \\ &\le \frac{C}{|x-z|^{n+1}}.\end{aligned}$$

Therefore, $\mathcal{K}$ is a (vector-valued) standard kernel, so it satisfies Hörmander conditions (4.5) and (4.6). However, we cannot directly apply the vector-valued Calderón–Zygmund theorem of Remark 4.15(*iii*) just yet because $X = \ell^\infty(\mathbb{Q}_+)$ is not separable. To fix this, consider, for any finite subset J of $\mathbb{Q}_+$, the operator

$$M_\Phi^J f(x) = \sup_{r\in J} \left| \int_{\mathbb{R}^n} \Phi_r(x-z) f(z)\, dz \right|.$$

Obviously, the analysis above can be carried out with $X = \ell^\infty(J)$ in place of $\ell^\infty(\mathbb{Q}_+)$. The advantage now is that $\ell^\infty(J)$ is a separable Banach space. Then the vector-valued Calderón–Zygmund theorem applies to M_Φ^J, giving boundedness with constants independent of J. To pass to M_Φ, it is enough to observe that, by the continuity of $r \longmapsto (\Phi_r * f)(x)$ we can always choose a sequence J_k such that $M_\Phi^{J_k} f(x) \to M_\Phi f(x)$ for every x, as $k \to \infty$. Therefore, we conclude that T is of weak-type $(1,1)$:

$$|\{x \in \mathbb{R}^n : |M_\Phi f(x)| > \lambda\}| = |\{x \in \mathbb{R}^n : \|Tf(x)\|_X > \lambda\}| \le \frac{C}{\lambda}\|f\|_{L^1(\mathbb{R}^n)}$$

and bounded from $L^p(\mathbb{R}^n)$ to $L^p(\mathbb{R}^n; X)$:

$$\|M_\Phi f\|_{L^p(\mathbb{R}^n)}^p = \int_{\mathbb{R}^n} \|Tf(x)\|_X^p\, dx \le C\|f\|_{L^p(\mathbb{R}^n)}^p.$$

Example 4.17 (Maximal functions in metric spaces)**.** Let (Ω, d, μ) be a metric measure space, that is, Ω is a metric space with metric d and measure μ. Assume that the set of continuous functions over Ω with bounded support is dense in $L^1(\Omega, d\mu)$. Furthermore, suppose that the metric balls $B_r(x)$ of radius $r > 0$ centered at $x \in \Omega$ are measurable and satisfy the doubling property $\mu(B_{2r}(x)) \le c\mu(B_r(x)) < \infty$ for some constant $c > 1$, for all r and x. Consider the maximal function

$$M_\Omega f(x) = \sup_{r>0} \frac{1}{\mu(B_r(x))} \int_{B_r(x)} |f|\, d\mu.$$

Then it can be proved that M_Ω is bounded on $L^p(\Omega, d\mu)$, $1 < p \leq \infty$, and satisfies a weak-type $(1, 1)$ estimate: there is a constant $C > 0$ depending on (Ω, d, μ) such that for any $t > 0$ and $f \in L^1(\Omega, d\mu)$,

$$\mu(\{x \in \Omega : Mf(x) > t\}) \leq \frac{C}{t} \int_\Omega |f| \, d\mu.$$

In fact, the result is valid when d is replaced by a quasi-metric.

4.4 THE BMO SPACE

As we saw in Example 4.14, Calderón–Zygmund operators are not bounded on L^∞. This means that if $f \in L^\infty$ then solutions to $-\Delta u = f$ do not in general satisfy $D^2u \in L^\infty$, namely, $u \notin C^{1,1}$, see also Example 3.18. The correct endpoint space for $p = \infty$ is the BMO space, which is larger than L^∞. In fact, we will show that if $-\Delta u = f$ and $f \in L^\infty$ then $D^2u \in BMO$.

Definition 4.18. Let $f \in L^1_{\text{loc}}(\mathbb{R}^n)$. For any cube $Q \subset \mathbb{R}^n$ we denote

$$f_Q = \frac{1}{|Q|} \int_Q f \, dx.$$

We say that f belongs to the space BMO (**bounded mean oscillation**) if

$$\|f\|_{BMO} = \sup_Q \frac{1}{|Q|} \int_Q |f - f_Q| \, dx < \infty.$$

The quantity $\|f\|_{BMO}$ is a seminorm: if f is constant then $\|f\|_{BMO} = 0$. By identifying functions in BMO as equal whenever they differ by a constant, one can check that $(BMO, \|\cdot\|_{BMO})$ becomes a Banach space.

Lemma 4.19. *For any $f \in BMO$ we have*

$$\frac{1}{2}\|f\|_{BMO} \leq \sup_Q \inf_{a \in \mathbb{R}} \frac{1}{|Q|} \int_Q |f(x) - a| \, dx \leq \|f\|_{BMO}.$$

Therefore, $f \in BMO$ if and only if there is a constant $C > 0$ such that for any cube Q there exists a constant a_Q, that may depend on Q, such that

$$\frac{1}{|Q|} \int_Q |f(x) - a_Q| \, dx \leq C.$$

Proof. The second inequality is immediate. For the first one,

$$\int_Q |f - f_Q| \, dx \leq \int_Q |f - a| \, dx + \int_Q |a - f_Q| \, dx \leq 2 \int_Q |f - a| \, dx.$$

Divide both sides by $2|Q|$, take the infimum over a and the supremum over Q. □

Clearly, $L^\infty(\mathbb{R}^n) \subset BMO$, with $\|f\|_{BMO} \leq 2\|f\|_{L^\infty(\mathbb{R}^n)}$. But, as we will see after the following main two results, there are functions in BMO that are not bounded.

Theorem 4.20 (Calderón–Zygmund operators on L^∞_c). *Let T be a Calderón–Zymgund operator associated with a standard kernel $\mathcal{K}(x,z)$. If $f \in L^\infty_c(\mathbb{R}^n)$, then $Tf \in BMO$ and there exists a constant $C > 0$ independent of f such that*

$$\|Tf\|_{BMO} \leq C\|f\|_{L^\infty(\mathbb{R}^n)}.$$

Proof. The following is a typical technique to estimate BMO norms. Let Q be a cube with center z_Q. Let Q^* be the cube centered at z_Q whose side length is twice the side length of Q. We decompose

$$f = f\chi_{Q^*} + (f - f\chi_{Q^*}) = f\chi_{Q^*} + f\chi_{(Q^*)^c} = f_1 + f_2.$$

Since f_2 has compact support and $z_Q \notin \operatorname{supp}(f_2)$, we can write

$$a_Q = Tf_2(z_Q) = \int_{\mathbb{R}^n} \mathcal{K}(z_Q, z) f_2(z)\,dz.$$

Then, by applying the L^2 boundedness of T and using the smoothness of $\mathcal{K}$ with respect to the first variable as we did in (4.4),

$$\begin{aligned}
\frac{1}{|Q|}\int_Q |Tf - a_Q|\,dx &\leq \frac{1}{|Q|}\int_Q |Tf_1|\,dx + \frac{1}{|Q|}\int_Q |Tf_2(x) - Tf_2(z_Q)|\,dx \\
&\leq \left(\frac{1}{|Q|}\int_Q |Tf_1|^2\,dx\right)^{1/2} + \frac{1}{|Q|}\int_Q \left|\int_{(Q^*)^c} (K(x,z) - K(z_Q,z))f(z)\,dz\right| dx \\
&\leq \left(\frac{C}{|Q|}\int_{Q^*} |f|^2\,dx\right)^{1/2} + \frac{\|f\|_{L^\infty(\mathbb{R}^n)}}{|Q|}\int_Q\int_{(Q^*)^c} |K(x,z) - K(z_Q,z)|\,dz\,dx \\
&\leq C\|f\|_{L^\infty(\mathbb{R}^n)}.
\end{aligned}$$

□

The immediate consequence of Theorem 4.20 is the following BMO estimate for the Poisson problem with bounded right hand side.

Corollary 4.21. *Let f be a bounded function with support contained in Ω. Let $u = (-\Delta)^{-1}f = K * f$. Then $D^2u \in BMO$ and*

$$\|D^2u\|_{BMO} \leq C\|f\|_{L^\infty(\Omega)}.$$

In particular, $R_2 : L^\infty_c(\mathbb{R}^n) \to BMO$ is well defined, linear and bounded.

Example 4.22. In Example 4.14 we computed the Hilbert transform of $g(x) = \chi_{(0,1)}(x)$. Since g is bounded with compact support and the Hilbert transform is a Calderón–Zygmund operator, Theorem 4.20 gives that

$$Hg(x) = \frac{1}{\pi}\log\left|\frac{x}{x-1}\right|$$

is an unbounded function in BMO.

4.5 THE JOHN–NIRENBERG INEQUALITY

This section is devoted to the proof of the John–Nirenberg inequality, which establishes the exponential integrability of functions in BMO. In particular, by Corollary 4.21, it applies to D^2u, where $u = (-\Delta)^{-1}f$ and $f \in L_c^\infty(\Omega)$. A consequence of the John–Nirenberg inequality given in Corollary 4.24 below will be very important in Chapter 6 to close a gap in the proof of the Harnack inequality for weak solutions to divergence form equations, see Section 6.3.

Theorem 4.23 (John–Nirenberg inequality). *There exists a constant $C_n > 0$ depending only on n such that, for any $f \in BMO$, any cube Q and $\lambda > 0$,*

$$|\{x \in Q : |f(x) - f_Q| > \lambda\}| \leq \sqrt{2}e^{-C_n\lambda/\|f\|_{BMO}}|Q|. \tag{4.7}$$

Proof. We can assume that $\|f\|_{BMO} = 1$. Otherwise we consider $g = f/\|f\|_{BMO}$, for which $g_Q = f_Q/\|f\|_{BMO}$ and $\|g\|_{BMO} = 1$. Applying (4.7) to g at level $\lambda/\|f\|_{BMO}$,

$$|\{x \in Q : |g(x) - g_Q| > \lambda/\|f\|_{BMO}\}| \leq \sqrt{2}e^{-C_n\lambda/\|f\|_{BMO}}|Q|$$

and the inequality for f follows.

We saw in the proof of the Calderón–Zygmund theorem (Theorem 4.8) that the Calderón–Zygmund decomposition (Lemma 4.12) is the key tool for proving the weak-type $(1,1)$ estimate. Indeed, the decomposition is crucial to control the measure of superlevel sets, and this is what we also need to do to prove (4.7).

Fix any cube Q. It is enough to prove the following.

Claim. For each $N \geq 1$ there is a family of disjoint cubes $\{Q_{N,j}\}_{j=1}^\infty$ such that

$$|f(x) - f_Q| \leq N2^{n+1} \qquad \text{for all } x \in Q \setminus \bigcup_{j=1}^\infty Q_{N,j} \tag{4.8}$$

and

$$\sum_{j=1}^\infty |Q_{N,j}| \leq \frac{1}{2^N}|Q|. \tag{4.9}$$

Assume for the moment that the claim is true. To prove the John–Nirenberg inequality we consider two cases. Suppose first that $\lambda \geq 2^{n+1}$. Let $N \geq 1$ be such that $N2^{n+1} \leq \lambda < (N+1)2^{n+1}$. Then, by (4.8) and (4.9),

$$\begin{aligned}
|\{x \in Q : |f(x) - f_Q| > \lambda\}| &\leq |\{x \in Q : |f(x) - f_Q| > N2^{n+1}\}| \\
&\leq \left|\bigcup_{j=1}^\infty Q_{N,j}\right| \\
&\leq \sum_{j=1}^\infty |Q_{N,j}| \\
&\leq \frac{1}{2^N}|Q| \\
&= e^{-N\log 2}|Q| \leq e^{-C_n\lambda}|Q|
\end{aligned}$$

where we have chosen $C_n = \log 2/2^{n+2}$. Notice that indeed we have $C_n\lambda \le N \log 2$, because this is equivalent to $\lambda/2^{n+1} \le 2N$, which again is true because $\lambda/2^{n+1} < N+1$ and $N \ge 1$. Once C_n has been chosen, we consider the second case. Suppose that $\lambda < 2^{n+1}$. Then

$$C_n\lambda = \frac{\log 2}{2^{n+2}}\lambda < \frac{\log 2}{2} = \log\sqrt{2}.$$

Hence, it trivially follows that

$$|\{x \in Q : |f(x) - f_Q| > \lambda\}| \le |Q| \le \sqrt{2}e^{-C_n\lambda}|Q|.$$

Let us now turn to the proof of the claim. As anticipated, we need to apply the Calderón–Zygmund decomposition given in Lemma 4.12. But we will use it in successive ways. Recall that

$$\frac{1}{|Q|}\int_Q |f(x) - f_Q|\,dx \le 1. \tag{4.10}$$

We form the Calderón–Zygmund decomposition of $f - f_Q$ on Q at height 2. This is done exactly as in the usual Calderón–Zygmund decomposition on $\mathbb{R}^n$ except that we begin by bisecting the sides of Q to form 2^n equal cubes and repeating this for all new cubes. We obtain a family of disjoint cubes $\{Q_{1,j}\}_{j=1}^{\infty}$ such that

$$2 < \frac{1}{|Q_{1,j}|}\int_{Q_{1,j}} |f(x) - f_Q|\,dx \le 2^{n+1}$$

and

$$|f(x) - f_Q| \le 2 \le 2^n \qquad \text{for all } x \in Q \setminus \bigcup_{j=1}^{\infty} Q_{1,j}, \tag{4.11}$$

see (vi) and (v) in Lemma 4.12, respectively. In particular,

$$|f_{Q_{1,j}} - f_Q| = \left|\frac{1}{|Q_{1,j}|}\int_{Q_{1,j}} (f(x) - f_Q)\,dx\right| \le 2^{n+1}.$$

Moreover, by Lemma 4.12(iv) and (4.10),

$$\sum_{j=1}^{\infty} |Q_{1,j}| \le \frac{1}{2}\int_Q |f - f_Q|\,dx \le \frac{1}{2}|Q|. \tag{4.12}$$

On each cube $Q_{1,j}$, $j \ge 1$, we form the Calderón–Zygmund decomposition of $f - f_{Q_{1,j}}$ at height 2. The average of this function over the cube $Q_{1,j}$ is at most 1. Then we obtain a family of cubes $\{Q_{1,j,k}\}_{k=1}^{\infty}$ such that (in a parallel way as before)

$$|f_{Q_{1,j,k}} - f_{Q_{1,j}}| \le 2^{n+1}$$

$$|f(x) - f_{Q_{1,j}}| \le 2 \qquad \text{for all } x \in Q_{1,j} \setminus \bigcup_{k=1}^{\infty} Q_{1,j,k}$$

$$\sum_{k=1}^{\infty} |Q_{1,j,k}| \le \frac{1}{2}|Q_{1,j}|. \tag{4.13}$$

We gather together all the cubes $\{Q_{1,j,k}\}_{k=1}^{\infty}$ corresponding to all of the cubes $\{Q_j\}_{j=1}^{\infty}$ and relabel this family as $\{Q_{2,j}\}_{j=1}^{\infty}$. Then, from (4.13) and (4.12),

$$\sum_{j=1}^{\infty} |Q_{2,j}| \leq \frac{1}{4}|Q| \tag{4.14}$$

and for $x \in Q \setminus \bigcup_{j=1}^{\infty} Q_{2,j}$,

$$|f(x) - f_Q| \leq |f(x) - f_{Q_{1,j}}| + |f_{Q_{1,j}} - f_Q| \leq 2 + 2^{n+1} \leq 2 \cdot 2^{n+1}. \tag{4.15}$$

We repeat this process indefinitely, and the claim holds: for (4.8), see (4.11) and (4.15); and for (4.9), see (4.12) and (4.14). □

We give several important consequences from the John–Nirenberg inequality.

Corollary 4.24. *If $1 < p < \infty$ then*

$$\|f\|_{BMO,p} = \sup_Q \left(\frac{1}{|Q|} \int_Q |f(x) - f_Q|^p \, dx \right)^{1/p}$$

is a norm in BMO equivalent to $\|f\|_{BMO}$. Moreover, there exists $0 < \theta < 1$ depending only on n such that for any $f \in BMO$ with $\|f\|_{BMO} \leq 1$,

$$\sup_Q \frac{1}{|Q|} \int_Q e^{\theta |f(x) - f_Q|} \, dx \leq 2 + 2\sqrt{2}.$$

In particular, any $f \in BMO$ with $\|f\|_{BMO} \leq 1$ satisfies

$$\sup_Q \left(\frac{1}{|Q|} \int_Q e^{\theta f(x)} \, dx \right) \left(\frac{1}{|Q|} \int_Q e^{-\theta f(x)} \, dx \right) \leq 12 + 8\sqrt{2}. \tag{4.16}$$

Proof. By Hölder's inequality, $\|f\|_{BMO} \leq \|f\|_{BMO,p}$. By (4.7) in Theorem 4.23,

$$\begin{aligned}
\frac{1}{|Q|} \int_Q |f(x) - f_Q|^p \, dx &= \frac{1}{|Q|} \int_Q \int_0^{|f(x) - f_Q|} p\lambda^{p-1} \, d\lambda \, dx \\
&= \frac{1}{|Q|} \int_0^{\infty} p\lambda^{p-1} |\{x \in Q : |f(x) - f_Q| > \lambda\}| \, d\lambda \\
&\leq \sqrt{2} \int_0^{\infty} p\lambda^{p-1} e^{-C_n \lambda / \|f\|_{BMO}} \, d\lambda \\
&= p\sqrt{2} C_n^{-p} \|f\|_{BMO}^p \int_0^{\infty} r^{p-1} e^{-r} \, dr \\
&= p\sqrt{2} C_n^{-p} \Gamma(p) \|f\|_{BMO}^p
\end{aligned}$$

where Γ denotes the Gamma function. The equivalence of norms is proved.

Let $0 < \theta < 1$. By using the previous estimate and $\|f\|_{BMO} \leq 1$,

$$\begin{aligned}\frac{1}{|Q|}\int_Q e^{\theta|f(x)-f_Q|}\,dx &= \frac{1}{|Q|}\sum_{k=0}^{\infty}\frac{\theta^k}{k!}\int_Q |f(x)-f_Q|^k\,dx \\ &\leq 1+\theta+\sqrt{2}\sum_{k=2}^{\infty}\frac{\theta^k}{k!}kC_n^{-k}\Gamma(k)\|f\|_{BMO}^k \\ &\leq 2+\sqrt{2}\sum_{k=0}^{\infty}\left(\frac{\theta}{C_n}\right)^k \\ &\leq 2+\sqrt{2}\sum_{k=0}^{\infty}\left(\frac{1}{2}\right)^k = 2+2\sqrt{2}\end{aligned}$$

as soon as θ is chosen so that $\theta/C_n < 1/2$.

As $e^{\pm\theta(f(x)-f_Q)} \leq e^{\theta|f(x)-f_Q|}$, the inequality just proved implies that

$$\frac{1}{|Q|}e^{\mp\theta f_Q}\int_Q e^{\pm\theta f(x)}\,dx \leq 2+2\sqrt{2}.$$

By multiplying both inequalities we reach (4.16). □

4.6 PRINCIPAL VALUE REPRESENTATION OF SECOND DERIVATIVES

We know from Chapter 3 that if $f \in C_c^\infty(\mathbb{R}^n)$ and $u = (-\Delta)^{-1}f = K * f$ then $u \in C^1(\mathbb{R}^n)$ and

$$\partial_j u(x) = \int_{\mathbb{R}^n}\partial_j K(x-z)f(z)\,dz = \int_{\mathbb{R}^n}\partial_j K(z)f(x-z)\,dz.$$

Since f is smooth with compact support, for any $x \in \mathbb{R}^n$ we can differentiate with respect to x again inside of the integral and apply the divergence theorem to get

$$\begin{aligned}\partial_{ij}u(x) &= \int_{\mathbb{R}^n}\partial_j K(z)\partial_{x_i}f(x-z)\,dz = -\int_{\mathbb{R}^n}\partial_j K(z)\partial_{z_i}f(x-z)\,dz \\ &= -\lim_{\varepsilon\to 0}\int_{|z|>\varepsilon}\partial_j K(z)\partial_{z_i}f(x-z)\,dz \\ &= \lim_{\varepsilon\to 0}\left[\int_{|z|>\varepsilon}\partial_{ij}K(z)f(x-z)\,dz - \int_{|z|=\varepsilon}f(x-z)\partial_j K(z)\nu_i\,dS\right] \qquad (4.17) \\ &= \lim_{\varepsilon\to 0}\left[\int_{|x-z|>\varepsilon}\partial_{ij}K(x-z)f(z)\,dz - \frac{\Gamma(n/2)}{2\pi^{n/2}}\int_{|z|=\varepsilon}f(x-z)\frac{z_j}{|z|^n}\frac{z_i}{\varepsilon}\,dS\right] \\ &= \lim_{\varepsilon\to 0}\left[\int_{|x-z|>\varepsilon}\partial_{ij}K(x-z)f(z)\,dz - \frac{\Gamma(n/2)}{2\pi^{n/2}\varepsilon^{n+1}}\int_{|z|=\varepsilon}f(x-z)z_j z_i\,dS\right].\end{aligned}$$

Let us prove that

$$\lim_{\varepsilon\to 0}\frac{\Gamma(n/2)}{2\pi^{n/2}\varepsilon^{n+1}}\int_{|z|=\varepsilon}f(x-z)z_j z_i\,dS = f(x)\frac{\delta_{ij}}{n}. \qquad (4.18)$$

For this, we add and subtract $f(x)$ to the integrand and notice that

$$\frac{1}{\varepsilon^{n+1}}\int_{|z|=\varepsilon}|f(x-z)-f(x)||z_iz_j|\,dS \le \frac{\|\nabla f\|_{L^\infty(\mathbb{R}^n)}}{\varepsilon^{n+1}}\int_{|z|=\varepsilon}|z|^3\,dS = C\varepsilon \to 0$$

as $\varepsilon \to 0$. On the other hand, by (3.19),

$$\begin{aligned}\frac{\Gamma(n/2)}{2\pi^{n/2}\varepsilon^{n+1}}\int_{|z|=\varepsilon} z_jz_i\,dS &= \frac{\Gamma(n/2)\varepsilon^2\varepsilon^{n-1}}{2\pi^{n/2}\varepsilon^{n+1}}\int_{|z|=1} z_jz_i\,dS\\ &= \frac{\Gamma(n/2)}{2\pi^{n/2}}\delta_{ij}\int_{|z|=1}(z_1)^2\,dS\\ &= \frac{\Gamma(n/2)}{2\Gamma(n/2+1)}\delta_{ij} = \frac{1}{n}\delta_{ij}.\end{aligned}$$

Hence, (4.18) is true. Using this in (4.17), we find the following pointwise principal value representation for $D^2u(x)$:

$$\partial_{ij}u(x) = \lim_{\varepsilon\to 0}\int_{|x-z|>\varepsilon}\partial_{ij}K(x-z)f(z)\,dz - f(x)\frac{\delta_{ij}}{n} \tag{4.19}$$

whenever $f \in C_c^\infty(\mathbb{R}^n)$. In particular, (4.19) shows that $-\Delta u = f$ in $\mathbb{R}^n$ and that the limit as $\varepsilon \to 0$ of the truncated integrals

$$T_\varepsilon f(x) = \int_{|x-y|>\varepsilon}\partial_{ij}K(x-z)f(z)\,dz$$

converges at every point $x \in \mathbb{R}^n$, whenever $f \in C_c^\infty(\mathbb{R}^n)$. Our aim is to extend the representation (4.19) to the case when $f \in L_c^p(\mathbb{R}^n)$, $1 < p < \infty$.

Theorem 4.25 (Pointwise principal value representation formula for D^2u). *Let $f \in L^p(\mathbb{R}^n)$, $1 < p < \infty$, with support contained in Ω and $u = (-\Delta)^{-1}f = K * f$. Then*

$$\partial_{ij}u(x) = \lim_{\varepsilon\to 0}\int_{|x-z|>\varepsilon}\partial_{ij}K(x-z)f(z)\,dz - f(x)\frac{\delta_{ij}}{n}$$

in $L^p(\mathbb{R}^n)$ and for a.e. $x \in \mathbb{R}^n$.

It is important to stress that Theorem 4.25 requires a nontrivial proof, that we present now. Let f and u be as in the hypotheses of Theorem 4.25. We know that $-\Delta u \in L^p(\Omega)$ and $-\Delta u = f$ a.e. in Ω. By the Calderón–Zygmund theorem, $D^2u \in L^p(\mathbb{R}^n)$ and

$$\|D^2u\|_{L^p(\mathbb{R}^n)} \le C\|f\|_{L^p(\Omega)}. \tag{4.20}$$

Moreover, for $x \notin \operatorname{supp}(f)$, we can clearly represent

$$\begin{aligned}\partial_{ij}u(x) &= \int_{\mathbb{R}^n}\partial_{ij}K(x-z)f(z)\,dz\\ &= \lim_{\varepsilon\to 0}\int_{|x-z|>\varepsilon}\partial_{ij}K(x-z)f(z)\,dz - f(x)\frac{\delta_{ij}}{n}.\end{aligned}$$

Let us show that this representation of $\partial_{ij}u(x)$ also holds for a.e. $x \in \Omega$. To this end, let $f_k \in C_c^\infty(\Omega)$ such that $f_k \to f$ in $L^p(\Omega)$ and let $u_k = (-\Delta)^{-1}f_k$. By (4.20) we have that $D^2u_k \to D^2u$ in $L^p(\Omega)$. Suppose that the following two statements hold.

(*a*) The limit

$$\lim_{\varepsilon\to 0} T_\varepsilon f(x) = \lim_{\varepsilon\to 0} \int_{|x-z|>\varepsilon} \partial_{ij} K(x-z) f(z)\,dz$$

exists for a.e. $x \in \mathbb{R}^n$, for every $f \in L^p(\mathbb{R}^n)$, $1 < p < \infty$.

(*b*) The maximal operator

$$T_* f(x) = \sup_{\varepsilon>0} |T_\varepsilon f(x)| = \sup_{\varepsilon>0} \left| \int_{|x-z|>\varepsilon} \partial_{ij} K(x-z) f(z)\,dz \right|$$

is bounded on $L^p(\mathbb{R}^n)$, $1 < p < \infty$.

Then (*a*) gives that

$$\lim_{\varepsilon\to 0} T_\varepsilon f(x) = \lim_{\varepsilon\to 0} \int_{|x-z|>\varepsilon} \partial_{ij} K(x-z) f(z)\,dz$$

exists and is well defined for a.e. $x \in \Omega$ and for our particular $f \in L^p(\mathbb{R}^n)$ with support contained in Ω. Next, (*b*) implies that

$$\begin{aligned}\left\| \lim_{\varepsilon\to 0} T_\varepsilon f_k - \lim_{\varepsilon\to 0} T_\varepsilon f \right\|_{L^p(\Omega)} &= \left\| \lim_{\varepsilon\to 0} T_\varepsilon (f_k - f) \right\|_{L^p(\Omega)} \\ &\le \left\| \sup_{\varepsilon>0} |T_\varepsilon (f_k - f)| \right\|_{L^p(\Omega)} \\ &= \|T^*(f_k - f)\|_{L^p(\Omega)} \le C\|f_k - f\|_{L^p(\Omega)} \to 0\end{aligned}$$

as $k \to \infty$. In other words,

$$\partial_{ij} u_k(x) = \lim_{\varepsilon\to 0} \int_{|x-z|>\varepsilon} \partial_{ij} K(x-z) f_k(z)\,dz - f_k(x)\frac{\delta_{ij}}{n}$$

converges in $L^p(\Omega)$ to

$$\lim_{\varepsilon\to 0} \int_{|x-z|>\varepsilon} \partial_{ij} K(x-z) f(z)\,dz - f(x)\frac{\delta_{ij}}{n}.$$

By uniqueness of the limits in $L^p(\Omega)$, we have that

$$\partial_{ij} u(x) = \lim_{\varepsilon\to 0} \int_{|x-z|>\varepsilon} \partial_{ij} K(x-z) f(z)\,dz - f(x)\frac{\delta_{ij}}{n}$$

as functions in $L^p(\Omega)$ and, as a consequence, for a.e. $x \in \Omega$. Therefore, under the assumption that (*a*) and (*b*) hold, the proof of Theorem 4.25 is completed.

We are only left to prove (*a*) and (*b*). We will see that (*a*) is a consequence of (*b*). Now, (*b*) is given by the following result, which is valid for general Calderón–Zygmund operators (and can be extended to the vector-valued setting as well).

Theorem 4.26. *Let T be a Calderón–Zygmund operator associated with a standard kernel $\mathcal{K}(x,z)$. Define the maximal operator*

$$T_* f(x) = \sup_{\varepsilon>0} |T_\varepsilon f(x)| = \sup_{\varepsilon>0} \left| \int_{|x-z|>\varepsilon} \mathcal{K}(x,z) f(z)\,dz \right|.$$

Then T_* is bounded on $L^p(\mathbb{R}^n)$ for $1 < p < \infty$.

Notice that the truncated operator $T_\varepsilon f(x)$, $\varepsilon > 0$, of a Calderón–Zygmund operator T is well defined for any $x \in \mathbb{R}^n$ whenever $f \in L^p(\mathbb{R}^n)$, $1 \leq p < \infty$, because $\chi_{|x-z|>\varepsilon}|\mathcal{K}(x,z)| \leq C\chi_{|x-z|>\varepsilon}|x-z|^{-n}$ is in $L^q(\mathbb{R}^n)$ for $1 < q \leq \infty$.

Clearly, Theorem 4.26 is an immediate consequence of the following result and the L^p boundedness of M and T for $1 < p < \infty$.

Lemma 4.27 (Cotlar). *Let T be a Calderón–Zygmund operator associated with a standard kernel $\mathcal{K}(x,z)$. Then, for any $f \in C_c^\infty(\mathbb{R}^n)$,*

$$T_* f(x) \leq C(M(|Tf|)(x) + Mf(x)) \qquad \text{for } x \in \mathbb{R}^n$$

where $C > 0$ is a constant independent of f, and M is the Hardy–Littlewood maximal function.

Proof. For simplicity, we set $x = 0$. The proof for any other point follows the same lines. It is enough to prove that

$$|T_\varepsilon f(0)| \leq C(M(|Tf|)(0) + Mf(0)) \tag{4.21}$$

where $C > 0$ is a constant independent of $\varepsilon > 0$. Let us fix $\varepsilon > 0$. If $T_\varepsilon f(0) = 0$, there is nothing to prove. We then show that for any $0 < \lambda < |T_\varepsilon f(0)|$ we have

$$\lambda \leq C(M(|Tf|)(0) + Mf(0)) \tag{4.22}$$

with $C > 0$ independent of ε, from which (4.21) follows.

For (4.22), we begin by decomposing

$$f = f\chi_{B_\varepsilon} + f\chi_{(B_\varepsilon)^c} = f_1 + f_2.$$

Since $0 \notin \operatorname{supp}(f_2)$,

$$Tf_2(0) = \int_{\mathbb{R}^n} \mathcal{K}(0,z) f_2(z)\, dz = \int_{|z|>\varepsilon} \mathcal{K}(0,z) f(z)\, dz = T_\varepsilon f(0).$$

Now, consider any $x \in B_{\varepsilon/2}$. By the linearity of T, we can bound

$$|T_\varepsilon f(0)| \leq |Tf_2(0) - Tf_2(x)| + |Tf(x)| + |Tf_1(x)|.$$

We estimate the first term above by noticing that $x \notin \operatorname{supp}(f_2)$ and using the size estimate of $\nabla_x \mathcal{K}$:

$$\begin{aligned}
|Tf_2(0) - Tf_2(x)| &\leq \int_{|z|>\varepsilon} |\mathcal{K}(0,z) - \mathcal{K}(x,z)||f(z)|\,dz \\
&\leq C|x| \int_{|z|>\varepsilon} \frac{|f(z)|}{|z|^{n+1}}\,dz \\
&= C|x| \sum_{k=0}^{\infty} \int_{2^k\varepsilon<|z|\leq 2^{k+1}\varepsilon} \frac{|f(z)|}{|z|^{n+1}}\,dz \\
&\leq C\varepsilon \sum_{k=0}^{\infty} \frac{1}{(2^k\varepsilon)^{n+1}} \int_{|z|\leq 2^{k+1}\varepsilon} |f(z)|\,dz \\
&= C \sum_{k=0}^{\infty} \frac{1}{2^k} \left(\frac{1}{|B_{2^{k+1}\varepsilon}|} \int_{B_{2^{k+1}\varepsilon}} |f(z)|\,dz \right) \\
&\leq CMf(0) \sum_{k=0}^{\infty} \frac{1}{2^k} = cMf(0).
\end{aligned}$$

Therefore

$$|T_\varepsilon f(0)| \leq cMf(0) + |Tf(x)| + |Tf_1(x)| \qquad \text{for any } x \in B_{\varepsilon/2}. \tag{4.23}$$

Now, let $\lambda < |T_\varepsilon f(0)|$. Then, by (4.23), $cMf(0) > \lambda/3$, or $|Tf(x)| > \lambda/3$, or $|Tf_1(x)| > \lambda/3$, for any $x \in B_{\varepsilon/2}$. If $\lambda < 3cMf(0)$, then we are done as (4.22) is trivial. If not, we can split

$$B_{\varepsilon/2} = \{x \in B_{\varepsilon/2} : |Tf(x)| > \lambda/3\} \cup \{x \in B_{\varepsilon/2} : |Tf_1(x)| > \lambda/3\}.$$

By the Chebyshev and weak-type $(1,1)$ inequalities,

$$\begin{aligned}
|B_{\varepsilon/2}| &\leq |\{x \in B_{\varepsilon/2} : |Tf(x)| > \lambda/3\}| + |\{x \in B_{\varepsilon/2} : |Tf_1(x)| > \lambda/3\}| \\
&\leq \frac{3}{\lambda} \int_{B_{\varepsilon/2}} |Tf(x)|\,dx + \frac{3C}{\lambda} \int_{B_\varepsilon} |f(x)|\,dx \\
&\leq \frac{3}{\lambda} |B_{\varepsilon/2}| M(Tf)(0) + \frac{3C}{\lambda} 2^n |B_{\varepsilon/2}| Mf(0)
\end{aligned}$$

and this is (4.22). □

As we anticipated, (a) will be a consequence of (b) and the following real analysis result, whose proof is omitted.

Lemma 4.28. *Let $\{T_t\}$ be a family of linear operators on $L^p(X,\mu)$, where (X,μ) is a measure space and $1 \leq p \leq \infty$. Define the maximal operator*

$$T_* f(x) = \sup_t |T_t f(x)|.$$

If T_* is of weak-type (p,p) then the set

$$\left\{ f \in L^p(X,\mu) : \lim_{t \to t_0} T_t f(x) \text{ exists a.e.} \right\}$$

is closed in $L^p(X,\mu)$.

Therefore, if T is a Calderón–Zygmund operator, Theorem 4.26 implies that T_* is of weak-type (p,p), $1 < p < \infty$, and so the set of functions $f \in L^p(\mathbb{R}^n)$ for which the limit

$$\lim_{\varepsilon \to 0} \int_{|z|>\varepsilon} \mathcal{K}(x,z) f(z)\, dz$$

exists a.e. is closed in $L^p(\mathbb{R}^n)$. We apply this to $\mathcal{K}(x,z) = D^2K(x-z)$, which is a standard kernel, to get (a). Indeed, we now know that the set of functions $f \in L^p(\mathbb{R}^n)$ for which the limit

$$\lim_{\varepsilon \to 0} \int_{|z|>\varepsilon} \partial_{ij} K(x-z) f(z)\, dz$$

exists a.e. is closed in $L^p(\mathbb{R}^n)$. We also know that such limit exists for every function $f \in C_c^\infty(\mathbb{R}^n)$. Therefore, the set where the limit exists a.e. contains $C_c^\infty(\mathbb{R}^n)$ and, since Lemma 4.28 says that such a set is closed in $L^p(\mathbb{R}^n)$, it is the whole $L^p(\mathbb{R}^n)$, $1 < p < \infty$. This concludes the proof of (a).

NOTES ON CHAPTER 4

The Calderón–Zygmund theorem and decomposition were originally proved in the famous paper by Caderón–Zygmund [18]. The proofs of the Marcinkiewicz interpolation lemma and Lemma 4.28, can be seen, for instance, in Duoandikoetxea [30]. Properties of the Hilbert transform and its relation to complex analysis and the conjugate function on the torus, as well as more about the Hardy–Littlewood maximal function, including other proofs of the weak-type $(1,1)$ inequality, can be found, for instance, in Duoandikoetxea [30] or Stein [63, 64].

The vector-valued theory has its origins in Benedek–Calderón–Panzone [2]. However, it was fully developed and applied to new examples by Rubio de Francia–Ruiz–Torrea [54], see also Torrea [73]. The boundedness of the maximal operator M_Φ when $\Phi \in L^1(\mathbb{R}^n)$ satisfies a Hörmander condition was originally shown by Zó in [75]. The proof we give here is taken from [54, 73]. For Example 4.17, see Calderón [17].

Then John–Nirenberg inequality of Theorem 4.23 was proved in [46], and we follow their paper and Duoandikoetxea [30].

The a.e. pointwise formula for $D^2u(x)$ in Theorem 4.25 also holds when $p = 1$. Our presentation of Theorem 4.26 follows the lines of Duoandikoetxea [30], where one can also find a proof for the endpoint case $p = 1$.

II

Divergence form Equations

CHAPTER 5

The De Giorgi theorem

In this chapter we prove the De Giorgi theorem, which says that weak solutions to divergence form elliptic equations with bounded measurable coefficients are locally bounded and Hölder continuous. The De Giorgi method is divided into two steps: $L^2 \implies L^\infty$ and $L^\infty \implies C^\alpha$. We will give two proofs of the second step: the original one by De Giorgi, which is geometric in nature, and an alternative proof by Moser based on a critical density estimate.

5.1 THE DE GIORGI THEOREM

We consider divergence form equations

$$\operatorname{div}(a(x)\nabla u) = 0 \qquad \text{in } \Omega$$

where Ω is a bounded domain of $\mathbb{R}^n$, $n \geq 2$. The coefficients $a(x) = (a^{ij}(x))_{i,j=1}^n$ are bounded, measurable functions in Ω with $\|a\|_{L^\infty(\Omega)} \leq \Lambda$ for some $\Lambda > 0$, symmetric $a^{ij}(x) = a^{ji}(x)$, and uniformly elliptic: there exists a constant $\lambda > 0$ such that

$$a^{ij}(x)\xi_i\xi_j \geq \lambda|\xi|^2$$

for all $\xi \in \mathbb{R}^n$ and a.e. $x \in \Omega$.

We say that $u \in H^1(\Omega)$ is a **weak subsolution** to the equation in Ω, written

$$\operatorname{div}(a(x)\nabla u) \geq 0 \qquad \text{in } \Omega$$

if

$$\int_\Omega a(x)\nabla u \nabla\varphi\, dx \leq 0$$

for every $\varphi \in C_c^1(\Omega)$ such that $\varphi \geq 0$. We say that $u \in H^1(\Omega)$ is a **weak supersolution** in Ω if $-u$ is a subsolution in Ω. Finally, $u \in H^1(\Omega)$ is a **weak solution** in Ω if it is both a subsolution and a supersolution. By density, equivalent definitions are obtained if we take $\varphi \in H_0^1(\Omega)$ in the definitions above. If u is a solution in Ω then, for any $\varphi \in H_0^1(\Omega)$,

$$\int_\Omega a(x)\nabla u \nabla\varphi\, dx = 0.$$

DOI: 10.1201/9781032679853-5

irrespectively of the sign of φ. Indeed, we can write $\varphi = \varphi^+ - \varphi^-$, where $\varphi^+ = \max(\varphi, 0) \geq 0$ and $\varphi^- = -\min(\varphi, 0) \geq 0$, and $\nabla\varphi = \nabla\varphi^+ - \nabla\varphi^-$.

For the sake of simplicity we will set $\Omega = B_1$ or B_2. The general case of a bounded domain Ω can always be deduced from this one via translation and rescaling, see Corollary 2.15(1) in Chapter 2 or Corollary 5.3 below for examples of this technique.

Theorem 5.1 (De Giorgi). *There exists $0 < \alpha < 1$ depending only on n, λ and Λ such that if $u \in H^1(B_1)$ is a solution to $\operatorname{div}(a(x)\nabla u) = 0$ in B_1 then $u \in C^{0,\alpha}(B_{1/2})$ and*

$$\|u\|_{C^{0,\alpha}(B_{1/2})} \leq C\|u\|_{L^2(B_1)}$$

where $C > 0$ is a constant depending only on n, λ and Λ.

The proof is split into two steps:

Step I: from L^2 to L^∞. Here we show that the weak solution $u \in H^1(B_1)$ is actually locally bounded. In fact, it is sufficient to prove that u^+ is bounded above by $C\|u\|_{L^2(B_1)}$ as we can always flip the picture by considering $-u$. The key is to obtain a recurrence relation for successive truncations of u by an interplay between the Sobolev and energy inequalities.

Step II: from L^∞ to C^α. Once we know that u is locally bounded, we show that $u \in C^\alpha(0)$. We present two proofs. The first one is the original proof of De Giorgi, and it is based on the $L^2 \implies L^\infty$ lemma and an isoperimetric inequality for Sobolev functions to gain a decay in the oscillation of u. De Giorgi's proof is very geometrical in nature. The second one, due to Moser, follows from a so-called critical density estimate, whose proof relies on the fact that the negative part of the logarithm of any supersolution is uniformly in $H^1(B_1)$.

From now on, and for the rest of this chapter, when we say that u is a (sub/super)solution in Ω, we mean that $u \in H^1(\Omega)$ is a weak (sub/super)solution to $\operatorname{div}(a(x)\nabla u) = 0$ in Ω in the sense defined at the beginning of this section.

5.2 L^2 IMPLIES L^∞

In this section we focus on Step I and prove the following main result.

Lemma 5.2 ($L^2 \implies L^\infty$). *There exists $\delta_0 > 0$ depending only on n and Λ/λ such that for any subsolution u in B_1, if*

$$\|u^+\|_{L^2(B_1)} \leq \delta_0$$

then

$$\sup_{B_{1/2}} u^+ \leq 1/2.$$

The $L^2 \implies L^\infty$ lemma says that if the positive part u^+ of a subsolution u has very small mass in B_1 (or the subsolution u itself is negative in a large part of B_1) then in $B_{1/2}$ it cannot go very high, say, it cannot cross $1/2$. Notice that $\delta_0 = \delta_0(n, \Lambda/\lambda) > 0$ is a universal quantity, namely, it depends on the structural

conditions of the equation but not on u. We should compare Lemma 5.2 with the case of subharmonic functions. If u is subharmonic in B_1 then, for any $x \in B_{1/2}$, $B_{1/4}(x) \subset B_1$ and, by the sub-mean value property and Hölder's inequality,

$$u(x) \leq \frac{1}{|B_{1/4}(x)|} \int_{B_{1/4}(x)} u(z)\, dz \leq C_n \|u\|_{L^2(B_1)}.$$

Hence, if $\|u\|_{L^2(B_1)}$ is small enough (depending only on n) then $u \leq 1/2$ in $B_{1/2}$.

As we mentioned before, the general case of solutions on bounded domains Ω can be deduced from the result in B_1.

Corollary 5.3. *There exists a constant $C > 0$ depending on n and Λ/λ such that for any solution u in Ω and for any subdomain $\Omega' \subset\subset \Omega$ we have $u \in L^\infty(\Omega')$ with*

$$\|u\|_{L^\infty(\Omega')} \leq \frac{C}{\operatorname{dist}(\Omega', \partial\Omega)^{n/2}} \|u\|_{L^2(\Omega)}.$$

Proof. Let $\rho = \operatorname{dist}(\Omega', \partial\Omega) > 0$. Let $\delta_0 > 0$ be as in Lemma 5.2. For any $x \in \Omega'$, we look at (a normalization of) u in the ball $B_\rho(x) \subset \Omega$, that is, we consider

$$\tilde{u}(z) = \delta_0 \frac{\rho^{n/2}}{\|u\|_{L^2(\Omega)}} u(x + \rho z) \qquad \text{for } z \in B_1.$$

Then

$$u(z) = \frac{\|u\|_{L^2(\Omega)}}{\delta_0 \rho^{n/2}} \tilde{u}((z - x)/\rho) \qquad \text{for } z \in \Omega,\ |z - x| < \rho$$

and

$$\begin{aligned}\|\tilde{u}\|_{L^2(B_1)} &= \delta_0 \frac{\rho^{n/2}}{\|u\|_{L^2(\Omega)}} \left(\int_{B_1} |u(x + \rho z)|^2\, dz \right)^{1/2} \\ &= \delta_0 \frac{1}{\|u\|_{L^2(\Omega)}} \left(\int_{|z-x|<\rho} |u(z)|^2\, dz \right)^{1/2} \leq \delta_0.\end{aligned}$$

For any $\tilde{\varphi} \in H_0^1(B_1)$ we define $\varphi(z) = \tilde{\varphi}((z - x)/\rho)$, for $z \in \Omega$ such that $|z - x| < \rho$. Then $\varphi \in H_0^1(B_\rho(x)) \subset H_0^1(\Omega)$ and $\tilde{\varphi}(z) = \varphi(x + \rho z)$ for any $z \in B_1$. We also let $\tilde{a}(z) = a(x + \rho z)$, for any $z \in B_1$. Then $\tilde{a}$ is bounded, measurable, symmetric and uniformly elliptic in B_1 with the same ellipticity constants as those of a. Moreover,

$$\begin{aligned}\int_{B_1} \tilde{a}(z) \nabla \tilde{u}(z) \nabla \tilde{\varphi}(z)\, dz &= c_{\delta_0,\rho,u} \int_{|z|<1} a(x + \rho z)(\nabla u)(x + \rho z)(\nabla \varphi)(x + \rho z)\, dz \\ &= c_{\delta_0,\rho,u} \int_{|z-x|<\rho} a(z) \nabla u(z) \nabla \varphi(z)\, dz = 0.\end{aligned}$$

Hence Lemma 5.2 implies that

$$\tilde{u} \leq \tilde{u}^+ \leq 1/2 \qquad \text{in } B_{1/2}.$$

By applying the same result to $-u$ we get $\min(\tilde{u}, 0) \geq -1/2$ in $B_{1/2}$. Thus

$$|\tilde{u}(z)| \leq 1/2 \qquad \text{for } z \in B_{1/2}, \text{ for any } x \in \Omega'.$$

In other words,

$$|u(x+\rho z)| = \frac{\|u\|_{L^2(\Omega)}}{\delta_0 \rho^{n/2}} |\tilde{u}(z)| \le \frac{\|u\|_{L^2(\Omega)}}{2\delta_0 \rho^{n/2}}$$

for any $|z| < 1/2$, for all $x \in \Omega'$. In particular, $\|u\|_{L^\infty(\Omega')} \le \frac{1}{2\delta_0 \rho^{n/2}} \|u\|_{L^2(\Omega)}$. □

The proof of Lemma 5.2 is based on the interaction between two competing inequalities applied to truncations v of u: the Sobolev inequality, that controls an $L^{2+\varepsilon}$ norm of v in terms of the L^2 norm of ∇v, and the energy inequality, that controls the L^2 norm of ∇v in terms of the L^2 norm of v.

Lemma 5.4 (Sobolev inequality). *There exists $p > 2$ depending only on n such that for any $v \in H_0^1(B_1)$,*

$$\|v\|_{L^p(B_1)} \le C \|\nabla v\|_{L^2(B_1)}$$

where $C > 0$ is a constant depending only on n.

Proof. It is enough to prove the inequality for $v \in C_c^\infty(B_1)$. For any $x \in B_1$ and any direction ω such that $|\omega| = 1$,

$$v(x) = -\int_0^\infty \frac{d}{dr}(v(x+r\omega))\, dr.$$

Integrating over all $\omega \in \partial B_1$ and using polar coordinates $z - x = r\omega$,

$$\begin{aligned} v(x) &= -\frac{1}{|\partial B_1|} \int_0^\infty \int_{\partial B_1} \frac{d}{dr}(v(x+r\omega))\, dS_\omega\, dr \\ &= -\frac{1}{|\partial B_1|} \int_0^\infty \int_{\partial B_1} \frac{r\omega}{r^n} \cdot \nabla v(x+r\omega)\, dS_\omega\, r^{n-1}\, dr \\ &= \frac{1}{|\partial B_1|} \int_{B_1} \frac{x-z}{|x-z|^n} \cdot \nabla v(z)\, dz. \end{aligned}$$

Then

$$|v(x)| \le C_n \int_{B_1} \frac{|\nabla v(z)|}{|x-z|^{n-1}}\, dy = F * |\nabla v|(x).$$

Now $F(z) = C_n |z|^{-(n-1)} \in L^r(B_1)$ for $1 \le r < n/(n-1)$ and $|\nabla v| \in L^2(B_1)$. By Young's convolution inequality, $v \in L^p(B_1)$ for $1/p = 1/r + 1/2 - 1 > 1 - 1/n - 1/2 = (n-2)/(2n)$ and, in particular, for any $2 < p < 2n/(n-2)$, with the estimate

$$\|v\|_{L^p(B_1)} \le C_n \|F\|_{L^r(B_1)} \|\nabla v\|_{L^2(B_1)} \le C_n \|\nabla v\|_{L^2(B_1)}.$$

□

Remark 5.5. Our proof of Lemma 5.4 does not provide the sharp Sobolev exponent $p = 2^* = \frac{2n}{n-2}$, $n \ge 3$, for which it is known that

$$\|v\|_{L^{2^*}(B_1)} \le C \|\nabla v\|_{L^2(B_1)}$$

for every $v \in H_0^1(B_1)$. Nevertheless, we do not need to know the sharp exponent $p > 2$. The only important aspect in the proof of the $L^2 \implies L^\infty$ lemma is that we gain some integrability $p = 2 + \varepsilon > 2$. Lemma 5.4 for $p = 2^*$ requires another proof. The one we presented here is quite elementary.

Lemma 5.6 (Energy inequality for truncations). *Let u be a subsolution in B_1. For any constant c, consider the truncation*

$$v = (u-c)^+ = \max(0, u-c) \geq 0.$$

Then, for any $\varphi \in C_c^1(B_1)$,

$$\int_{B_1} |\nabla(\varphi v)|^2 \, dx \leq C \sup_{B_1} |\nabla \varphi|^2 \int_{B_1 \cap \mathrm{supp}(\varphi)} v^2 \, dx$$

where $C > 0$ is a constant depending only on Λ/λ.

Proof. Let us first recall that if $u \in H^1(B_1)$ then $v \in H^1(B_1)$ and

$$\nabla v = \chi_{\{u-c>0\}} \nabla(u-c) = \chi_{\{v>0\}} \nabla u. \tag{5.1}$$

By applying the numerical inequality $(A+B)^2 \leq 2A^2 + 2B^2$,

$$\int_{B_1} |\nabla(\varphi v)|^2 \, dx \leq 2 \sup_{B_1} |\nabla \varphi|^2 \int_{B_1 \cap \mathrm{supp}(\varphi)} v^2 \, dx + 2 \int_{B_1} |\varphi \nabla v|^2 \, dx.$$

To bound the second term, we choose $\varphi^2 v \geq 0$ as a test function. Then, by (5.1), the ellipticity and boundedness of $a(x)$ and the Cauchy inequality $|AB| \leq \varepsilon A^2 + \frac{1}{4\varepsilon} B^2$, valid for any $\varepsilon > 0$,

$$\begin{aligned}
0 &\geq \int_{B_1} a(x) \nabla u \nabla(\varphi^2 v) \, dx \\
&= \int_{B_1} a(x)(\varphi \nabla u)(\varphi \nabla v) \, dx + 2 \int_{B_1} a(x)(\varphi \nabla u)(v \nabla \varphi) \, dx \\
&= \int_{B_1} a(x)(\varphi \nabla v)(\varphi \nabla v) \, dx + 2 \int_{B_1} a(x)(\varphi \nabla v)(v \nabla \varphi) \, dx \\
&\geq \lambda \int_{B_1} |\varphi \nabla v|^2 \, dx - 2\Lambda \left[\varepsilon \int_{B_1} |\varphi \nabla v|^2 \, dx + \frac{1}{4\varepsilon} \int_{B_1} |v \nabla \varphi|^2 \, dx \right].
\end{aligned}$$

Let $\varepsilon > 0$ such that $2\Lambda\varepsilon = \lambda/2$. Then, by passing the last term above to the left hand side and dividing by $\lambda/2$,

$$\int_{B_1} |\varphi \nabla v|^2 \, dx \leq \left(\frac{2\Lambda}{\lambda} \right)^2 \int_{B_1} |v \nabla \varphi|^2 \, dx \leq C \sup_{B_1} |\nabla \varphi|^2 \int_{B_1 \cap \mathrm{supp}(\varphi)} v^2 \, dx.$$

□

Before entering into the proof of Lemma 5.2 we present a geometric analogue due to Caffarelli that avoids some of the technicalities and highlights the main ideas.

Let $\Omega \subset \mathbb{R}^n$ be a smooth domain. We assume that $S = \partial\Omega \cap B_1$, the boundary of Ω in B_1, is a minimizer of area in the sense that the boundary $\tilde{S} = \partial\tilde{\Omega} \cap B_1$ of any smooth perturbation $\tilde{\Omega}$ of Ω inside of B_1 has larger area:

$$\mathrm{Area}(\tilde{S}) \geq \mathrm{Area}(S).$$

The geometric analogue of the $L^2 \implies L^\infty$ lemma is the following.

Lemma 5.7 (S has no tentacles). *There exists $\delta_0 > 0$ depending only on n such that if*

$$\mathrm{Vol}(\Omega \cap B_1) \leq \delta_0$$

then

$$\mathrm{Vol}(\Omega \cap B_{1/2}) = 0.$$

In other words, if the volume $\mathrm{Vol}(\Omega \cap B_1)$ or mass of Ω in B_1 is very small (analogue to $\|u^+\|_{L^2(B_1)} \leq \delta_0$) then Ω never reaches the origin (analogue to u never reaching $+\infty$). In fact, Ω stays outside of $B_{1/2}$ (analogue to $u^+ \leq 1/2$). The proof is an interaction between two quantities that compete: volume that is controlled by area in a nonhomogeneous way because of the isoperimetric inequality and the minimality condition (analogue to the Sobolev inequality and the equation), and area that is controlled by volume (analogue to the energy inequality).

Proof of Lemma 5.7. We consider a sequence of dyadic balls $B_k \searrow B_{1/2}$ as

$$B_k = B_{r_k} = B_{1/2+1/2^k} \qquad k \geq 1.$$

The idea is to find a nonlinear recurrence relation for the mass of Ω in B_k

$$V_k = \mathrm{Vol}(\Omega \cap B_k)$$

that will imply that $V_k \to 0$ as $k \to \infty$, provided $V_1 = \mathrm{Vol}(\Omega \cap B_1) \leq \delta_0$ is sufficiently small. In particular, Ω never crosses $B_{1/2}$. Let us call

$$A_r = \mathrm{Area}(\Omega \cap \partial B_r).$$

Area controls volume. We show that A_r controls V_k for any $r \geq r_k = 1/2 + 1/2^k$ in a nonhomogeneous way. Indeed, by the isoperimetric inequality and the minimality of $S = \partial\Omega \cap B_1$, for any $r \geq 1/2 + 1/2^k$,

$$\begin{aligned} V_k &\leq V_r \\ &\leq C_n \,\mathrm{Area}(\partial(\Omega \cap B_r))^{n/(n-1)} \\ &= C_n \big[A_r + \mathrm{Area}(\partial\Omega \cap B_r)\big]^{n/(n-1)} \\ &\leq C_n (2A_r)^{n/(n-1)}. \end{aligned}$$

Volume controls area. We show that V_k controls A_r for some slice at level r living inside of the ring $B_k \setminus B_{k+1}$ in a homogeneous way. By Cavalieri's principle,

$$\begin{aligned} V_k &\geq V_k - V_{k+1} \\ &= \int_{1/2+1/2^{k+1}}^{1/2+1/2^k} A_r \, dr \\ &\geq \left(\frac{1}{2^k} - \frac{1}{2^{k+1}} \right) \inf_{r_{k+1} \leq r \leq r_k} A_r \\ &= \frac{1}{2^{k+1}} \inf_{r_{k+1} \leq r \leq r_k} A_r. \end{aligned}$$

By pasting together both estimates,

$$\begin{aligned} V_{k+1} &\le C_n \inf_{r\ge r_{k+1}} (2A_r)^{n/(n-1)} \\ &\le C_n \inf_{r_{k+1}\le r\le r_k} (2A_r)^{n/(n-1)} \le C_n (2^{n/(n-1)})^{k+2} V_k^{n/(n-1)}. \end{aligned}$$

This is a nonlinear recurrence relation. By choosing $V_1 \le \delta_0$ with δ_0 depending only on n we conclude that $V_k \to 0$ as $k \to \infty$ (see Lemma 5.8). □

Lemma 5.8 (Nonlinear recurrence relation). *Let V_k, $k \ge 1$, be a sequence of non-negative numbers linked by the recurrence inequalities*

$$V_{k+1} \le cb^k V_k^{1+\delta}$$

for some constants $c, b > 1$ and $\delta > 0$. If

$$V_1 < c^{-1/\delta} b^{-1/\delta^2}$$

then $V_k \to 0$ as $k \to \infty$.

Proof. We have

$$\begin{aligned} V_{k+1} &\le cb^k V_k^{1+\delta} \\ &\le cb^k (cb^{k-1} V_{k-1}^{1+\delta})^{1+\delta} \\ &= c^{1+(1+\delta)} b^{k+(k-1)(1+\delta)} V_{k-1}^{(1+\delta)^2} \\ &\le c^{1+(1+\delta)} b^{k+(k-1)(1+\delta)} (cb^{k-2} V_{k-2}^{1+\delta})^{(1+\delta)^2} \\ &= c^{1+(1+\delta)+(1+\delta)^2} b^{k+(k-1)(1+\delta)+(k-2)(1+\delta)^2} V_{k-2}^{(1+\delta)^3} \\ &\le c^{S_1(k)} b^{S_2(k)} V_1^{(1+\delta)^k} \end{aligned}$$

where

$$S_1(k) = 1 + (1+\delta) + (1+\delta)^2 + \cdots + (1+\delta)^{k-1} = \frac{(1+\delta)^k - 1}{\delta} \le \frac{(1+\delta)^k}{\delta}$$

and

$$S_2(k) = k + (k-1)(1+\delta) + (k-2)(1+\delta)^2 + \cdots + 2(1+\delta)^{k-2} + (1+\delta)^{k-1}.$$

To estimate $S_2(k)$, we divide it by $(1+\delta)$ and subtract it from $S_2(k)$ to get

$$S_2(k)\left(1 - \frac{1}{1+\delta}\right) = \frac{(1+\delta)^k - 1}{\delta} - \frac{k}{1+\delta}.$$

Therefore

$$S_2(k) = (1+\delta)\frac{(1+\delta)^k - 1}{\delta^2} - \frac{k}{\delta} \le \frac{(1+\delta)^{k+1}}{\delta^2}.$$

Hence, for all $k \ge 1$,

$$V_{k+1} \le b^{(1+\delta)/\delta^2} (c^{1/\delta} b^{1/\delta^2} V_1)^{(1+\delta)^k}.$$

Since $c^{1/\delta} b^{1/\delta^2} V_1 < 1$, we get $V_k \to \infty$ as $k \to \infty$. □

Let us now prove the $L^2 \implies L^\infty$ lemma. In the geometric analogue, the origin becomes $+\infty$, $\|u^+\|_{L^2}$ plays the role of volume and $\|\nabla u^+\|_{L^2}$ that of area. Apart from the truncation in height, we need to truncate also in space to stay away from ∂B_1.

Proof of Lemma 5.2. Let u be a subsolution. We are going to consider a sequence of truncations in space and height of u of the form $\varphi_k u_k$.

Space truncations. We take φ_k to be a sequence of smooth cutoffs converging to $\chi_{B_{1/2}}$. More precisely, we let $\varphi_0 = 1$ in B_1 and, for any $k \geq 1$, we take $\varphi_k \in C_c^1(B_k)$, $0 \leq \varphi_k \leq 1$, such that

$$\varphi_k = 1 \text{ in } B_{1/2+1/2^{k+1}} = B_{k+1}$$

$$\varphi_k = 0 \text{ outside of } B_{1/2+1/2^k} = B_k$$

and

$$|\nabla \varphi_k| \leq C2^k$$

for some constant $C > 0$. Notice that

$$\varphi_k \equiv 1 \qquad \text{in } \operatorname{supp}(\varphi_{k+1}).$$

Height truncations. We define u_k to be a sequence of monotone truncations of u converging to $(u - 1/2)^+$. More precisely, for any $k \geq 0$, we let

$$u_k = \left[u - \frac{1}{2}\left(1 - \frac{1}{2^k}\right)\right]^+.$$

Notice that $u_0 = u^+$,

$$u_k \geq u_{k+1}$$

and

$$u_k > \frac{1}{2} \cdot \frac{1}{2^{k+1}} \qquad \text{in } \{u_{k+1} > 0\}.$$

Therefore,

$$\text{if} \quad (\varphi_{k+1} u_{k+1})(x) > 0 \quad \text{then} \quad (\varphi_k u_k)(x) > \frac{1}{2} \cdot \frac{1}{2^{k+1}}.$$

We then look at

$$V_0 = \|u^+\|^2_{L^2(B_1)} = \|\varphi_0 u_0\|^2_{L^2(B_1)}$$

and

$$V_k = \|\varphi_k u_k\|^2_{L^2(B_1)} = \int_{B_1} (\varphi_k u_k)^2 \, dx \qquad k \geq 1.$$

We will find a nonlinear recurrence relation as in Lemma 5.8 that will show that if $V_0 = \|u^+\|^2_{L^2(B_1)} \leq \delta_0$, where $\delta_0 > 0$ depends only on n and Λ/λ, then $V_k \to 0$ as $k \to \infty$. In other words,

$$\int_{B_{1/2}} [(u - 1/2)^+]^2 \, dx = 0$$

which means that $u \leq 1/2$ in $B_{1/2}$. For the nonlinear recurrence relation, we apply Hölder's inequality with exponent $p/2 > 1$, where $p > 2$ is as in Lemma 5.4, the

Sobolev inequality, the energy estimate from Lemma 5.6 with $v = u_{k+1}$ and the Chebyshev inequality, to get, for some $0 < \delta < 1$,

$$
\begin{aligned}
V_{k+1} &= \int_{B_1} (\varphi_{k+1} u_{k+1})^2 \, dx \\
&\le \left[\int_{B_1} (\varphi_{k+1} u_{k+1})^p \, dx \right]^{2/p} \cdot |\{\varphi_{k+1} u_{k+1} > 0\}|^\delta \\
&\le C \int_{B_1} |\nabla(\varphi_{k+1} u_{k+1})|^2 \, dx \cdot |\{\varphi_k u_k > 2^{-(k+2)}\}|^\delta \\
&\le C \|\nabla \varphi_{k+1}\|^2_{L^\infty(B_1)} \left(\int_{B_1 \cap \operatorname{supp}(\varphi_{k+1})} u_{k+1}^2 \, dx \right) \cdot |\{\varphi_k u_k > 2^{-(k+2)}\}|^\delta \\
&\le C 4^k \left(\int_{B_1} (\varphi_k u_k)^2 \, dx \right) \cdot \left[4^{k+2} \int_{B_1} (\varphi_k u_k)^2 \, dx \right]^\delta \\
&\le C 4^{2k} V_k^{1+\delta}.
\end{aligned}
$$

□

5.3 L^∞ IMPLIES C^α: DE GIORGI'S GEOMETRIC PROOF

As we saw in Corollary 5.3, De Giorgi's $L^2 \implies L^\infty$ result, Lemma 5.2, implies that solutions are locally bounded by their L^2 norm. Hence, from now on, we can work with bounded solutions. This section is devoted to Step II, namely, to prove that bounded solutions are Hölder continuous.

The main result we show is the following De Giorgi's oscillation decay lemma.

Lemma 5.9 (De Giorgi oscillation decay). *For any $\mu > 0$ there exists $0 < \gamma < 1$ depending only on n, Λ/λ and μ such that if u is a subsolution in B_2 with $u \le 1$ in B_2 and*

$$|\{u \le 0\} \cap B_1\}| \ge \mu$$

then

$$\sup_{B_{1/2}} u \le 1 - \gamma.$$

Lemma 5.9 says that if $u \le 1$ is a (sub)solution in B_2 that is "far from 1" on a set of nontrivial measure inside of B_1, then it cannot get too close to 1 in $B_{1/2}$. In fact, u stays below $1 - \gamma$ where $\gamma << 1$ depends on how much negative u is in B_1 (the proof will show that $\gamma(\mu) \to 0$ as $\mu \to 0$). In other words, bounded solutions gain a decay in oscillation.

Before presenting the proof of Lemma 5.9, let us see how it implies the Hölder regularity of solutions, completing De Giorgi's proof of the $L^\infty \implies C^\alpha$ step. For that, it is enough to have the following oscillation decay, which follows directly from Lemma 5.9. Recall that $\operatorname{osc}_\Omega u = \sup_\Omega u - \inf_\Omega u$.

Theorem 5.10 (Oscillation decay). *There exists $0 < \sigma < 1$ depending only on n and Λ/λ such that if u is a bounded solution in B_2 then*

$$\operatorname*{osc}_{B_{1/2}} u \le \sigma \operatorname*{osc}_{B_2} u. \tag{5.2}$$

Proof. Let u be a bounded solution in B_2. Consider

$$\tilde{u}(x) = \frac{2}{\operatorname{osc}_{B_2} u}\Big(u(x) - \frac{1}{2}\Big(\sup_{B_2} u + \inf_{B_2} u\Big)\Big).$$

Then $\tilde{u}$ is a solution in B_2 such that

$$-1 \leq \tilde{u} \leq 1 \qquad \text{in } B_2.$$

Moreover,

$$\operatorname*{osc}_{B_{1/2}} \tilde{u} = \frac{2}{\operatorname{osc}_{B_2} u} \operatorname*{osc}_{B_{1/2}} u.$$

Suppose that $\tilde{u}$ is "mostly negative" in B_1, that is, it satisfies

$$|\{\tilde{u} \leq 0\} \cap B_1| \geq \frac{1}{2}|B_1|.$$

Then, by Lemma 5.9,

$$\operatorname*{osc}_{B_{1/2}} \tilde{u} \leq \sup_{B_{1/2}} \tilde{u} + 1 \leq 2 - \gamma.$$

Hence

$$\operatorname*{osc}_{B_{1/2}} u = \frac{1}{2}\big(\operatorname*{osc}_{B_2} u\big)\big(\operatorname*{osc}_{B_{1/2}} \tilde{u}\big) \leq (1 - \gamma/2) \operatorname*{osc}_{B_2} u$$

and $\sigma = 1 - \gamma/2 \in (0,1)$ depends only on n and Λ/λ. In the case when $\tilde{u}$ is mostly positive,

$$|\{\tilde{u} \leq 0\} \cap B_1| \leq \frac{1}{2}|B_1|$$

we obtain the same conclusion by flipping the picture, that is, by considering $-\tilde{u}$ and using that $\operatorname{osc}(-\tilde{u}) = \operatorname{osc} \tilde{u}$. □

Corollary 5.11 (Hölder continuity). *There exist $0 < \alpha < 1$ and a constant $C > 0$ depending only on n and Λ/λ such that if u is a solution in B_2 then $u \in C^\alpha$ at the origin with*

$$|u(x) - u(0)| \leq C\|u\|_{L^2(B_2)}|x|^\alpha \qquad \textit{for every } |x| < 1. \tag{5.3}$$

As a consequence, any solution u in Ω is locally Hölder continuous: for any subdomain $\Omega' \subset\subset \Omega$ there exists a constant $C > 0$ depending on n, Λ/λ and $\operatorname{dist}(\partial\Omega', \Omega)$ *such that*

$$\|u\|_{C^{0,\alpha}(\Omega')} \leq C\|u\|_{L^2(\Omega)}.$$

Proof. To prove (5.3), we zoom out u from $B_{1/4^k}$ to B_2 and apply Theorem 5.10. For this we consider, for any $k \geq 0$,

$$\tilde{u}(x) = u(x/4^{k+1/2}) \qquad \text{for } |x| < 2.$$

Then, if we define $\tilde{a}(x) = a(x/4^{k+1/2})$ for $|x| < 2$, we see that $\tilde{a}$ has the same ellipticity constants than a and that $\tilde{u} \in H^1(B_2)$ is a solution to

$$\operatorname{div}(\tilde{a}(x)\nabla\tilde{u}) = 0 \qquad \text{in } B_2.$$

By Theorem 5.10,

$$\operatorname*{osc}_{B_{1/4^{k+1}}} u = \operatorname*{osc}_{B_{1/2}} \tilde{u} \le \sigma \operatorname*{osc}_{B_2} \tilde{u} = \sigma \operatorname*{osc}_{B_{1/4^k}} u.$$

By iteration of this inequality and the local boundedness of solutions, for any $k \ge 0$,

$$\operatorname*{osc}_{B_{1/4^k}} u \le \sigma^k \operatorname*{osc}_{B_1} u \le 2\sigma^k \|u\|_{L^\infty(B_1)} \le C\sigma^k \|u\|_{L^2(B_2)}$$

where $C > 0$ depends only on n and Λ/λ. Notice that we can assume that $\sigma > 1/4$. To obtain the Hölder regularity at the origin, we first choose $0 < \alpha < 1$ such that

$$\sigma = \left(\frac{1}{4}\right)^\alpha.$$

Indeed, this is equivalent to setting

$$\alpha = -\frac{\log \sigma}{\log 4} \in (0, 1).$$

Then α depends only on σ, which by Theorem 5.10 depends only in n and Λ/λ. Next, for any $|x| < 1$ there exists $k \ge 0$ such that $\frac{1}{4^{k+1}} \le |x| \le \frac{1}{4^k}$, so that

$$\begin{aligned}
|u(x) - u(0)| &\le \operatorname*{osc}_{B_{1/4^k}} u \\
&\le C\sigma^k \|u\|_{L^2(B_2)} \\
&= C\left(\frac{1}{4^k}\right)^\alpha \|u\|_{L^2(B_2)} \\
&= 4^\alpha C\left(\frac{1}{4^{k+1}}\right)^\alpha \|u\|_{L^2(B_2)} \\
&\le C|x|^\alpha \|u\|_{L^2(B_2)}
\end{aligned}$$

where $C > 0$ depends only on n and Λ/λ.

For the general case of a solution u in Ω one can use an argument based on translation and rescaling of (5.3) as in the proof of Corollary 5.3. □

Corollary 5.12 (Liouville theorem). *If u is a bounded solution in $\mathbb{R}^n$ then u is constant.*

Proof. Let $M = \sup_{\mathbb{R}^n} u$ and $m = \inf_{\mathbb{R}^n} u$. Then, by passing to the limit in r in the oscillation decay inequality

$$\operatorname*{osc}_{B_{r/2}} u \le \sigma \operatorname*{osc}_{B_r} u \qquad \text{where } 0 < \sigma < 1$$

(which is easily obtained from (5.2) by rescaling) we get

$$M - m \le \sigma(M - m).$$

Hence, $M = m$ and u is constant. □

The idea for the proof of De Giorgi's oscillation decay in Lemma 5.9 is based on the following observation. Let δ_0 be as in the $L^2 \implies L^\infty$ result, Lemma 5.2. Suppose that, for some reason, we know that u is "very negative" in B_1, namely,

$$|\{u \le 0\} \cap B_1| \ge |B_1| - \delta_0^2.$$

Then we would be done. Indeed, since $u \le 1$ in B_2,

$$\begin{aligned} \|u^+\|_{L^2(B_1)}^2 &= \int_{B_1} (u^+)^2\, dx \\ &\le |\{u^+ > 0\} \cap B_1| \\ &= |B_1| - |\{u \le 0\} \cap B_1| \\ &\le \delta_0^2. \end{aligned}$$

Then, by Lemma 5.2,

$$u^+ \le 1/2 = 1 - 1/2 \qquad \text{in } B_{1/2}$$

so $\gamma = 1/2$. The problem with this reasoning is that we cannot ensure from the beginning that

$$|\{u \le 0\} \cap B_1| \ge |B_1| - \delta_0^2$$

but rather that

$$|\{u \le 0\} \cap B_1| \ge \frac{|B_1|}{2}$$

(in case the opposite inequality were true we would work with $-u$). We must "fill the gap" between these two inequalities. As a matter of fact, Lemma 5.9 says that if u is only "a little bit negative" in B_1, namely,

$$|\{u \le 0\} \cap B_1| \ge \mu > 0.$$

then we gain a universal decay of u in $B_{1/2}$.

The main tool to "fill the gap" in the proof of Lemma 5.9 is the following isoperimetric inequality of De Giorgi. It basically says that functions in H^1 "need room" to go from being negative to positive. In other words, functions in H^1 cannot jump from 0 to 1/2 "instantaneously," that is, cannot have jump discontinuities. More importantly, the De Giorgi isoperimetric inequality is fundamental because it gives a quantitative estimate on the size of the transition set for w in terms of the sets where $w \le 0$ and $w \ge 1/2$.

Lemma 5.13 (De Giorgi isoperimetric lemma). *There exists a constant $C_n > 0$ depending only on n such that the following holds. Let $w \in H^1(B_1)$ such that*

$$\int_{B_1} |\nabla w^+|^2\, dx \le C_0$$

for some constant $C_0 > 0$. Let

$$|A| = |\{w \le 0\} \cap B_1|$$

$$|E| = |\{w \ge 1/2\} \cap B_1|$$

$$|D| = |\{0 < w < 1/2\} \cap B_1|.$$

Then

$$C_0|D| \geq C_n(|E||A|^{(n-1)/n})^2.$$

Proof. Let

$$\overline{w} = \sup(0, \inf(w, 1/2)).$$

Note that

$$\nabla\overline{w} = \chi_{\{0<w<1/2\}}\nabla w^+$$

and

$$|A| = |\{\overline{w} = 0\} \cap B_1|$$
$$|E| = |\{\overline{w} = 1/2\} \cap B_1|$$
$$|D| = |\{0 < \overline{w} < 1/2\} \cap B_1|.$$

Let $x \in A$ and $z \in E$. Then, by the change of variables $s = t|x - z|$, denoting $e = \frac{z-x}{|x-z|} \in \partial B_1$ and extending $\overline{w} = 0$ outside of B_1,

$$\begin{aligned}
\frac{1}{2} &= \overline{w}(z) - \overline{w}(x) \\
&= \int_0^1 \frac{d}{dt}(\overline{w}(x + t(z-x)))\, dt \\
&= \int_0^1 \nabla\overline{w}(x + t(z-x)) \cdot (z-x)\, dt \\
&= \int_0^{|x-z|} \nabla\overline{w}(x + se) \cdot e\, ds \\
&\leq \int_0^\infty |\nabla\overline{w}(x + se)|\, ds.
\end{aligned}$$

Now we integrate this inequality in $z \in E$, apply polar coordinates for $z - x$ and use that $\nabla\overline{w}(x + se)$ does not depend on the radial direction in B_1 because $e \in \partial B_1$:

$$\begin{aligned}
\frac{|E|}{2} &\leq \int_E \int_0^\infty |\nabla\overline{w}(x + se)|\, ds\, dz \\
&\leq \int_{B_1} \int_0^\infty |\nabla\overline{w}(x + se)|\, ds\, dz \\
&= \int_0^1 \int_{\partial B_1} \int_0^\infty |\nabla\overline{w}(x + s\omega)|\, ds\, dS_\omega\, r^{n-1}\, dr \\
&\leq \int_{\partial B_1} \int_0^\infty |\nabla\overline{w}(x + s\omega)|\, ds\, dS_\omega \\
&= \int_{\partial B_1} \int_0^\infty \frac{|\nabla\overline{w}(x + s\omega)|}{s^{n-1}} s^{n-1}\, ds\, dS_\omega \\
&= \int_{B_1} \frac{|\nabla\overline{w}(z)|}{|x-z|^{n-1}}\, dz.
\end{aligned}$$

By integrating over all $x \in A$ and using that $\nabla\overline{w} = \chi_D \nabla w^+$,

$$\frac{|A||E|}{2} \leq \int_D |\nabla w^+(z)| \left[\int_A \frac{1}{|x-z|^{n-1}}\, dx\right] dz. \tag{5.4}$$

The integral in x inside the brackets is maximized by the ball B centered at z and having the same measure as A, for which its radius has to be $R = c_n|A|^{1/n}$. Thus

$$\int_A \frac{1}{|x-z|^{n-1}}\,dx \le \int_{B_R(z)} \frac{1}{|x-z|^{n-1}}\,dx = C_n|A|^{1/n}.$$

We conclude by using this in (5.4) and applying Hölder's inequality:

$$\frac{|A||E|}{2} \le C_n|A|^{1/n}\int_D |\nabla w^+(z)|\,dz \le C_n|A|^{1/n}C_0^{1/2}|D|^{1/2}.$$

□

Remark 5.14 (Another version of De Giorgi's isoperimetric inequality). Under the assumptions of Lemma 5.13, one can also prove that

$$C_0|D| \ge C_n(|E||A|)^2.$$

Indeed, let $\overline{w}$ be as in the proof of Lemma 5.13. Then, for any $x \in E$ and $z \in A$, $1/2 = \overline{w}(x) - \overline{w}(z)$. Integrating this identity over E and A, adding and subtracting the average $\overline{w}_{B_1}$ of $\overline{w}$ over the ball B_1, and applying the Poincaré and Hölder inequalities,

$$\begin{aligned}|E||A| &\le 2\int_A\int_E |\overline{w}(x) - \overline{w}(z)|\,dx\,dz\\ &\le 4|B_1|\int_{B_1} |\overline{w}(y) - \overline{w}_{B_1}|\,dy\\ &\le C\int_D |\nabla w^+|\,dy \le CC_0^{1/2}|D|^{1/2}.\end{aligned}$$

Either this version of the De Giorgi isoperimetric inequality or the one in Lemma 5.13 can be used for the application to the proof of Lemma 5.9. However, De Giorgi's proof is much more geometric in nature.

Proof of Lemma 5.9. For $k \ge 0$ we consider the sequence of rescaled translations

$$w_k = 2^k\left[u - \left(1 - \frac{1}{2^k}\right)\right] = 2^k u_k.$$

We want to show that there exists $k_0 \ge 0$ depending only on n, Λ/λ and μ such that

$$\int_{B_1} [(w_{k_0+1})^+]^2\,dx \le \delta_0^2 \tag{5.5}$$

where $\delta_0 > 0$ is as in the $L^2 \implies L^\infty$ lemma. Indeed, if (5.5) were true then, since w_{k_0+1} is a subsolution, by Lemma 5.2,

$$w_{k_0+1} \le (w_{k_0+1})^+ \le 1/2 \qquad \text{in } B_{1/2}.$$

Thus

$$2^{k_0+1}\left[u - \left(1 - \frac{1}{2^{k_0+1}}\right)\right] \le 1/2 \qquad \text{in } B_{1/2}$$

which implies that

$$u \le 1 - 1/2^{k_0+2}$$

so that $\gamma = 1/2^{k_0+2}$ and we would be done.

We prove (5.5) by using the De Giorgi isoperimetric inequality of Lemma 5.13. Since $u \le 1$ in B_2, we have

$$w_k \le 2^k(1 - (1 - 2^{-k})) = 1 \qquad \text{in } B_2.$$

Therefore, if we choose $\varphi \in C_c^1(B_2)$ such that $\varphi = 1$ in B_1 and $|\nabla\varphi| \le 4$ in the energy inequality for truncations, Lemma 5.6,

$$\begin{aligned}\int_{B_1} |\nabla(w_k)^+|^2\,dx &\le \int_{B_2} |\nabla[\varphi(w_k)^+]|^2\,dx \\ &\le C_{\Lambda/\lambda}\int_{B_2} [(w_k)^+]^2\,dx \\ &\le C_{\Lambda/\lambda}|B_2| =: C_0.\end{aligned}$$

In other words, $\nabla(w_k)^+$ are all uniformly bounded in $L^2(B_1)$. Then we can apply De Giorgi's isoperimetric inequality to get

$$C_0|D_k| \ge C_n(|E_k||A_k|^{(n-1)/n})^2$$

where the sets A_k, E_k and D_k are defined as in Lemma 5.13 with w_k in place of w. We can bound $|A_k|$ from below uniformly in k:

$$\begin{aligned}|A_k| &= |\{w_k \le 0\} \cap B_1| \\ &= |\{u \le 1 - 2^{-k}\} \cap B_1| \\ &\ge |\{u \le 0\} \cap B_1| = \mu > 0.\end{aligned}$$

This gives

$$|D_k| \ge \tau|E_k|^2$$

where $\tau > 0$ depends only on n, Λ/λ and μ. Next we bound $|E_k|$ using that $w_{k+1} \le 1$:

$$\begin{aligned}|E_k| &= |\{w_k \ge 1/2\} \cap B_1| \\ &= |\{w_{k+1} \ge 0\} \cap B_1| \ge \int_{B_1} [(w_{k+1})^+]^2\,dx.\end{aligned}$$

Therefore, for every $k \ge 0$ such that

$$\int_{B_1} [(w_{k+1})^+]^2\,dx > \delta_0 \tag{5.6}$$

we get

$$|D_k| = |\{0 < w_k < 1/2\} \cap B_1| \ge \beta > 0$$

where $\beta = \tau\delta_0^2$ depends only on n, Λ/λ and μ. In other words, every time (5.6) occurs, the transition set D_k contains a fixed nontrivial portion of B_1 of universal

size at least $\beta > 0$. But this cannot happen for all k because the sets D_k are subsets of B_1. Indeed, let us iterate to find that, for every $k \geq 0$ for which (5.6) is true,

$$
\begin{aligned}
|B_1| &\geq |\{w_k \leq 0\} \cap B_1| \\
&= |\{w_{k-1} \leq 1/2\} \cap B_1| \\
&\geq |\{w_{k-1} \leq 0\} \cap B_1| + |\{0 < w_{k-1} < 1/2\} \cap B_1| \\
&\geq |\{w_{k-1} \leq 0\} \cap B_1| + \beta \\
&\geq |\{w_{k-2} \leq 0\} \cap B_1| + 2\beta \\
&\geq |\{w_0 \leq 0\} \cap B_1| + k\beta \\
&= |\{u \leq 0\} \cap B_1| + k\beta \geq \mu + k\beta.
\end{aligned}
\tag{5.7}
$$

Now (5.7) fails to be true after some finite $k_0 > 0$ that depends only on n, Λ/λ and μ. At this k_0 we necessarily have that (5.6) does not hold, so we must have (5.5). □

5.4 L^∞ IMPLIES C^α: MOSER'S CRITICAL DENSITY PROOF

Recall once again that the De Giorgi $L^2 \implies L^\infty$ lemma gives that if u is a subsolution in B_2 then there exists a constant $C > 0$ depending only on n and Λ/λ such that

$$\sup_{B_{1/2}} u^+ \leq C\|u^+\|_{L^2(B_1)}.$$

Here we present a different way of proving that solutions are locally Hölder continuous that is due to Moser. It is based on the local boundedness of subsolutions and the so-called critical density estimate for supersolutions (Theorem 5.16).

We begin with a general fact about Sobolev functions, which can be seen as a measure-type Poincaré inequality.

Lemma 5.15 (Fabes' lemma). *For any $\varepsilon > 0$ there exists a constant $C > 0$ depending only on n and ε such that if $v \in H^1(B_1)$ satisfies*

$$|\{x \in B_1 : v = 0\}| \geq \varepsilon |B_1|$$

then

$$\int_{B_1} v^2\,dx \leq C \int_{B_1} |\nabla v|^2\,dx.$$

Proof. We prove the result by contradiction. Suppose that there exists a sequence $\{v_k\}_{k=1}^\infty \subset H^1(B_1)$ such that

$$|\{x \in B_1 : v_k = 0\}| \geq \varepsilon |B_1| \quad \text{and} \quad \int_{B_1} v_k^2\,dx = 1$$

for all $k \geq 1$ but

$$\int_{B_1} |\nabla v_k|^2\,dx \to 0 \qquad \text{as } k \to \infty.$$

Then $\{v_k\}_{k=1}^\infty$ is a bounded sequence in $H^1(B_1)$. Therefore, there exist a subsequence, that we still denote by $\{v_k\}_{k=1}^\infty$, and $v_\infty \in H^1(B_1)$ such that $v_k \to v_\infty$ weakly in

$H^1(B_1)$ as $k \to \infty$. Since $H^1(B_1)$ is compactly embedded in $L^2(B_1)$, there exists a further subsequence, that we keep on denoting by $\{v_k\}_{k=1}^\infty$, such that $v_k \to v_\infty$ strongly in $L^2(B_1)$. In particular, $\|v_\infty\|_{L^2(B_1)} = 1$ and

$$\int_{B_1} |\nabla v_\infty|^2 \, dx \leq \liminf_{k\to\infty} \int_{B_1} |\nabla v_k|^2 \, dx = 0.$$

Hence v_∞ is a nonzero constant. Thus,

$$\begin{aligned} 0 &= \lim_{k\to\infty} \int_{B_1} |v_k - v_\infty|^2 \, dx \\ &\geq \lim_{k\to\infty} \int_{\{x\in B_1 : v_k = 0\}} |v_k - v_\infty|^2 \, dx \\ &\geq (v_\infty)^2 \cdot \inf_{k\geq 1} |\{x \in B_1 : v_k = 0\}| \\ &\geq (v_\infty)^2 \cdot \varepsilon |B_1| > 0 \end{aligned}$$

which is a contradiction. □

Theorem 5.16 (Critical density estimate). *For any $\varepsilon > 0$ there exists $0 < c < 1$ depending only on n, Λ/λ and ε such that if u is a nonnegative supersolution in B_2 and*

$$|\{x \in B_1 : u \geq 1\}| \geq \varepsilon |B_1|,$$

then

$$\inf_{B_{1/2}} u \geq c.$$

The critical density estimate states that if a nonnegative supersolution u is greater than 1 in some portion of B_1, even if it is a very small set, then in $B_{1/2}$ it has to stay uniformly away from 0. In some sense, we can say that the value $u = 1$ somewhere in B_1 pulls up the whole supersolution away from 0 everywhere in $B_{1/2}$.

Proof of Theorem 5.16. We may assume that $u \geq \delta$ in B_2 for some small $\delta > 0$. Indeed, for the general case we consider the supersolution $u + \delta > 0$ in B_2. Then

$$|\{x \in B_1 : (u + \delta) \geq 1\}| \geq |\{x \in B_1 : u \geq 1\}| \geq \varepsilon |B_1|$$

and so

$$\inf_{B_{1/2}} (u + \delta) = \inf_{B_{1/2}} u + \delta \geq c.$$

Since $\delta > 0$ was arbitrary the result for u follows.

The idea is to translate the hypothesis on the measure of the set where $u \geq 1$ to the measure of the set where $v = (\log u)^-$ vanishes and apply Fabes' lemma. Then the key points are to prove that v is a subsolution (so as to apply the $L^2 \implies L^\infty$ lemma to v) and to bound $\|\nabla v\|_{L^2}$ universally.

Let $\Phi(t) = (\log t)^- = -\min(0, \log t)$. We begin by proving that

$$v = \Phi(u) = (\log u)^-$$

is a subsolution in B_2. Clearly, $0 \le v \le \log \delta^{-1}$, so that $v \in L^2(B_2)$. Notice that $\Phi(t)$ is convex, nonincreasing and Lipschitz in $t \in [\delta, \infty)$, with $\operatorname{supp}(\Phi) = [\delta, 1]$, and differentiable in $[\delta, \infty)$ except at $t = 1$. The Lipschitz continuity of Φ ensures that $\nabla(\Phi(u)) = \Phi'(u)\nabla u \in L^2(B_2)$. Therefore, $v \in H^1(B_2)$. To verify the subsolution property we use a mollification of Φ. Take $\zeta \in C_c^\infty((-1,1))$, $\zeta \ge 0$ even, with $\int \zeta = 1$. Consider the dilations defined by $\zeta_\tau(z) = \frac{1}{\tau}\zeta(\frac{z}{\tau})$ for $\tau > 0$. Let $\Phi_\tau(t) = (\zeta_\tau * \Phi)(t)$. Then $\Phi_\tau \in C_c^\infty((\delta-\tau, 1+\tau))$. Recall that locally Lipschitz functions are differentiable a.e. with locally bounded derivative (Rademacher's theorem) and convex functions are second-order differentiable a.e. with nonnegative second derivative (Aleksandrov's theorem). Since Φ is nonincreasing and Lipschitz in $[\delta, \infty)$, we have $\Phi'_\tau = \zeta_\tau * \Phi' \le 0$. Similarly, since Φ is convex, $\Phi''_\tau = \zeta_\tau * \Phi'' \ge 0$. Let $\varphi \in C_c^1(B_2)$ such that $\varphi \ge 0$. Then $\Phi'_\tau(u)\varphi \in L^2(B_2)$ and $\nabla(\Phi'_\tau(u)\varphi) = \Phi'_\tau(u)\nabla\varphi + \Phi''_\tau(u)\varphi\nabla u \in L^2(B_2)$. Therefore, $\Phi'_\tau(u)\varphi$ is an admissible nonpositive test function. Using this, the fact that u is a supersolution and ellipticity,

$$\begin{aligned}\int_{B_2} a(x)\nabla(\Phi_\tau(u))\nabla\varphi\,dx &= \int_{B_2} a(x)\Phi'_\tau(u)\nabla u\nabla\varphi\,dx\\ &= \int_{B_2} a(x)\nabla u\nabla(\Phi'_\tau(u)\varphi)\,dx - \int_{B_2}(a(x)\nabla u\nabla u)\varphi\Phi''_\tau(u)\,dx\\ &\le 0 - \lambda\int_{B_2}|\nabla u|^2\varphi\Phi''_\tau(u)\,dx \le 0.\end{aligned}$$

In other words, $\Phi_\tau(u)$ is a subsolution. Now $\Phi'_\tau(u) \to \Phi'(u)$ a.e. and $\Phi'_\tau(u)$ is bounded in B_1, uniformly in τ. Therefore, by the dominated convergence theorem,

$$\begin{aligned}\int_{B_2} a(x)\nabla(\Phi(u))\nabla\varphi\,dx &= \int_{B_2} a(x)\Phi'(u)\nabla u\nabla\varphi\,dx\\ &= \lim_{\tau\to 0}\int_{B_2} a(x)\Phi'_\tau(u)\nabla u\nabla\varphi\,dx \le 0\end{aligned}$$

and $v = \Phi(u) = (\log u)^-$ is a subsolution.

We continue the proof by observing that

$$|\{x \in B_1 : v = 0\}| = |\{x \in B_1 : u \ge 1\}| \ge \varepsilon|B_1|.$$

By the $L^2 \implies L^\infty$ and Fabes' lemmas applied to the subsolution v,

$$\sup_{B_{1/2}} v \le C\|v\|_{L^2(B_1)} \le C\|\nabla v\|_{L^2(B_1)}.$$

If we prove that the right hand side $\|\nabla v\|_{L^2(B_1)}$ is universally bounded (independently of δ) then we can conclude. Indeed, we get

$$\sup_{B_{1/2}} v = \sup_{B_{1/2}}(\log u)^- \le C$$

which gives $\log u \ge \min(0, \log u) \ge -C$ in $B_{1/2}$, so that

$$\inf_{B_{1/2}} u \ge e^{-C} = c > 0.$$

Let us estimate $\|\nabla v\|_{L^2(B_1)}$. We have $\nabla v = -\chi_{\{u<1\}}\frac{1}{u}\nabla u$. Let $\eta \in C^1_c(B_2)$ and

$$\varphi = \frac{1}{u}\eta^2 > 0.$$

Since $u \geq \delta$, φ is a valid nonnegative test function. Using that u is a supersolution,

$$\begin{aligned}0 &\leq \int_{B_2} a(x)\nabla u \nabla\varphi\, dx \\ &= 2\int_{B_2} \eta a(x)\left(\frac{1}{u}\nabla u\right)\nabla\eta\, dx - \int_{B_2}\eta^2 a(x)\left(\frac{1}{u}\nabla u\right)\left(\frac{1}{u}\nabla u\right) dx.\end{aligned}$$

This implies, by ellipticity, the boundedness of $a(x)$ and the Cauchy inequality with $\varepsilon > 0$, that

$$\begin{aligned}\lambda\int_{B_2}\eta^2\left|\frac{1}{u}\nabla u\right|^2 dx &\leq \int_{B_2}\eta^2 a(x)\left(\frac{1}{u}\nabla u\right)\left(\frac{1}{u}\nabla u\right) dx \\ &\leq 2\int_{B_2}\eta a(x)\left(\frac{1}{u}\nabla u\right)\nabla\eta\, dx \\ &\leq \varepsilon\int_{B_2}\eta^2\left|\frac{1}{u}\nabla u\right|^2 dx + \frac{\Lambda^2}{4\varepsilon}\int_{B_2}|\nabla\eta|^2\, dx.\end{aligned}$$

By choosing $\varepsilon = \lambda/2$,

$$\int_{B_2}\eta^2|\nabla \log u|^2\, dx \leq \left(\frac{\Lambda}{\lambda}\right)^2\int_{B_2}|\nabla\eta|^2\, dx.$$

Therefore, if we fix η such that $\eta = 1$ in B_1, we get

$$\int_{B_1}|\nabla v|^2\, dx \leq \int_{B_2}\eta^2|\nabla\log u|^2\, dx \leq C(n, \Lambda/\lambda)$$

as desired. □

Corollary 5.17 (Critical density implies oscillation decay). *There exists $0 < \sigma < 1$ depending only on n and Λ/λ such that if u is a bounded solution in B_2 then*

$$\operatorname*{osc}_{B_{1/2}} u \leq \sigma \operatorname*{osc}_{B_2} u.$$

In particular, u is locally Hölder continuous in B_2.

Proof. For $r > 0$, let

$$M(r) = \sup_{B_r} u \qquad \text{and} \qquad m(r) = \inf_{B_r} u.$$

Consider the two nonnegative solutions

$$\frac{u - m(2)}{M(2) - m(2)} \qquad \text{and} \qquad \frac{M(2) - u}{M(2) - m(2)}$$

in B_2. Note that

$$u \geq \frac{1}{2}(M(2)+m(2)) \qquad \text{if and only if} \qquad \frac{u-m(2)}{M(2)-m(2)} \geq \frac{1}{2}$$

and

$$u \leq \frac{1}{2}(M(2)+m(2)) \qquad \text{if and only if} \qquad \frac{M(2)-u}{M(2)-m(2)} \geq \frac{1}{2}.$$

We consider two separate cases. In the first one, suppose that

$$\left|\left\{x \in B_1 : \frac{2(u-m(2))}{M(2)-m(2)} \geq 1\right\}\right| \geq \frac{1}{2}|B_1|.$$

Apply the critical density estimate of Theorem 5.16 to the nonnegative solution $\frac{2(u-m(2))}{M(2)-m(2)}$ with $\varepsilon = 1/2$ to get that, for some $0 < c < 1$,

$$\inf_{B_{1/2}} \frac{(u-m(2))}{M(2)-m(2)} \geq c.$$

In other words,

$$m(1/2) = \inf_{B_{1/2}} u \geq m(2) + c(M(2)-m(2)).$$

This implies that

$$\operatorname*{osc}_{B_{1/2}} u = M(1/2) - m(1/2) \leq M(2) - m(1/2) \leq (1-c)(M(2)-m(2)) = \sigma \operatorname*{osc}_{B_2} u.$$

For the second case, assume that

$$\left|\left\{x \in B_1 : \frac{2(M(2)-u)}{M(2)-m(2)} \geq 1\right\}\right| \geq \frac{1}{2}|B_1|.$$

Applying the critical density estimate to $\frac{2(M(2)-u)}{M(2)-m(2)}$,

$$\inf_{B_{1/2}} \frac{(M(2)-u)}{M(2)-m(2)} \geq c.$$

In other words,

$$M(1/2) = \sup_{B_{1/2}} u \leq M(2) - c(M(2)-m(2))$$

which implies that

$$\operatorname*{osc}_{B_{1/2}} u = M(1/2) - m(1/2) \leq M(1/2) - m(2) \leq (1-c)(M(2)-m(2)) = \sigma \operatorname*{osc}_{B_2} u.$$

The local Hölder continuity is proved as in Corollary 5.11. □

NOTES ON CHAPTER 5

The $L^2 \implies L^\infty$ and $L^\infty \implies C^\alpha$ results, including De Giorgi's isoperimetric inequality, were originally proved by Ennio De Giorgi in [24], see also [25] for an account on how the De Giorgi theorem solved Hilbert's 19th problem.

De Giorgi's method was extended to more general equations with lower-order terms and nonzero right hand side by Stampacchia [62].

Our presentation is based on unpublished lectures notes by Caffarelli [6] and the paper by Caffarelli–Vasseur [16].

The geometric analogue of the $L^2 \implies L^\infty$ lemma in Section 5.2 is due to Caffarelli and can be found in [25, Section 2.8].

The proof of Hölder regularity via the critical density estimate was done by Moser in [52].

The theorems by Rademacher for Lipschitz functions and Aleksandrov for convex functions can be seen in Evans–Gariepi [33].

CHAPTER 6

The Moser theorem

In this chapter we present Moser's Harnack inequality for nonnegative weak solutions to divergence form elliptic equations. Moser's method consists in using energy estimates for powers of u and Sobolev embeddings to first prove local upper and lower bounds for positive solutions. Then a gap must be closed by means of the Poincaré inequality, the energy estimate for $\log u$ and the John–Nirenberg inequality for BMO functions. In particular, one obtains local boundedness of subsolutions and the weak Harnack inequality for nonnegative supersolutions. We finally show how the Harnack inequality implies the local Hölder continuity of solutions.

6.1 THE MOSER THEOREM

As in Chapter 5, we consider weak (sub/super)solutions to uniformly elliptic equations in divergence form

$$\operatorname{div}(a(x)\nabla u) = 0 \qquad \text{in } \Omega$$

where Ω is a bounded domain of $\mathbb{R}^n$, $n \geq 2$, the matrix of coefficients $a(x) = (a^{ij}(x))_{i,j=1}^n$ is symmetric, bounded by $\Lambda > 0$, that is, $\|a\|_{L^\infty(\Omega)} \leq \Lambda$, and uniformly elliptic: $a^{ij}(x)\xi_i\xi_j \geq \lambda|\xi|^2$ for some $\lambda > 0$, for all $\xi \in \mathbb{R}^n$ and a.e. $x \in \Omega$. The reader should recall the basic notions introduced in Section 5.1.

Throughout this chapter, when we say that u is a (sub/super)solution in Ω, we mean that $u \in H^1(\Omega)$ is a weak (sub/super)solution to $\operatorname{div}(a(x)\nabla u) = 0$ in Ω.

Our objective is to prove Moser's Harnack inequality.

Theorem 6.1 (Moser's Harnack inequality). *For any subdomain $\Omega' \subset\subset \Omega$ there exists a constant $C_H > 0$ depending only on n, Λ/λ and $\operatorname{dist}(\Omega', \partial\Omega)$ such that for any solution $u \in H^1(\Omega)$ to*

$$\begin{cases} \operatorname{div}(a(x)\nabla u) = 0 & \text{in } \Omega \\ u \geq 0 & \text{in } \Omega \end{cases}$$

we have

$$\sup_{\Omega'} u \leq C_H \inf_{\Omega'} u.$$

Recall that in Section 2.3.3 we showed the Harnack inequality for harmonic functions using the mean value property. We will prove Theorem 6.1 for cubes, with

DOI: 10.1201/9781032679853-6

$\Omega = Q_4 = (-4,4)^n$ and $\Omega' = Q_1 = (-1,1)^n$. Since cubes and balls are equivalent in $\mathbb{R}^n$, Theorem 6.1 can also be stated for Euclidean balls. The general case can be obtained through a covering argument as in the proof of Theorem 2.16. Furthermore, by considering $u + \delta$ for some $\delta > 0$, it is enough to work with positive solutions.

The Harnack inequality together with the local boundedness of solutions implies Hölder regularity of solutions. We prove this in Section 6.4.

Another direct corollary of the Harnack inequality is the strong maximum principle: if u is a nonnegative solution in Ω and there is $x_0 \in \Omega$ such that $u(x_0) = 0$ then $0 \le \sup_{\Omega'} u \le C_H \inf_{\Omega'} u = 0$ for any $\Omega' \subset\subset \Omega$, so that $u \equiv 0$ in Ω.

Moser's strategy of proof begins by observing that one can characterize the supremum and the infimum of u as limits of integrals of u^p as $p \to \pm\infty$. The following real analysis result makes this precise.

Lemma 6.2. *Let $|u|^r \in L^1(\Omega)$ for some $r \in \mathbb{R}$. Define*

$$\Phi(p) = \left(\int_\Omega |u|^p \, dx \right)^{1/p}.$$

Then

$$\lim_{p\to\infty} \Phi(p) = \sup_\Omega |u| \qquad \textit{and} \qquad \lim_{p\to-\infty} \Phi(p) = \inf_\Omega |u|.$$

Proof. Let $0 \le t < \sup_\Omega |u|$. By definition,

$$0 < |A| = |\{x \in \Omega : |u(x)| > t\}| \le |\Omega| < \infty.$$

Then

$$\Phi(p) \ge \left(\int_A |u|^p \, dx \right)^{1/p} \ge t|A|^{1/p}$$

which implies that

$$\liminf_{p\to\infty} \Phi(p) \ge t.$$

Since t was arbitrary,

$$\liminf_{p\to\infty} \Phi(p) \ge \sup_\Omega |u|.$$

For the reverse inequality, if $r < p < \infty$ then

$$\Phi(p) = \left(\int_\Omega |u|^{p-r} |u|^r \, dx \right)^{1/p} \le \left(\sup_\Omega |u| \right)^{1-r/p} \left(\int_\Omega |u|^r \, dx \right)^{1/p}.$$

Since $|u|^r \in L^1(\Omega)$, we can take the limit as $p \to \infty$ to get

$$\limsup_{p\to\infty} \Phi(p) \le \sup_\Omega |u|.$$

For the second limit,

$$\lim_{p\to-\infty} \Phi(p) = \lim_{|p|\to\infty} \left[\int_\Omega \left(\frac{1}{|u|} \right)^{|p|} dx \right]^{-1/|p|} = \frac{1}{\sup_\Omega \frac{1}{|u|}} = \inf_\Omega |u|.$$

□

Let u be a positive solution in Q_4. For $p \in \mathbb{R}$, $p \neq 0$, and $0 < r < 4$, let

$$\Phi(p, r) = \left(\int_{Q_r} u^p \, dx \right)^{1/p}.$$

In view of Lemma 6.2, the Harnack inequality takes the form

$$\Phi(\infty, 1) \leq C_H \Phi(-\infty, 1).$$

The proof goes as follows. By solely using the energy estimate for positive and negative powers of u and the Sobolev inequality, we first show that, for any $p > 0$, the upper bound for the supremum (or local boundedness)

$$\Phi(\infty, 1) = \sup_{Q_1} u \leq C \left(\int_{Q_3} u^p \, dx \right)^{1/p} = C\Phi(p, 3) \tag{6.1}$$

and the lower bound for the infimum

$$\Phi(-p, 3) = \left(\int_{Q_3} u^{-p} \, dx \right)^{-1/p} \leq C \inf_{Q_1} u = C\Phi(-\infty, 1) \tag{6.2}$$

hold. In fact, we initially prove (6.1) for any $p \geq 2$ and then use a real analysis lemma to decrease the exponent all the way to $p > 0$. The proof of (6.2) is more direct. Now the gap between $\Phi(p, 3)$ and $\Phi(-p, 3)$ above cannot be directly closed because $p = 0$ is not allowed. Rather, we show that there exists some $p_0 > 0$ (which may be smaller than 2) for which

$$\Phi(p_0, 3) \leq C\Phi(-p_0, 3).$$

Then we can finally connect (6.1) and (6.2). To find this p_0, we actually work with the correct limiting value of u^p when $p \to 0$, which is not u^0 but

$$\log u = \lim_{p \to 0} \frac{u^p - 1}{p}.$$

In fact, $\log u$ is the adequate homogeneity 0 power of u. We use the Poincaré inequality and the energy estimate for $\log u$ to get that $\log u$ is in BMO. Finally, the p_0 that closes the gap is obtained by applying a consequence of the John–Nirenberg inequality.

Since we need to deal with integrals of u^p, for p both positive and negative, and of $\log u$ (corresponding to $p = 0$), we need energy estimates for powers of u. In order to have well defined integrals in the case of subsolutions, we must consider truncations of u and then pass to the limit.

For $m > 0$, we use the notation

$$\overline{u} = \begin{cases} u & \text{if } u < m \\ m & \text{if } u \geq m. \end{cases} \tag{6.3}$$

Lemma 6.3 (Energy estimates for powers of u and for $\log u$). *There exists a constant $C > 0$ depending only on Λ/λ such that the following statements hold. Fix any $\eta \in C_c^1(\Omega)$.*

(a) If $u > 0$ *is a subsolution in* Ω *and*

$$v = \overline{u}^{k-1}u \qquad \text{for } k \geq 1,$$

then

$$\int_\Omega \eta^2|\nabla v|^2\,dx \leq C(2k-1)\int_\Omega |\nabla\eta|^2 v^2\,dx. \tag{6.4}$$

(b) If $u > 0$ *is a supersolution in* Ω *and*

$$v = u^k \qquad \text{for } k < 1/2,$$

then

$$\int_\Omega \eta^2|\nabla v|^2\,dx \leq C\Big(\frac{2k}{1-2k}\Big)^2\int_\Omega |\nabla\eta|^2 v^2\,dx. \tag{6.5}$$

(c) If $u > 0$ *is a supersolution in* Ω *and*

$$v = \log u,$$

then

$$\int_\Omega \eta^2|\nabla v|^2\,dx \leq C\int_\Omega |\nabla\eta|^2\,dx. \tag{6.6}$$

Proof. The idea is to choose appropriate test functions.

We begin with (a). We assume that the right hand side in (6.4) is finite. If not, there is nothing to prove. Since $\nabla\overline{u} = \chi_{u<m}\nabla u$, we have $u\nabla\overline{u} = \overline{u}\nabla\overline{u}$ and

$$\nabla v = \overline{u}^{k-1}[(k-1)\nabla\overline{u} + \nabla u].$$

Without loss of generality, it can be assumed that $u \geq \delta > 0$ in Ω. In the general case, consider the subsolution $u_\delta = u + \delta > 0$. Applying Fatou's lemma and (6.4) to $v_\delta = (\overline{u_\delta})^{k-1}u_\delta$,

$$\begin{aligned}
\int_\Omega \eta^2|\nabla v|^2\,dx &= \int_\Omega \eta^2\overline{u}^{2k-2}|(k-1)\nabla\overline{u} + \nabla u|^2\,dx\\
&\leq \liminf_{\delta\to 0}\int_\Omega \eta^2\overline{u_\delta}^{2k-2}|(k-1)\nabla\overline{u_\delta} + \nabla u_\delta|^2\,dx\\
&= \liminf_{\delta\to 0}\int_\Omega \eta^2|\nabla v_\delta|^2\,dx\\
&\leq C(2k-1)\liminf_{\delta\to 0}\int_\Omega |\nabla\eta|^2 v_\delta^2\,dx\\
&= C(2k-1)\int_\Omega |\nabla\eta|^2 v^2\,dx.
\end{aligned}$$

Let

$$\varphi = \eta^2\overline{u}^{2k-2}u \geq 0.$$

Since $\delta \leq \overline{u} \leq m$ and $u\nabla\overline{u} = \overline{u}\nabla\overline{u}$, we get $\varphi \in H_0^1(\Omega)$. Then, using ellipticity, the boundedness of the coefficients and the Cauchy inequality with $\varepsilon = \lambda/(4\Lambda)$, that is, $AB \leq \frac{\lambda}{4\Lambda}A^2 + \frac{\Lambda}{\lambda}B^2$,

$$\begin{aligned}
0 &\geq \int_\Omega a(x)\nabla u\nabla\varphi\,dx \\
&= (2k-2)\int_\Omega \eta^2\overline{u}^{2k-2}a(x)\nabla\overline{u}\nabla\overline{u}\,dx + \int_\Omega \eta^2\overline{u}^{2k-2}a(x)\nabla u\nabla u\,dx \\
&\quad + 2\int_\Omega \eta\overline{u}^{2k-2}ua(x)\nabla u\nabla\eta\,dx \\
&\geq (2k-2)\lambda\int_\Omega \eta^2\overline{u}^{2k-2}|\nabla\overline{u}|^2\,dx + \lambda\int_\Omega \eta^2\overline{u}^{2k-2}|\nabla u|^2\,dx \\
&\quad - 2\Lambda\int_\Omega (\eta\overline{u}^{k-1}|\nabla u|)(\overline{u}^{k-1}u|\nabla\eta|)\,dx \\
&\geq (2k-2)\lambda\int_\Omega \eta^2\overline{u}^{2k-2}|\nabla\overline{u}|^2\,dx + \frac{\lambda}{2}\int_\Omega \eta^2\overline{u}^{2k-2}|\nabla u|^2\,dx \\
&\quad - \frac{2\Lambda^2}{\lambda}\int_\Omega \overline{u}^{2k-2}u^2|\nabla\eta|^2\,dx.
\end{aligned}$$

We pass the last term to the left hand side to get

$$\int_\Omega \eta^2\overline{u}^{2k-2}\left[(2k-2)|\nabla\overline{u}|^2 + |\nabla u|^2\right]dx \leq \left(\frac{2\Lambda}{\lambda}\right)^2\int_\Omega |\nabla\eta|^2v^2\,dx.$$

Using that $\nabla\overline{u}\nabla u = |\nabla\overline{u}|^2$, one can easily check the pointwise inequality

$$|\nabla v|^2 \leq (2k-1)\overline{u}^{2k-2}\left[(2k-2)|\nabla\overline{u}|^2 + |\nabla u|^2\right].$$

This and the previous estimate give (a) with $C = (2\Lambda/\lambda)^2$.

Consider next (b). We assume that the right hand side in (6.5) is finite and that $k \neq 0$, otherwise there is nothing to prove. Without loss of generality, $u \geq \delta > 0$ in Ω. Indeed, for the general case, by applying Fatou's lemma and (6.5) to $v_\delta = (u+\delta)^k$,

$$\begin{aligned}
\int_\Omega \eta^2|\nabla v|^2\,dx &= \int_\Omega \eta^2|ku^{k-1}\nabla u|^2\,dx \\
&\leq \liminf_{\delta\to 0}\int_\Omega \eta^2|k(u+\delta)^{k-1}\nabla u|^2\,dx \\
&= \liminf_{\delta\to 0}\int_\Omega \eta^2|\nabla v_\delta|^2\,dx \\
&\leq C\left(\frac{2k}{1-2k}\right)^2\liminf_{\delta\to 0}\int_\Omega |\nabla\eta|^2(u+\delta)^{2k}\,dx \\
&= C\left(\frac{2k}{1-2k}\right)^2\int_\Omega |\nabla\eta|^2v^2\,dx.
\end{aligned}$$

Let

$$\varphi = \eta^2u^{2k-1} \geq 0.$$

Note that $2k-1<0$. Since $u \geq \delta$ and $\nabla u \in L^2(\Omega)$, we get $\varphi \in H_0^1(\Omega)$. Then

$$\begin{aligned} 0 &\leq \int_\Omega a(x)\nabla u \nabla\varphi\,dx \\ &= (2k-1)\int_\Omega \eta^2 u^{2k-2} a(x)\nabla u\nabla u\,dx + 2\int_\Omega \eta u^{2k-1}a(x)\nabla u\nabla\eta\,dx. \end{aligned}$$

By ellipticity, the boundedness of the coefficients and the Cauchy inequality with $\varepsilon = \frac{\lambda(1-2k)}{4\Lambda}$, that is, $AB \leq \frac{\lambda(1-2k)}{4\Lambda}A^2 + \frac{\Lambda}{\lambda(1-2k)}B^2$,

$$\begin{aligned} \lambda(1-2k)\int_\Omega \eta^2 u^{2k-2}|\nabla u|^2\,dx &\leq (1-2k)\int_\Omega \eta^2 u^{2k-2}a(x)\nabla u\nabla u\,dx \\ &\leq 2\Lambda\int_\Omega (\eta u^{k-1}|\nabla u|)(u^k|\nabla\eta|)\,dx \\ &\leq \frac{\lambda(1-2k)}{2}\int_\Omega \eta^2 u^{2k-2}|\nabla u|^2\,dx \\ &\quad + \frac{2\Lambda^2}{\lambda(1-2k)}\int_\Omega u^{2k}|\nabla\eta|^2\,dx. \end{aligned}$$

After rearranging and using that $\nabla v = ku^{k-1}\nabla u$ we arrive at (6.5) with $C=(\Lambda/\lambda)^2$.

For (c), as before, we can assume that $u \geq \delta > 0$. Indeed, if $v_\delta = \log(u+\delta)$ then, by Fatou's lemma,

$$\begin{aligned} \int_\Omega \eta^2|\nabla v|^2\,dx &= \int_\Omega \eta^2\left|\frac{\nabla u}{u}\right|^2 dx \\ &\leq \liminf_{\delta\to 0}\int_\Omega \eta^2\left|\frac{\nabla u}{u+\delta}\right|^2 dx \\ &= \liminf_{\delta\to 0}\int_\Omega \eta^2|\nabla v_\delta|^2\,dx \leq C\int_\Omega |\nabla\eta|^2\,dx. \end{aligned}$$

The proof proceeds as we did for $(\log u)^-$ in the proof of Theorem 5.16. But for the reader's convenience we give the details here. Let

$$\varphi = \frac{1}{u}\eta^2 \geq 0.$$

Due to the lower bound $u \geq \delta$, we have $\varphi \in H_0^1(\Omega)$ and $\nabla v = \frac{\nabla u}{u} \in L^2(\Omega)$. Then

$$0 \leq \int_\Omega a(x)\nabla u\nabla\varphi\,dx = -\int_\Omega \eta^2 a(x)\nabla v\nabla v\,dx + 2\int_\Omega a(x)(\eta\nabla v)\nabla\eta\,dx.$$

By ellipticity, the boundedness of the coefficients and the Cauchy inequality with $\varepsilon = \lambda/(4\Lambda)$, that is, $AB \leq \frac{\lambda}{4\Lambda}A^2 + \frac{\Lambda}{\lambda}B^2$,

$$\begin{aligned} \lambda\int_\Omega \eta^2|\nabla v|^2\,dx &\leq \int_\Omega \eta^2 a(x)\nabla v\nabla v\,dx \\ &\leq 2\Lambda\int_\Omega |\eta\nabla v||\nabla\eta|\,dx \\ &\leq \frac{\lambda}{2}\int_\Omega \eta^2|\nabla v|^2\,dx + \frac{2\Lambda^2}{\lambda}\int_\Omega |\nabla\eta|^2\,dx. \end{aligned}$$

The integrals are all finite so, after rearranging, we get (6.6) with $C=(2\Lambda/\lambda)^2$. □

Remark 6.4. If u is a *bounded* positive subsolution, then the truncation $\overline{u}$ is not needed in Lemma 6.3(a) and one can prove that $v = u^k$ satisfies (6.4) for all $k > 1/2$. The proof follows what we did for (b) using the test function $\varphi = \eta^2 u^{2k-1}$.

The second ingredient is the Sobolev embedding.

Lemma 6.5 (Sobolev embedding). *For any $w \in H_0^1(Q_r)$, $r > 0$,*

$$\left(\int_{Q_r} w^{2\gamma}\,dx\right)^{1/(2\gamma)} \leq C\left(\int_{Q_r} |\nabla w|^2\,dx\right)^{1/2}$$

where $\gamma = \frac{n}{n-2} > 1$ if $n \geq 3$, or $\gamma > 1$ if $n = 2$, and $C > 0$ depends only on γ.

We next combine the energy estimate for powers and the Sobolev embedding.

Corollary 6.6 (Local improvement of integrability). *Let $\gamma > 1$ be as in Lemma 6.5. There exists a constant $C > 0$ depending only on Λ/λ and γ such that the following holds. If $u > 0$ is a subsolution in Q_4 and $k \geq 1$, or if $u > 0$ is a supersolution in Q_4 and $k < 1/2$ then, for any $0 < r < R < 4$,*

$$\left(\int_{Q_r} (u^k)^{2\gamma}\,dx\right)^{1/(2\gamma)} \leq \frac{C_{2k}}{R-r}\left(\int_{Q_R} (u^k)^2\,dx\right)^{1/2} \tag{6.7}$$

where

$$C_{2k} = \begin{cases} C(2k-1)^{1/2} & \text{if } k \geq 1 \\ C\frac{2|k|}{1-2k} & \text{if } k < 1/2.. \end{cases}$$

In particular, if $u > 0$ is a subsolution and $p = 2k \geq 2$, or if $u > 0$ is a supersolution and $0 < p = 2k < 1$,

$$\Phi(\gamma p, r) \leq \left(\frac{C_p}{R-r}\right)^{2/p} \Phi(p, R). \tag{6.8}$$

If $u > 0$ is a supersolution and $p = 2k < 0$,

$$\Phi(p, R) \leq \left(\frac{C}{R-r}\right)^{2/|p|} \Phi(\gamma p, r). \tag{6.9}$$

Proof. Recall (6.3). As in Lemma 6.3, let $v = \overline{u}^{k-1}u \leq u^k$ if $u > 0$ is a subsolution and $k \geq 1$, or $v = u^k$ if $u > 0$ is a supersolution and $k < 1/2$. Assume that the right hand side in (6.7) is finite, otherwise there is nothing to prove. Then $v \in L^2(Q_R)$. Let $\eta \in C_c^1(Q_R)$, $\eta \geq 0$, such that $\eta = 1$ in Q_r and $|\nabla \eta| \leq 2/(R-r)$. Using this η in Lemma 6.3,

$$\int_{Q_R} |\eta \nabla v|^2\,dx \leq C\int_{Q_R} |\nabla \eta|^2 v^2\,dx \leq C\frac{4}{(R-r)^2}\int_{Q_R} v^2\,dx. \tag{6.10}$$

In particular, $\eta v \in H_0^1(Q_R)$. By the Sobolev embedding from Lemma 6.5 and (6.10),

$$\begin{aligned}
\left(\int_{Q_r} v^{2\gamma}\,dx\right)^{1/(2\gamma)} &\le \left(\int_{Q_R} (\eta v)^{2\gamma}\,dx\right)^{1/(2\gamma)} \\
&\le C_\gamma\left(\int_{Q_R} |\nabla(\eta v)|^2\,dx\right)^{1/2} \\
&\le C\left(\frac{4}{(R-r)^2}\int_{Q_R} v^2\,dx + \int_{Q_R} |\eta\nabla v|^2\,dx\right)^{1/2} \\
&\le \frac{C_{2k}}{R-r}\int_{Q_R} v^2\,dx \le \frac{C_{2k}}{R-r}\int_{Q_R} (u^k)^2\,dx.
\end{aligned}$$

This is (6.7) for supersolutions. For subsolutions, take $m \to \infty$ in the left hand side above and use the monotone convergence theorem to get (6.7). The constant $C_{2k} > 0$ can be tracked down from Lemmas 6.3 and 6.5. Finally, (6.8) and (6.9) follow by letting $p = 2k$ in (6.7) and, for the case of (6.9), noticing that $\frac{2|k|}{1-2k} \le 1$ if $k < 0$. □

6.2 UPPER AND LOWER BOUNDS

In this section we implement Moser's iteration technique which consists on recursively applying the initial integrability of powers of u we obtained in Corollary 6.6 and conclude by means of Lemma 6.2. With this, we get the local boundedness of positive solutions (6.1) for all $p \ge 2$ and the lower bound for the infimum (6.2) for all $p > 0$. A rescaling and the following real analysis lemma are used to show that (6.1) actually also holds for all $0 < p < 2$.

Lemma 6.7. *Let $\phi(r) \ge 0$ be a bounded function defined on an interval I. Suppose that for any $r, R \in I$ with $r < R$, we have*

$$\phi(r) \le \sigma\phi(R) + \frac{C}{(R-r)^\beta}$$

where $0 < \sigma < 1$ and $C, \beta > 0$. Then there exists $c > 0$ depending only on σ and β such that for any $r, R \in I$ with $r < R$,

$$\phi(r) \le c\frac{C}{(R-r)^\beta}.$$

Proof. Given any $0 < \varepsilon < 1$, we consider the increasing sequence of points

$$r_0 = r \qquad \text{and} \qquad r_{j+1} = r_j + (R-r)(1-\varepsilon)\varepsilon^j, \text{ for } j \ge 0.$$

Then

$$r_{j+1} = r + (R-r)(1-\varepsilon)\sum_{i=0}^{j} \varepsilon^j \to R \qquad \text{as } j \to \infty.$$

We iterate the inequality in the hypothesis to get

$$\begin{aligned}\phi(r) = \phi(r_0) &\leq \sigma\phi(r_1) + \frac{C}{(r_1 - r_0)^\beta} \\ &\leq \sigma\left[\sigma\phi(r_2) + \frac{C}{(r_2 - r_1)^\beta}\right] + \frac{C}{((1-\varepsilon)(R-r))^\beta} \\ &= \sigma^2\phi(r_2) + \frac{C}{((1-\varepsilon)(R-r))^\beta}\left[1 + \frac{\sigma}{\varepsilon^\beta}\right] \\ &\leq \sigma^j\phi(r_j) + \frac{C}{((1-\varepsilon)(R-r))^\beta}\left[1 + \frac{\sigma}{\varepsilon^\beta} + \cdots + \left(\frac{\sigma}{\varepsilon^\beta}\right)^j\right].\end{aligned}$$

We choose $\varepsilon > 0$ such that $\sigma/\varepsilon^\beta = 1/2$. Then, as $j \to \infty$, the series above converges to 2 and $\sigma^j\phi(r_j) \to 0$ because ϕ is bounded. The conclusion follows. □

First, we consider positive subsolutions. We iterate (6.8) and also use Lemma 6.7.

Theorem 6.8 (Upper bound). *If $u > 0$ is a subsolution in Q_4 then, for any $p > 0$,*

$$\sup_{Q_1} u \leq C\left(\int_{Q_2} u^p\, dx\right)^{1/p}$$

where $C > 0$ depends only on n, Λ/λ and p.

Proof. Fix $p \geq 2$. Let $\gamma > 1$ be as in the Sobolev embedding, Lemma 6.5. We define

$$p_j = \gamma^j p \qquad \text{and} \qquad r_j = 1 + \frac{1}{2^j}$$

for $j \geq 0$. Then, $p_j \geq 2$ and $r_j - r_{j+1} = 1/2^{j+1}$. By (6.8) in Corollary 6.6 applied with $r = r_{j+1}$ and $R = r_j$,

$$\begin{aligned}\Phi(p_{j+1}, r_{j+1}) &= \Phi(\gamma p_j, r_{j+1}) \\ &\leq \left(2^{j+1}C(p_j - 1)^{1/2}\right)^{2/p_j}\Phi(p_j, r_j) \\ &\leq C^{j/\gamma^j}\Phi(p_j, r_j)\end{aligned}$$

where $C > 0$ depends on Λ/λ, γ and p, but not on j. We then iterate this to obtain, for any $j \geq 0$,

$$\Phi(p_{j+1}, r_{j+1}) \leq C^{j/\gamma^j + \cdots + 1/\gamma + 1}\Phi(p_0, r_0) \leq C^{S+1}\Phi(p, 2)$$

where $S = \sum_{j=1}^\infty \frac{j}{\gamma^j}$ is a convergent series because of the ratio test and the fact that $\gamma > 1$. We conclude by noticing that, by Lemma 6.2,

$$\limsup_{j\to\infty} \Phi(p_{j+1}, r_{j+1}) \geq \limsup_{j\to\infty} \Phi(p_{j+1}, 1) = \Phi(\infty, 1).$$

Next, we need to push the value of $p \geq 2$ all the way to $p > 0$. For this, let us begin by observing that, by using a translation and rescaling argument as we did in the proof of Corollary 5.3, we can easily show that, for any $0 < r < R < 4$,

$$\sup_{Q_r} u \leq \frac{C}{(R-r)^{n/2}} \left(\int_{Q_R} u^2 \, dx \right)^{1/2}.$$

Let $0 < p < 2$. By writing $u^2 = u^{2-p} u^p$ in this estimate and applying Young's inequality $AB \leq \frac{1}{q} A^q + \frac{1}{q'} B^{q'}$ with $q = 1/(1-p/2) > 1$ and $q' = 2/p > 1$,

$$\begin{aligned} \sup_{Q_r} u &\leq \left(\sup_{Q_R} u \right)^{1-p/2} \frac{C}{(R-r)^{n/2}} \left(\int_{Q_R} u^p \, dx \right)^{1/2} \\ &\leq (1-p/2) \sup_{Q_R} u + \frac{C}{(R-r)^{n/p}} \left(\int_{Q_R} u^p \, dx \right)^{1/p}. \end{aligned}$$

Let us call $\phi(r) = \sup_{Q_r} u$ and $\sigma = 1 - p/2 \in (0,1)$. Then, $\phi \geq 0$ is bounded and the inequality above says that for any $0 < r < R < 4$,

$$\phi(r) \leq \sigma \phi(R) + \frac{C}{(R-r)^{n/p}} \left(\int_{Q_R} u^p \, dx \right)^{1/p}.$$

By Lemma 6.7 with $\beta = n/p$,

$$\phi(r) \leq c \frac{C}{(R-r)^{n/p}} \left(\int_{Q_R} u^p \, dx \right)^{1/p}.$$

We conclude by letting $r = 1$ and $R = 2$. □

Next, we look at positive supersolutions.

Theorem 6.9 (Lower bound). *If $u > 0$ is a supersolution in Q_4 then, for any $p > 0$,*

$$\left(\int_{Q_3} u^{-p} \, dx \right)^{-1/p} \leq C \inf_{Q_1} u$$

where $C > 0$ depends only on n, Λ/λ and p.

Proof. Fix $p > 0$. Let $\gamma > 1$ be as in the Sobolev embedding, Lemma 6.5. We define

$$r_j = 2 + \frac{1}{2^j}$$

for $j \geq 0$. Then $r_j - r_{j+1} = 1/2^{j+1}$. We successively apply (6.9) in Corollary 6.6 starting from $j = 0$ to get that for any $j \geq 0$,

$$\begin{aligned} \Phi(-p, 3) &\leq (2C)^{2/p} \Phi(-\gamma p, r_1) \\ &\leq (2C)^{2/p} (2^2 C)^{2/(\gamma p)} \Phi(-\gamma^2 p, r_2) \\ &\leq (2C)^{2/p} (2^2 C)^{2/(\gamma p)} (2^3 C)^{2/(\gamma^2 p)} \Phi(-\gamma^3 p, r_3) \\ &\leq C^{(j+1)/\gamma^j + j/\gamma^{j-1} + \cdots + 2/\gamma + 1} \Phi(-\gamma^{j+1} p, r_{j+1}) \\ &\leq C^S \Phi(-\gamma^{j+1} p, 2) \end{aligned}$$

where $S = \sum_{j=0}^{\infty} \frac{j+1}{\gamma^j}$ is a convergent series because of the ratio test and the fact that $\gamma > 1$. We conclude by using Lemma 6.2, which ensures that

$$\lim_{j\to\infty} \Phi(-\gamma^{j+1}p, 2) = \inf_{Q_2} u \le \inf_{Q_1} u.$$

□

From Theorems 6.8 and 6.9 we immediately get the following result.

Corollary 6.10. *For any $p > 0$, there exists $C > 0$ depending only on n, Λ/λ and p such that for any positive solution u in Q_4,*

$$\sup_{Q_1} u \le C\left(\int_{Q_3} u^p\,dx\right)^{1/p}$$

and

$$\left(\int_{Q_3} u^{-p}\,dx\right)^{-1/p} \le C \inf_{Q_1} u.$$

We have the $L^2 \implies L^\infty$ result (Corollary 5.3), this time via Moser's technique.

Corollary 6.11 (Local boundedness of solutions). *If u is a solution in Q_4 then*

$$\sup_{Q_1} u^+ \le C\left(\int_{Q_2} u^2\,dx\right)^{1/2}$$

where $C > 0$ is a constant depending only on n and Λ/λ. In particular, solutions are locally bounded.

Proof. We only need to check that u^+ is a subsolution in Q_4 and apply Theorem 6.8 with $p = 2$ to $u^+ + \delta$, $\delta > 0$, and let $\delta \to 0$. Define, for every $\varepsilon > 0$,

$$f_\varepsilon(u) = \begin{cases} (u^2+\varepsilon^2)^{1/2} - \varepsilon & \text{if } u > 0 \\ 0 & \text{if } u \le 0. \end{cases}$$

Then f_ε is differentiable in $\mathbb{R}$, with f'_ε bounded and nonnegative. Also, f'_ε is piecewise smooth, with $f''_\varepsilon \in L^\infty(\mathbb{R})$ and nonnegative. Therefore, if we take $\varphi \in C^1_c(Q_4)$, $\varphi \ge 0$, then $f'_\varepsilon(u)\varphi \in H^1_0(Q_4)$. Hence, using that u is a solution and ellipticity,

$$\begin{aligned}\int_{Q_4} a(x)\nabla(f_\varepsilon(u))\nabla\varphi\,dx &= \int_{Q_4} f'_\varepsilon(u)a(x)\nabla u\nabla\varphi\,dx \\ &= \int_{Q_4} a(x)\nabla(f'_\varepsilon(u)\varphi)\nabla u\,dx - \int_{Q_4} f''_\varepsilon(u)\varphi a(x)\nabla u\nabla u\,dx \\ &\le 0 - \lambda\int_{Q_4} f''_\varepsilon(u)\varphi|\nabla u|^2\,dx \le 0.\end{aligned}$$

Therefore, by the dominated convergence theorem,

$$\begin{aligned}\int_{Q_4} a(x)\nabla(u^+)\nabla\varphi\,dx &= \int_{Q_4} \chi_{u>0}a(x)\nabla u\nabla\varphi\,dx \\ &= \lim_{\varepsilon\to 0}\int_{Q_4} f'_\varepsilon(u)a(x)\nabla u\nabla\varphi\,dx \le 0\end{aligned}$$

and u^+ is a subsolution.

The local boundedness of solutions follows by noticing that $-u$ is also a solution, so u^- is locally bounded. □

6.3 CLOSING THE GAP

The following result closes the gap between (6.1) and (6.2), see Corollary 6.10.

Theorem 6.12. *There exist $p_0 > 0$ and $C > 0$ depending on n and Λ/λ such that for any positive supersolution u in Q_4,*

$$\Phi(p_0, 3) \le C\Phi(-p_0, 3). \tag{6.11}$$

Clearly, Corollary 6.10 and Theorem 6.12 give the Harnack inequality of Theorem 6.1 with $\Omega = Q_4$ and $\Omega' = Q_1$.

To prove Theorem 6.12 we need the Poincaré inequality for cubes and a consequence of the John–Nirenberg inequality for BMO functions. Recall the notions of average of a function over a cube Q and the BMO space in Definition 4.18.

Lemma 6.13 (Poincaré inequality for cubes). *There exists a constant $C > 0$ depending only on n such that for any $w \in H^1(Q_r)$ we have*

$$\int_{Q_r} |w - w_{Q_r}|^2\, dx \le Cr^2 \int_{Q_r} |\nabla w|^2\, dx.$$

We considered the space BMO in Section 4.4 and proved the John–Nirenberg inequality in Section 4.5. In those cases, the BMO space was defined for locally integrable functions over the whole space $\mathbb{R}^n$. For our purposes, we need a localized definition, replacing $\mathbb{R}^n$ by Q_3 and considering cubes contained in Q_3.

We say that a function $f \in L^1_{\mathrm{loc}}(Q_3)$ is in $BMO(Q_3)$ if

$$\|f\|_{BMO(Q_3)} = \sup_{Q \subset Q_3} \frac{1}{|Q|} \int_Q |f - f_Q|\, dx < \infty.$$

The reader can easily verify that everything we did in Section 4.5 holds true for this localized BMO space. In particular, the proof of the John–Nirenberg inequality (Theorem 4.23) already uses a localization of the Calderón–Zygmund decomposition. Thus, we obtain the estimate (4.16) in Corollary 4.24 that we need here.

Lemma 6.14 (Consequence of the John–Nirenberg inequality). *There is $0 < \theta < 1$ depending only on n such that for any $f \in BMO(Q_3)$ with $\|f\|_{BMO(Q_3)} \le 1$,*

$$\sup_{Q \subset Q_3} \left(\frac{1}{|Q|} \int_Q e^{\theta f(x)}\, dx \right) \left(\frac{1}{|Q|} \int_Q e^{-\theta f(x)}\, dx \right) \le 12 + 8\sqrt{2}. \tag{6.12}$$

We are ready to prove Theorem 6.12. Here is where we use Lemma 6.3(c).

Proof of Theorem 6.12. Suppose that we have proved (6.11) when $u \ge \delta > 0$ in Q_4. Then the general case follows because, for any $\delta > 0$,

$$\left(\int_{Q_3} u^{p_0}\, dx \right)^{1/p_0} \le \left(\int_{Q_3} (u + \delta)^{p_0}\, dx \right)^{1/p_0}$$

and, by the monotone convergence theorem,

$$\left(\int_{Q_3} \frac{1}{(u+\delta)^{p_0}}\,dx\right)^{-1/p_0} \to \left(\int_{Q_3} \frac{1}{u^{p_0}}\,dx\right)^{-1/p_0}$$

as $\delta \to 0$. Therefore, without loss of generality, $u \geq \delta > 0$ in Q_4, and the right hand side of (6.11) is finite.

Let $v = \log u$. Since $|v| \leq |\log \delta|$ for $\delta \leq u \leq 1$, and $|v| \leq u - 1$ for $u > 1$, we get $v \in L^2(Q_4)$. Moreover, $|\nabla v| = \frac{|\nabla u|}{u} \leq \frac{|\nabla u|}{\delta} \in L^2(Q_4)$. Hence, $v \in H^1(Q_4)$. By the Hölder and Poincaré inequalities (see Lemma 6.13), for any cube $Q_r = Q_r(z) \subset Q_3$,

$$\begin{aligned}\frac{1}{|Q_r|}\int_{Q_r} |v - v_{Q_r}|\,dx &\leq \left(\frac{1}{|Q_r|}\int_{Q_r} |v - v_{Q_r}|^2\,dx\right)^{1/2} \\ &\leq C\left(\frac{r^2}{|Q_r|}\int_{Q_r} |\nabla v|^2\,dx\right)^{1/2}\end{aligned}$$

for $C > 0$ depending only on n. Notice that if $Q_r = Q_r(z) \subset Q_3$ then $Q_{4r/3} = Q_{4r/3}(z) \subset Q_4$. Choose $\eta \in C_c^1(Q_{4r/3})$, $\eta \geq 0$, such that $\eta = 1$ in Q_r and $|\nabla \eta| \leq C/r$. Then (6.6) in Lemma 6.3 with $\Omega = Q_4$ implies that

$$\int_{Q_r} |\nabla v|^2\,dx \leq \int_{Q_{4r/3}} \eta^2 |\nabla v|^2\,dx \leq C\int_{Q_{4r/3}} |\nabla \eta|^2\,dx \leq \frac{C}{r^2}|Q_r|$$

where $C > 0$ depends only on n and Λ/λ. Therefore, from these two estimates,

$$\frac{1}{|Q_r|}\int_{Q_r} |v - v_{Q_r}|\,dx \leq D$$

for all cubes $Q_r = Q_r(z) \subset Q_3$, where D depends only on n and Λ/λ. In other words, $f = D^{-1}v = D^{-1}\log u \in BMO(Q_3)$ and $\|f\|_{BMO(Q_3)} \leq 1$. Let $0 < \theta < 1$ be as in Lemma 6.14. Since $e^{\pm\theta f} = u^{\pm\theta/D} = u^{\pm p_0}$, where $p_0 = \theta/D > 0$, by (6.12),

$$\left(\frac{1}{|Q_3|}\int_{Q_3} u^{p_0}\,dx\right)\left(\frac{1}{|Q_3|}\int_{Q_3} u^{-p_0}\,dx\right) \leq 12 + 8\sqrt{2} = c$$

and (6.11) holds for $p_0 = \theta/D > 0$, which depends only on n and Λ/λ. □

By Hölder's inequality, if $0 < p < p_0$ then $\Phi(p,3) \leq C_{n,p,p_0}\Phi(p_0,3)$. Therefore, Theorems 6.12 and 6.9 give that if u is a positive supersolution in Q_4 then $\Phi(p,3) \leq C\Phi(-\infty,1)$ for all $0 < p \leq p_0$. However, by iterating (6.8) as we did in the proof of Theorem 6.8 but only a finite number of times, we can increase the range of p_0 all the way to the Sobolev exponent γ of Lemma 6.5. The following result is the so-called weak Harnack inequality for positive supersolutions.

Corollary 6.15 (Weak Harnack inequality). *Let $\gamma > 1$ be the Sobolev exponent of Lemma 6.5. If u is a positive supersolution in Q_4 then, for any $0 < p < \gamma$,*

$$\left(\int_{Q_2} u^p\,dx\right)^{1/p} \leq C\inf_{Q_1} u$$

where $C > 0$ depends only on n, Λ/λ and p.

Proof. As we mentioned before, by Hölder's inequality and Theorems 6.12 and 6.9, it is enough to show that for any $0 < p_0 < p < \gamma$ we have

$$\left(\int_{Q_2} u^p \, dx\right)^{1/p} \leq C\left(\int_{Q_3} u^{p_0} \, dx\right)^{1/p_0}.$$

Since $0 < p/\gamma < 1$, (6.8) gives

$$\Phi(p, 2) = \Phi(\gamma(p/\gamma), 2) \leq C(n, \Lambda/\lambda, p, \varepsilon)\Phi(p/\gamma, 2 + \varepsilon)$$

for any $\varepsilon > 0$. We iterate this a finite number of times so that in the right hand side we continue to decrease the exponent and to enlarge the domain of integration so as to arrive at $\Phi(p_0, 3)$. For this, observe that there exists $\ell > 0$ depending only on n, Λ/λ and p such that $0 < p/\gamma^{\ell+1} \leq p_0 < p/\gamma^{\ell} < 1$ and there is $\varepsilon > 0$ depending on ℓ such that $3 = 2 + (\ell + 1)\varepsilon$. Therefore, after ℓ iterations,

$$\Phi(p, 2) \leq C(n, \Lambda/\lambda, p)\Phi(p/\gamma^{\ell+1}, 2 + (\ell + 1)\varepsilon) \leq C\Phi(p_0, 3)$$

where in the second estimate we applied Hölder's inequality. □

6.4 HARNACK INEQUALITY IMPLIES HÖLDER REGULARITY

In this section we prove that the local boundedness of solutions (Corollary 6.11) and the Harnack inequality (Theorem 6.1) imply the Hölder continuity of solutions.

Let u be a solution in Q_4 and define, for $r \leq 2$,

$$M(r) = \sup_{Q_r} u \qquad \text{and} \qquad m(r) = \inf_{Q_r} u.$$

Since u is locally bounded, $M(r) < \infty$ and $m(r) > -\infty$. By linearity,

$$M(2) - u \qquad \text{and} \qquad u - m(2)$$

are nonnegative solutions in Q_2. Then, by the Harnack inequality,

$$\begin{aligned} M(2) - m(1/2) &= \sup_{Q_{1/2}}(M(2) - u) \\ &\leq C_H \inf_{Q_{1/2}}(M(2) - u) = C_H(M(2) - M(1/2)) \end{aligned}$$

and

$$\begin{aligned} M(1/2) - m(2) &= \sup_{Q_{1/2}}(u - m(2)) \\ &\leq C_H \inf_{Q_{1/2}}(u - m(2)) = C_H(m(1/2) - m(2)). \end{aligned}$$

Observe that $C_H > 1$ unless u is constant. After adding these two inequalities,

$$(1 + C_H)(M(1/2) - m(1/2)) \leq (C_H - 1)(M(2) - m(2)).$$

In other words, we get the oscillation decay

$$\operatorname*{osc}_{Q_{1/2}} u \leq \sigma \operatorname*{osc}_{Q_2} u$$

with $0 < \sigma = \frac{C_H - 1}{C_H + 1} < 1$, which depends only on n and Λ/λ. As we have seen in Section 5.3 (see Corollary 5.11), this implies the local Hölder continuity of u.

NOTES ON CHAPTER 6

The Harnack inequality for divergence form elliptic equations was proved by Moser in [53]. It was generalized to equations with lower-order terms and nonzero right hand sides by Stampacchia [62], see also [43, 44] for other presentations.

De Giorgi's arguments from Chapter 5 can be modified to give a different proof of Moser's theorem, even for more general equations, see Di Benedetto–Trudinger [26].

CHAPTER 7

Perturbation theory for divergence form equations

In this chapter we prove the Schauder and Calderón–Zygmund estimates for weak solutions to divergence form elliptic equations with continuous coefficients. The techniques we present here were introduced by Caffarelli.

7.1 SCHAUDER ESTIMATES

We consider divergence form elliptic operators

$$\operatorname{div}(a(x)\nabla u) = \partial_i(a^{ij}(x)\partial_j u) \qquad \text{in } B_1 \subset \mathbb{R}^n$$

for $n \geq 2$. As in Chapters 5 and 6, we assume that the coefficients $a(x) = (a^{ij}(x))_{i,j=1}^n$ are bounded functions in B_1 with $\|a\|_{L^\infty(B_1)} \leq \Lambda$ for some $\Lambda > 0$, symmetric $a^{ij}(x) = a^{ji}(x)$, and uniformly elliptic: there exists a constant $\lambda > 0$ such that $a^{ij}(x)\xi_i\xi_j \geq \lambda|\xi|^2$, for all $\xi \in \mathbb{R}^n$ and a.e. $x \in B_1$.

Given $f \in L^2(B_1)$ and $F = (f_1, \dots, f_n) \in [L^2(B_1)]^n$, we say that $u \in H^1(B_1)$ is a **weak solution** to

$$-\operatorname{div}(a(x)\nabla u) = f - \operatorname{div} F \qquad \text{in } B_1 \tag{7.1}$$

if

$$\int_{B_1} a(x)\nabla u \nabla\varphi\, dx = \int_{B_1} f\varphi\, dx + \int_\Omega F \cdot \nabla\varphi\, dx$$

for every $\varphi \in C_c^1(B_1)$. By density, the same holds for all $\varphi \in H_0^1(B_1)$.

Throughout this section we will say that u is a solution in B_1 if $u \in H^1(B_1)$ is a weak solution to (7.1) in B_1.

The main result is the following.

Theorem 7.1 (Schauder estimates for weak solutions). *Fix any $0 < \alpha < 1$. Assume that $a(x) \in C^{0,\alpha}(B_1)$, $F \in [C^{0,\alpha}(B_1)]^n$ and $f \in L^p(B_1)$ with $p = n/(1-\alpha) > n$. There exists a constant $C > 0$ depending on n, λ, Λ, α and $[a]_{C^{0,\alpha}(B_1)}$ such that if u is a solution in B_1, then $u \in C^{1,\alpha}(B_{1/2})$ with*

$$\|u\|_{C^{1,\alpha}(B_{1/2})} \leq C\left(\|u\|_{L^2(B_1)} + \|f\|_{L^p(B_1)} + \|F\|_{C^{0,\alpha}(B_1)}\right).$$

DOI: 10.1201/9781032679853-7

Using a translation and covering argument as we did in Chapter 5, Theorem 7.1 can also be established for equations on bounded domains Ω by replacing B_1 by Ω and $B_{1/2}$ by any subdomain $\Omega' \subset\subset \Omega$. In this case the constant C will also depend on $\mathrm{dist}(\Omega', \partial\Omega)$.

To prove the Schauder estimates, the following characterization of the $C^{1,\alpha}$ space in terms of L^2 averages due to Campanato will be needed.

Theorem 7.2 (Campanato's theorem). *Let Ω be a Lipschitz domain and $0 < \alpha < 1$. Denote by $\mathcal{P}_1$ the set of all linear polynomials of the form $\ell(x) = A + B \cdot x$, where $A \in \mathbb{R}$ and $B \in \mathbb{R}^n$. Let $u \in L^2(\Omega)$. Then $u \in C^{1,\alpha}(\overline{\Omega})$ if and only if there exists $D > 0$ such that*

$$\inf_{\ell \in \mathcal{P}_1} \frac{1}{|B_r|} \int_{B_r(z)\cap\Omega} |u(x) - \ell(x)|^2 \, dx \le D r^{2(1+\alpha)} \tag{7.2}$$

for all $z \in \overline{\Omega}$ and $0 < r \le \mathrm{diam}(\Omega)$. In this case, if D_* denotes the least constant D for which the inequality above holds, then the norm $\|u\|_{C^{1,\alpha}(\overline{\Omega})}$ is equivalent to $\|u\|_{L^2(\Omega)} + \|\nabla u\|_{L^2(\Omega)} + D_*^{1/2}$.

Therefore, to prove $C^{1,\alpha}$ estimates, it is enough to construct linear polynomials ℓ that approximate u at every scale in L^2 with a $C^{1,\alpha}$ rate at every point as in (7.2). For simplicity, we now consider the origin $z = 0$ in (7.2). To prove that $u \in C^{1,\alpha}(0)$ in this L^2-sense, it is enough to assume a Campanato-type oscillation decay for the coefficients and F, and a Morrey-type decay for f at the origin, see (7.3) and (7.4), respectively. The following is the main result we need to deduce Theorem 7.1.

Theorem 7.3 (Pointwise Schauder estimates for divergence form equations). *Let $0 < \alpha < 1$. There exists $\delta_0 > 0$ depending only on n and α such that if*

$$\frac{1}{|B_r|} \int_{B_r} \left(|a(x) - a(0)|^2 + |F(x) - F(0)|^2\right) dx \le \delta_0^2 r^{2\alpha} \tag{7.3}$$

and

$$\frac{1}{|B_r|} \int_{B_r} |f(x)|^2 \, dx \le \delta_0^2 r^{2\alpha-2} \tag{7.4}$$

for $0 < r < 1$, and u is a solution in B_1, then there exists a linear polynomial $\ell_\infty(x) = A_\infty + B_\infty \cdot x$ such that

$$\frac{1}{|B_r|} \int_{B_r} |u - \ell_\infty|^2 \, dx \le C \|u\|_{L^2(B_1)}^2 r^{2(1+\alpha)} \qquad \text{for } 0 < r < 1$$

and $|A_\infty| + |B_\infty| \le C\|u\|_{L^2(B_1)}$, where $C > 0$ is a constant depending only on n, λ, Λ and α.

We first prove Theorem 7.3 and then show how to derive Theorem 7.1 from it.

The proof of Theorem 7.3 we present here is based on a compactness argument that allows us to approximate solutions u by harmonic functions h at small scales. The reader should compare this with the method of maximum principle for the Laplacian in Section 3.7 and for nondivergence form equations in Section 11.1.

The first ingredient to implement the method of compactness is the Caccioppoli inequality. The following should be compared with the energy inequality for truncations in Lemma 5.6 and the energy estimates for powers in Lemma 6.3.

Lemma 7.4 (Caccioppoli inequality). *There exists a constant $C > 0$ depending only on λ and Λ such that for any solution u in B_1 and any $\eta \in C_c^1(B_1)$ we have*

$$\int_{B_1} \eta^2 |\nabla u|^2 \, dx \le C \Big[\int_{B_1} \eta^2 (f^2 + |F|^2 + u^2) \, dx + \int_{B_1} |\nabla \eta|^2 u^2 \, dx \Big].$$

In particular,

$$\int_{B_{1/2}} |\nabla u|^2 \, dx \le C \int_{B_1} (f^2 + |F|^2 + u^2) \, dx.$$

Proof. Take $\varphi = \eta^2 u \in H_0^1(B_1)$ as test function in (7.1). We have

$$\int_{B_1} \eta^2 a(x) \nabla u \nabla u \, dx = -2 \int_{B_1} a(x)(\eta \nabla u)(u \nabla \eta) \, dx + \int_{B_1} (f\eta)(\eta u) \, dx \\ + \int_{B_1} F \cdot (2u\eta\nabla\eta + \eta^2 \nabla u) \, dx.$$

We next apply ellipticity, the boundedness of the coefficients and the Cauchy inequality with $\varepsilon > 0$ to get

$$\lambda \int_{B_1} \eta^2 |\nabla u|^2 \, dx \le 2\Lambda\varepsilon \int_{B_1} \eta^2 |\nabla u|^2 \, dx + \frac{\Lambda}{2\varepsilon} \int_{B_1} |\nabla \eta|^2 u^2 \, dx \\ + \frac{1}{2} \int_{B_1} f^2 \eta^2 \, dx + \frac{1}{2} \int_{B_1} \eta^2 u^2 \, dx \\ + \frac{5}{4\varepsilon} \int_{B_1} |F|^2 \eta^2 \, dx + 2\varepsilon \int_{B_1} \eta^2 |\nabla u|^2 \, dx.$$

Now choose $\varepsilon > 0$ small depending on λ and Λ to conclude. For the second inequality, take $\eta \in C_c^1(B_1)$ such that $0 \le \eta \le 1$, $\eta = 1$ in $B_{1/2}$ and $|\nabla \eta| \le 4$. □

The Caccioppoli inequality gives control of local L^2 norms of ∇u. This is key in the application of the compactness argument that gives the first approximation lemma by harmonic functions. Recall that we denote by I the identity matrix.

Lemma 7.5 (Approximation lemma). *Given any $\varepsilon > 0$ there exists $\delta > 0$ depending only on ε such that for any solution u in B_1, if*

$$\frac{1}{|B_1|} \int_{B_1} u^2 \, dx \le 1$$

and

$$\frac{1}{|B_1|} \int_{B_1} (|a(x) - I|^2 + |F|^2 + f^2) \, dx < \delta^2,$$

then there exists a harmonic function h in $B_{1/2}$ such that

$$\frac{1}{|B_{1/2}|} \int_{B_{1/2}} |u - h|^2 \, dx < \varepsilon^2.$$

Proof. We argue by contradiction and compactness. Suppose that there exist $\varepsilon_0 > 0$, coefficients $a_k(x)$, right hand sides F_k and f_k and solutions u_k such that

$$\frac{1}{|B_1|}\int_{B_1} u_k^2\,dx \leq 1$$

and

$$\frac{1}{|B_1|}\int_{B_1} (|a_k(x) - I|^2 + |F_k|^2 + f_k^2)\,dx < \frac{1}{k^2}$$

for all $k \geq 1$, but for any harmonic function h in $B_{1/2}$,

$$\frac{1}{|B_{1/2}|}\int_{B_{1/2}} |u_k - h|^2\,dx \geq \varepsilon_0^2. \tag{7.5}$$

By the Caccioppoli inequality, Lemma 7.4,

$$\int_{B_{1/2}} |\nabla u_k|^2\,dx \leq C$$

for all $k \geq 1$. Therefore $\{u_k\}_{k=1}^{\infty}$ is a bounded sequence in $H^1(B_{1/2})$. Then, by the weak compactness of Hilbert spaces and the compact embedding $H^1(B_{1/2}) \subset\subset L^2(B_{1/2})$, there exist a subsequence, that we still denote by $\{u_k\}_{k=1}^{\infty}$, and $u_\infty \in H^1(B_{1/2})$ such that

$$u_k \to u_\infty \text{ weakly in } H^1(B_{1/2}) \text{ and strongly in } L^2(B_{1/2}).$$

Let us show that u_∞ is a weakly harmonic function in $B_{1/2}$. For any $\varphi \in C_c^1(B_{1/2})$,

$$\int_{B_{1/2}} a(x)\nabla u_k \nabla\varphi\,dx = \int_{B_{1/2}} (f_k\varphi + F_k \cdot \nabla\varphi)\,dx.$$

As $k \to \infty$, the right hand side vanishes because f_k and $|F_k|$ converge to 0 in $L^2(B_1)$. For the left hand side, notice that

$$\left|\int_{B_{1/2}} (a_k(x)\nabla u_k \nabla\varphi - \nabla u_\infty \nabla\varphi)\,dx\right| \leq \int_{B_{1/2}} |a_k(x) - I||\nabla u_k||\nabla\varphi|\,dx + \left|\int_{B_{1/2}} (\nabla u_k - \nabla u_\infty)\nabla\varphi\,dx\right|.$$

The first term in the right hand side above is bounded by

$$\|\nabla\varphi\|_{L^\infty(B_{1/2})}\|a_k - I\|_{L^2(B_1)}\|\nabla u_k\|_{L^2(B_{1/2})} \leq C/k.$$

The second term also vanishes in the limit because of the weak convergence of ∇u_k to ∇u_∞ in $L^2(B_{1/2})$. We conclude that

$$\int_{B_{1/2}} \nabla u_\infty \nabla\varphi\,dx = 0$$

and u_∞ is a harmonic function (see Section 2.2 for the equivalence between weakly harmonic and classical harmonic functions). This contradicts (7.5) for $h = u_\infty$ because u_k converges to u_∞ in $L^2(B_{1/2})$ □

Corollary 7.6 (Linear approximation at a fixed scale). *For any $0 < \alpha < 1$ there exist $0 < \rho < 1$ and $\delta > 0$ depending only on n and α such that for any solution u in B_1, if*

$$\frac{1}{|B_1|}\int_{B_1} u^2\,dx \le 1$$

and

$$\frac{1}{|B_1|}\int_{B_1} \left(|a(x) - I|^2 + |F|^2 + f^2\right) dx < \delta^2$$

then there exists a linear polynomial $\ell(x) = A + B\cdot x$ such that

$$\frac{1}{|B_\rho|}\int_{B_\rho} |u-\ell|^2\,dx < \rho^{2(1+\alpha)}$$

and $|A| + |B| \le C_0$, where $C_0 > 0$ is a constant depending only on n.

Proof. Let $0 < \varepsilon < 1$ to be fixed and let $\delta = \delta(\varepsilon) > 0$ be as in the approximation lemma, Lemma 7.5. Then there exists a harmonic function h in $B_{1/2}$ such that

$$\frac{1}{|B_{1/2}|}\int_{B_{1/2}} |u-h|^2\,dx < \varepsilon^2.$$

Notice that

$$\int_{B_{1/2}} |h|^2\,dx \le 2\int_{B_{1/2}} |h-u|^2\,dx + 2\int_{B_{1/2}} u^2\,dx \le 4C_n.$$

By interior derivative estimates for harmonic functions (see Theorem 2.11), for any multi-index γ such that $|\gamma| \le 2$ and any $x \in B_{1/4}$,

$$|D^\gamma h(x)| \le \|D^\gamma h\|_{L^\infty(B_{1/4})} \le C_{n,\gamma}\|h\|_{L^2(B_{1/2})} \le 2C_{n,\gamma}.$$

Hence, if we define

$$\ell(x) = A + B\cdot x \qquad \text{where } A = h(0) \text{ and } B = \nabla h(0)$$

then there exists a constant $C_0 > 0$ depending only on n such that $|A| + |B| \le C_0$ and for any $x \in B_{1/4}$,

$$|h(x) - \ell(x)| \le C_0|x|^2.$$

To conclude, take any $0 < \rho < 1/4$. Then

$$\begin{aligned}\frac{1}{|B_\rho|}\int_{B_\rho} |u-\ell|^2\,dx &\le \frac{2}{|B_\rho|}\int_{B_\rho} |u-h|^2\,dx + \frac{2}{|B_\rho|}\int_{B_\rho} |h-\ell|^2\,dx \\ &\le \frac{2}{|B_\rho|}|B_{1/2}|\varepsilon^2 + 2C_0^2\rho^4.\end{aligned}$$

First, choose ρ sufficiently small depending on n and α such that $2C_0^2\rho^4 < \frac{1}{2}\rho^{2(1+\alpha)}$. Second, fix $\varepsilon > 0$ small depending on n and α such that $\frac{2}{|B_\rho|}|B_{1/2}|\varepsilon^2 < \frac{1}{2}\rho^{2(1+\alpha)}$. Finally, set $\delta = \delta(\varepsilon)$. The conclusion thus follows. □

Proof of Theorem 7.3. Let $0 < \alpha < 1$. The idea is to apply Corollary 7.6 at every scale ρ^k, $k \geq 1$. For this, a few reductions must be performed. After an orthogonal change of variables (that depends on n, λ and Λ), we can assume that $a(0) = I$. Also,

$$\operatorname{div}(a(x)\nabla u) = f - \operatorname{div} F = f - \operatorname{div}(F - F(0))$$

so, without loss of generality, $F(0) = 0$. By considering $u/\|u\|_{L^2(\Omega)}$, it can also be assumed that $\|u\|_{L^2(B_1)} = 1$ and

$$\frac{1}{|B_1|}\int_{B_1} u^2\,dx \leq 1.$$

Similarly as in Step 2 of the proof of Theorem 3.19, it is enough to prove the following claim.

Claim. For every $k \geq 0$, there exist linear polynomials $\ell_k = A_k + B_k \cdot x$ such that

$$\frac{1}{|B_{\rho^k}|}\int_{B_{\rho^k}} |u - \ell_k|^2\,dx \leq (\rho^k)^{2(1+\alpha)}$$

and

$$|A_k - A_{k+1}| + \rho^k |B_k - B_{k+1}| \leq C_0 \rho^{k(1+\alpha)}$$

where $C_0 > 0$ is a constant depending only on n.

Suppose that the claim has been proved. As we did in the proof of Theorem 3.19, we find that there are $A_\infty \in \mathbb{R}$ and $B_\infty \in \mathbb{R}^n$, both bounded by a constant that depends only on n and α, such that if we define $\ell_\infty(x) = A_\infty + B_\infty \cdot x$ then $|\ell_k(x) - \ell_\infty(x)| \leq c_{\rho,\alpha,n}\rho^{k(1+\alpha)}$ whenever $|x| < \rho^k$. Then, for any $0 < r < 1$ such that $\rho^{k+1} \leq r < \rho^k$,

$$\begin{aligned}\frac{1}{|B_r|}\int_{B_r} |u - \ell_\infty|^2\,dx &\leq \frac{2}{\rho^n}\left(\frac{1}{|B_{\rho^k}|}\int_{B_{\rho^k}} |u - \ell_k|^2\,dx + \frac{1}{|B_{\rho^k}|}\int_{B_\rho^k} |\ell_k - \ell_\infty|^2\,dx\right)\\ &\leq \frac{2}{\rho^n}\left((\rho^k)^{2(1+\alpha)} + c_{\rho,\alpha,n}^2(\rho^k)^{2(1+\alpha)}\right)\\ &\leq C r^{2(1+\alpha)}\end{aligned}$$

where $C > 0$ depends only on n and α.

We are left to prove the claim, which is done by induction and rescaling. Let $0 < \rho < 1$, $\delta > 0$ and $C_0 > 0$ be as in Corollary 7.6. Set $\delta_0 = \delta/2 > 0$ and $\ell_0 = 0$. For $k = 0$, Corollary 7.6 says that, under our hypothesis

$$\frac{1}{|B_1|}\int_{B_1} (|a(x) - I|^2 + |F|^2 + f^2)\,dx \leq \delta_0^2 < \delta^2$$

there exists a linear polynomial ℓ_1 with coefficients bounded by C_0 such that

$$\frac{1}{|B_\rho|}\int_{B_\rho} |u - \ell_1|^2\,dx \leq \rho^{2(1+\alpha)}.$$

For the inductive step, assume that the claim is true for some $k \geq 0$. Observe that if ℓ is a linear polynomial and u is a solution, then $u - \ell$ satisfies

$$\begin{aligned}\operatorname{div}(a(x)\nabla(u-\ell)) &= f - \operatorname{div} F - \operatorname{div}(a(x)\nabla\ell)\\ &= f - \operatorname{div} F - \operatorname{div}((a(x)-I)\nabla\ell)\\ &= f - \operatorname{div} G\end{aligned}$$

where $G = F + (a(x) - I)$, $G(0) = 0$. Define

$$\tilde{u}(x) = \frac{(u-\ell_k)(\rho^k x)}{\rho^{k(1+\alpha)}}.$$

By looking at the equation

$$\operatorname{div}(a(x)\nabla u) = f - \operatorname{div} F \qquad \text{in } B_{\rho^k}$$

and zooming out to B_1, that is, by performing the change of variables $x = \rho^k z$, $z \in B_1$, we can readily see that $\tilde{u}$ satisfies

$$\operatorname{div}(\tilde{a}(x)\nabla\tilde{u}) = \tilde{f} - \operatorname{div}(\tilde{G}) \qquad \text{in } B_1 \tag{7.6}$$

where

$$\tilde{a}(x) = a(\rho^k x) \qquad \tilde{f}(x) = \frac{f(\rho^k x)}{\rho^{(\alpha-1)k}}$$

and

$$\tilde{G}(x) = \frac{F(\rho^k x)}{\rho^{\alpha k}} + \frac{a(\rho^k x) - I}{\rho^{\alpha k}}$$

with $\tilde{a}(0) = I$, $\tilde{G}(0) = 0$. By changing variables again and the induction hypothesis,

$$\begin{aligned}\frac{1}{|B_1|}\int_{B_1} \tilde{u}^2\,dx &= \frac{1}{\rho^{2k(1+\alpha)}|B_1|}\int_{B_1} |u(\rho^k x) - \ell_k(\rho^k x)|^2\,dx\\ &= \frac{1}{\rho^{2k(1+\alpha)}|B_{\rho^k}|}\int_{B_{\rho^k}} |u(z) - \ell_k(z)|^2\,dx \leq 1.\end{aligned}$$

Similarly, by hypothesis,

$$\begin{aligned}&\frac{1}{|B_1|}\int_{B_1} (|\tilde{a}(x) - I|^2 + |\tilde{G}|^2 + \tilde{f}^2)\,dx\\ &\quad\leq \frac{1}{|B_{\rho^k}|}\int_{B_{\rho^k}} \left(|a(z) - I|^2 + 2\frac{|F(z)|^2}{\rho^{2\alpha k}} + 2\frac{|a(z) - I|^2}{\rho^{2\alpha k}} + \frac{|f(z)|^2}{\rho^{2(\alpha-1)k}}\right) dz\\ &\quad< 4\delta_0^2 = \delta^2.\end{aligned}$$

Thus, $\tilde{u}$ falls under the hypotheses of Corollary 7.6. Therefore, there exists a linear polynomial $\ell(x) = A + B \cdot x$ with $|A| + |B| \leq C_0$ such that

$$\frac{1}{|B_\rho|}\int_{B_\rho} |\tilde{u} - \ell|^2\,dx < \rho^{2(1+\alpha)}.$$

Changing variables back, this gives

$$\frac{1}{|B_{\rho^{k+1}}|}\int_{B_{\rho^{k+1}}} |u(z)-\ell_{k+1}(z)|^2\,dz < (\rho^{(k+1)})^{2(1+\alpha)}$$

where $\ell_{k+1}(z)=\ell_k(z)+\rho^{k(1+\alpha)}\ell(z/\rho^k)$. The coefficients of ℓ_{k+1} satisfy the estimates in the claim. Indeed, setting $z=0$ in the identity $\ell_{k+1}(z)-\ell_k(z)=\rho^{k(1+\alpha)}\ell(z/\rho^k)$ we find

$$|A_{k+1}-A_k|=\rho^{k(1+\alpha)}|A|$$

and, by considering the gradient, $\nabla(\ell_{k+1}(z)-\ell_k(z))=\rho^{k(1+\alpha)}\nabla(\ell(z/\rho^k))$,

$$|B_{k+1}-B_k|=\rho^{k(1+\alpha)}\rho^{-k}|B|.$$

Since $|A|+|B|\le C_0$, the estimate follows. □

Proof of Theorem 7.1. By replacing u by

$$\frac{u(x)}{\|u\|_{L^2(B_1)}+(\|f\|_{L^p(B_1)}+\|F\|_{C^{0,\alpha}(B_1)})/c_0}$$

for some $0<c_0<1$ (that will be chosen later on depending only on n and α), we can assume that

$$\frac{1}{|B_1|}\int_{B_1} u^2\,dx\le 1$$

and

$$\|f\|_{L^p(B_1)}+\|F\|_{C^{0,\alpha}(B_1)}<c_0. \tag{7.7}$$

Under these hypotheses, we have to show that

$$\|u\|_{C^{1,\alpha}(B_{1/2})}\le C(n,\lambda,\Lambda,\alpha,[a]_{C^\alpha(B_1)}).$$

By Campanato's theorem (Theorem 7.2), we are left to control $\|\nabla u\|_{L^2(B_{1/2})}$ and the constant D in (7.2). By the Caccioppoli inequality (Lemma 7.4), the Hölder inequality applied to $f\in L^p(B_1)$ for $p=n/(1-\alpha)>2$ and (7.7),

$$\begin{aligned}\int_{B_{1/2}}|\nabla u|^2\,dx &\le C\int_{B_1}(u^2+f^2+|F|^2)\,dx\\ &\le C(1+\|f\|^2_{L^p(B_1)}+\|F\|^2_{L^\infty(B_1)})\le C(1+c_0^2)\end{aligned}$$

as desired. Next, we estimate D in (7.2) by looking at how u detaches from a linear polynomial at $z=0$ by means of Theorem 7.3. The same argument can be done at any other point $z\in B_{1/2}$ giving the same bound for D, and with this the proof will be completed.

Let $\delta_0>0$ depending only on n and α be as in Theorem 7.3. Since F is Hölder continuous, by (7.7), for any $0<r<1$,

$$\frac{1}{|B_r|}\int_{B_r}|F(x)-F(0)|^2\,dx\le [F]^2_{C^\alpha(B_1)}r^{2\alpha}<c_0^2r^{2\alpha}.$$

Similarly, by Hölder's inequality and (7.7),

$$\frac{1}{|B_r|}\int_{B_r}|f(x)|^2\,dx \le C_n\|f\|^2_{L^p(B_1)}r^{2\alpha-2} \le C_n c_0^2 r^{2\alpha-2}.$$

Now we choose c_0 such that $c_0^2 = \delta_0^2/(C_n+2)$, so that, for any $0<r<1$,

$$\frac{1}{|B_r|}\int_{B_r}|F(x)-F(0)|^2\,dx \le \frac{\delta_0^2}{2}r^{2\alpha}$$

and

$$\frac{1}{|B_r|}\int_{B_r}|f(x)|^2\,dx \le \delta_0^2 r^{2\alpha-2}.$$

Next, by the Hölder continuity of the coefficients,

$$\frac{1}{|B_r|}\int_{B_r}|a(x)-a(0)|^2\,dx \le [a]^2_{C^\alpha(B_1)}r^{2\alpha}.$$

We need to make a rescaling so that $[a]^2_{C^\alpha(B_1)}$ becomes smaller than $\delta_0^2/2$. For this, fix $r_0>0$ depending on $[a]_{C^\alpha(B_1)}$, α and δ_0 such that

$$[a]^2_{C^\alpha(B_1)}r_0^{2\alpha} < \frac{\delta_0^2}{2}.$$

Then, for all $0<r<r_0$,

$$\frac{1}{|B_r|}\int_{B_r}|a(x)-a(0)|^2\,dx \le [a]^2_{C^\alpha(B_1)}r^{2\alpha} < \frac{\delta_0^2}{2}.$$

We look at the equation for u in B_{r_0} and rescale it back to B_1 by means of the change of variables $x=r_0z$ to find that $\tilde{u}(x)=u(r_0x)/r_0^{1+\alpha}$ solves

$$\operatorname{div}(\tilde{a}(x)\nabla\tilde{u}) = \tilde{f} - \operatorname{div}(\tilde{F}) \qquad \text{in } B_1$$

where

$$\tilde{a}(x)=a(r_0x) \qquad \tilde{f}(x)=\frac{f(r_0x)}{r_0^{\alpha-1}} \qquad \text{and} \qquad \tilde{F}(x)=\frac{F(r_0x)}{r_0^{\alpha}}$$

for $x\in B_1$. For any $0<r<1$, by our choice of r_0,

$$\begin{aligned}\frac{1}{|B_r|}\int_{B_r}|\tilde{a}(x)-\tilde{a}(0)|^2\,dx &= \frac{1}{|B_{r_0r}|}\int_{B_{r_0r}}|a(z)-a(0)|^2\,dz\\ &\le [a]^2_{C^\alpha(B_1)}(r_0r)^{2\alpha} < \frac{\delta_0^2}{2}r^{2\alpha}.\end{aligned}$$

Moreover, by the choice of c_0 we made above, for any $0<r<1$,

$$\begin{aligned}\frac{1}{|B_r|}\int_{B_r}|\tilde{F}(x)-\tilde{F}(0)|^2\,dx &= \frac{1}{r_0^{2\alpha}|B_{r_0r}|}\int_{B_{r_0r}}|F(z)-F(0)|^2\,dz\\ &\le \frac{\delta_0^2}{2r_0^{2\alpha}}(r_0r)^{2\alpha} = \frac{\delta_0^2}{2}r^{2\alpha}\end{aligned}$$

where $\tilde{a}(x) = a(r_0 x)$. Now, by our choice of r_0,

$$|\tilde{a}(x) - I| = |a(r_0 x) - I| < \varepsilon \qquad \text{for all } x \in Q_1.$$

Then, Theorem 7.11 can be applied to $\tilde{u}$ to get

$$\|\nabla \tilde{u}\|_{L^p(Q_{1/4})} \leq C\|\tilde{u}\|_{L^2(Q_1)}.$$

Changing variables back,

$$\|\nabla u\|_{L^p(Q_{r_0/4})} \leq \frac{C}{r_0^{1-n/p+n/2}}\|u\|_{L^2(Q_{r_0})}.$$

Next, we cover $Q_{1/4}$ by a finite numbers of cubes of radius $r_0/4$. At each cube we apply the same orthogonal change of variables to make $a = I$ at its center and use the estimate we just proved. By adding together all these bounds,

$$\|\nabla u\|_{L^p(Q_{1/4})} \leq C\|u\|_{L^2(Q_1)}$$

where $C > 0$ depends on n, λ, Λ, p and the modulus of continuity of $a(x)$.

To obtain (7.9), we are left to control $\|u\|_{L^p(Q_{1/4})}$. Let $(u)_{Q_{1/4}}$ denote the average of u over the cube $Q_{1/4}$. By the Poincaré and Hölder inequalities,

$$\begin{aligned}
\|u\|_{L^p(Q_{1/4})} &\leq \|u - (u)_{Q_{1/4}}\|_{L^p(Q_{1/4})} + |(u)_{Q_{1/4}}| \\
&\leq C\|\nabla u\|_{L^p(Q_{1/4})} + C_n\|u\|_{L^2(Q_{1/4})} \\
&\leq C\|u\|_{L^2(Q_1)}
\end{aligned}$$

where $C > 0$ is as in the statement. □

The strategy for proving Theorem 7.11 is the following. Instead of trying to estimate $\nabla u(x)$ in a pointwise way, we will study the Hardy–Littlewood maximal operator M (see Examples 4.16 and 4.17 in Section 4.3) acting on $f = |\nabla u|^2$:

$$M(|\nabla u|^2)(x) = \sup_{Q \ni x} \frac{1}{|Q|} \int_Q |\nabla u(z)|^2 \, dz$$

where we have defined $\nabla u = 0$ outside of Q_1 and the supremum is taken over all cubes Q that contain the point x. There are various reasons for introducing the maximal function. First, by Lebesgue's differentiation theorem, we have the pointwise control $|\nabla u(x)|^2 \leq M(|\nabla u|^2)(x)$. Hence, if we can show that $M(|\nabla u|^2) \in L^{p/2}$, $p > 2$, then $\nabla u \in L^p$. Second, since the equation is in fact a first-order integral equation, the averages of $|\nabla u|^2$ can be easily controlled. A clear evidence of this is the Caccioppoli inequality in Lemma 7.4. To prove that $M(|\nabla u|^2) \in L^{p/2}$ we show that the distribution function

$$\mu(t) = |\{x : M(|\nabla u|^2) > t\}|$$

decays sufficiently fast as $t \to \infty$ and apply the following simple real analysis result to the function $g(x) := M(|\nabla u|^2)(x)$ with $q = p/2$.

Lemma 7.12. *Let $g \geq 0$ be a measurable function in a bounded domain Ω and define*

$$\mu_g(t) = |\{x \in \Omega : g(x) > t\}|.$$

Let $\eta > 0$ and $N > 1$ be constants. Then $g \in L^q(\Omega)$ for $0 < q < \infty$ if and only if

$$S = \sum_{k=1}^{\infty} N^{kq} \mu_g(N^k \eta) < \infty$$

and, in this case,

$$C^{-1} S \leq \|g\|^q_{L^q(\Omega)} \leq C(|\Omega| + S)$$

where $C > 0$ depends only on η, N and q.

Proof. Since, for $x \in \Omega$ and $\tau > 0$, $\chi_{(0,g(x)^q)}(\tau) = \chi_{\{g(x)^q > \tau\}}(x)$, by Fubini's theorem and the change of variables $\tau = t^q$,

$$\begin{aligned}
\int_\Omega g(x)^q \, dx &= \int_\Omega \int_0^\infty \chi_{(0,|g(x)|^q)}(\tau) \, d\tau \, dx \\
&= \int_0^\infty |\{x \in \Omega : g(x) > \tau^{1/q}\}| \, d\tau \\
&= q \int_0^\infty t^{q-1} |\{x \in \Omega : g(x) > t\}| \, dt \\
&= q \int_0^\eta t^{q-1} \mu_g(t) \, dt + q \sum_{k=0}^{\infty} \int_{N^k \eta}^{N^{k+1}\eta} t^{q-1} \mu_g(t) \, dt.
\end{aligned}$$

The conclusion and estimates easily follow by using that $0 \leq \mu_g(t) \leq |\Omega|$ and that $\mu_g(t)$ is a nonincreasing function of t. □

The advantage of Lemma 7.12 is that we only need to estimate the distribution function of $M(|\nabla u|^2)$ for discrete values of t as we have to study the summability of

$$\sum_{k=1}^{\infty} N^{kp/2} |\{x \in Q_{1/4} : M(|\nabla u|^2)(x) > N^k \eta\}|$$

for some $\eta > 0$ and $N > 1$. Now, the decay of each term of the series above is proved by iteration. The key is to show that if the coefficients $a(x)$ are close to the identity matrix I then ∇u can be approximated in average by the gradient ∇h of a harmonic function h. In this scenario, the Laplace equation acts as a "reference equation," namely, one for which all the estimates hold (that is, local Lipschitz estimates) and from which we want to transfer regularity. With these ingredients, we estimate

$$t < M(|\nabla u|^2) \leq 2M(|\nabla(u - h)|^2) + 2M(|\nabla h|^2).$$

By the weak-type $(1,1)$ inequality for M and the approximation of ∇u by ∇h, we see that $|\{x \in Q_{1/4} : M(|\nabla(u - h)|^2)(x) > t\}|$ can be made arbitrarily small and, if t is sufficiently large, $M(|\nabla h|^2) = 0$ because ∇h is bounded.

To implement these ideas, let us begin with the following approximation lemma. Compare its rather simple proof with the proof of Lemma 7.5.

Lemma 7.13 (Approximation lemma). *Fix $\varepsilon > 0$. Let u be a solution in Q_1 and consider any cube $Q \subset Q_1$. If*

$$\|a(x) - I\|_{L^\infty(Q_1)} < \varepsilon$$

then the solution h to

$$\begin{cases} \Delta h = 0 & \text{in } Q \\ h = u & \text{on } \partial Q \end{cases}$$

satisfies

$$\frac{1}{|Q|}\int_Q |\nabla h|^2\,dx \le \frac{1}{|Q|}\int_Q |\nabla u|^2\,dx \tag{7.10}$$

and

$$\frac{1}{|Q|}\int_Q |\nabla(u-h)|^2\,dx \le \varepsilon^2 \frac{1}{|Q|}\int_Q |\nabla u|^2\,dx. \tag{7.11}$$

Proof. Notice that h exists and is unique because Q is a bounded Lipschitz domain. Since $u - h \in H_0^1(Q_1)$ and h is weakly harmonic in Q_1, by Hölder's inequality,

$$\begin{aligned}\int_Q |\nabla h|^2\,dx &= \int_Q \nabla h \nabla(h-u)\,dx + \int_Q \nabla h \nabla u\,dx \\ &= \int_Q \nabla h \nabla u\,dx \\ &\le \left(\int_Q |\nabla h|^2\,dx\right)^{1/2}\left(\int_Q |\nabla u|^2\,dx\right)^{1/2}\end{aligned}$$

and (7.10) follows. Using again that $u - h \in H_0^1(Q_1)$ and h is weakly harmonic,

$$\int_Q |\nabla(u-h)|^2\,dx = \int_Q \nabla u \nabla(u-h)\,dx.$$

By introducing $a(x)$, using the equation for u, the hypothesis that $|a(x) - I| < \varepsilon$ and Hölder's inequality,

$$\begin{aligned}\int_Q |\nabla(u-h)|^2\,dx &= \int_Q (I - a(x))\nabla u \nabla(u-h)\,dx + \int_Q a(x)\nabla u\nabla(u-h)\,dx \\ &= \int_Q (I - a(x))\nabla u \nabla(u-h)\,dx \\ &\le \varepsilon\left(\int_{Q_1} |\nabla u|^2\,dx\right)^{1/2}\left(\int_{Q_1} |\nabla(u-h)|^2\,dx\right)^{1/2}\end{aligned}$$

and this gives (7.11). □

Next, we need the following L^∞ estimate for ∇h, which is easily proved by using the mean value property for $\partial_{x_i} h$ as in (2.19) and Hölder's inequality.

Lemma 7.14 (Regularity for reference equation). *There exists a constant $C_n > 0$ depending only on n such that for any harmonic function h in an open set Ω and any cube Q such that $2Q \subset \Omega$,*

$$\|\nabla h\|^2_{L^\infty(Q)} \le C_n \frac{1}{|2Q|}\int_{2Q} |\nabla h|^2\,dx. \tag{7.12}$$

It became evident in Chapter 4 that covering lemmas are the key tool to estimate distribution functions, see Sections 4.3 and 4.5. The following is a consequence of the Calderón–Zygmund decomposition (Lemma 4.12) that will also be used in Chapter 9 to prove the Harnack inequality for nondivergence form equations.

Lemma 7.15 (Calderón–Zygmund AB lemma). *Let $A \subset B \subset Q_1$ be measurable sets and $0 < \delta < 1$ such that*

(a) $|A| \le \delta |Q_1|$ and

(b) if $Q \subset Q_1$ is a dyadic cube such that

$$\frac{|A \cap Q|}{|Q|} > \delta,$$

then the predecessor $\widetilde{Q} \subset Q_1$ of Q satisfies

$$\widetilde{Q} \subset B.$$

Then

$$|A| \le \delta |B|.$$

Proof. We perform the Calderón–Zygmund decomposition to the function $f = \chi_A$. We have

$$\frac{|A \cap Q_1|}{|Q_1|} \le \delta.$$

We subdivide Q_1 into 2^n congruent dyadic cubes Q by splitting each of its sides in half. For any of these cubes Q we have two possibilities:

(1) either $\dfrac{|A \cap Q|}{|Q|} > \delta$,

(2) or $\dfrac{|A \cap Q|}{|Q|} \le \delta$.

We keep the cubes Q that satisfy (1) and subdivide the cubes Q for which (2) holds. We continue this process by keeping the subcubes that satisfy (1) and dividing again the subcubes falling into possibility (2). We obtain a disjoint family of dyadic cubes $\{Q^i\}_{i=1}^{\infty}$ such that

$$\frac{|A \cap Q^i|}{|Q^i|} > \delta.$$

Observe that Q_1 is not in this family.

We claim that, up to a set of measure zero,

$$A \subset \bigcup_{i=1}^{\infty} Q^i.$$

Indeed, if $x \notin \bigcup_{i=1}^{\infty} Q^i$ then x is in a nested infinite family of closed dyadic cubes Q_ℓ such that $\operatorname{diam}(Q_\ell) \to 0$ as $\ell \to \infty$ and

$$\frac{|A \cap Q_\ell|}{|Q_\ell|} \leq \delta < 1.$$

By the Lebesgue differentiation theorem applied to χ_A we get

$$\chi_A(x) \leq \delta < 1.$$

This means that $x \notin A$.

The family of predecessors $\{\widetilde{Q}^i\}_{i=1}^{\infty}$ can be relabeled so that all of them are pairwise disjoint. Indeed, since all the cubes in this construction are dyadic cubes, if one takes two predecessors then either one is inside the other or they are disjoint. We then discard those predecessors that are contained in a previous predecessor. Clearly,

$$\frac{|A \cap \widetilde{Q}^i|}{|\widetilde{Q}^i|} \leq \delta. \tag{7.13}$$

Now, since

$$\frac{|A \cap Q^i|}{|Q^i|} > \delta$$

for all i, by hypothesis (b), $\widetilde{Q}^i \subset B$ for all i, so

$$A \subset \bigcup_{i=1}^{\infty} Q^i \subset \bigcup_{i=1}^{\infty} \widetilde{Q}^i \subset B.$$

This and (7.13) imply that

$$|A| = \sum_{i=1}^{\infty} |A \cap \widetilde{Q}^i| \leq \delta \sum_{i=1}^{\infty} |\widetilde{Q}^i| = \delta \left| \bigcup_{i=1}^{\infty} \widetilde{Q}^i \right| \leq \delta |B|.$$

□

Now we are ready to establish the key lemma. It is important that in the following statement $\eta > 0$ is arbitrary, as this will allow us to iterate it.

Lemma 7.16 (Main lemma). *Let u be a solution in Q_1 and extend ∇u outside of Q_1 as $\nabla u = 0$. Set $N = \max\{6^n, 5C_n\}$, where $C_n > 0$ is as in (7.12). Given $\delta > 0$ there exists $\varepsilon = \varepsilon(\delta) > 0$ such that if*

$$\|a(x) - I\|_{L^\infty(Q_1)} < \varepsilon$$

and Q is a dyadic cube such that $Q \subset \widetilde{Q} \subset Q_{1/4}$ and for which

$$\frac{|\{M(|\nabla u|^2) > N\eta\} \cap Q|}{|Q|} > \delta$$

where $\eta > 0$, then

$$\widetilde{Q} \subset \{M(|\nabla u|^2) > \eta\}.$$

Proof. Fix $\delta > 0$ and let $Q \subset \widetilde{Q} \subset Q_{1/4}$ be as in the hypothesis. Suppose, for the sake of contradiction, that the conclusion is false. Then there is $x_0 \in \widetilde{Q} \subset Q_{1/4}$ such that $M(|\nabla u|^2)(x_0) \leq \eta$ and, in particular,

$$\frac{1}{|Q_0|}\int_{Q_0} |\nabla u|^2\, dx \leq \eta \tag{7.14}$$

for any cube Q_0 containing x_0. Let h be the solution to

$$\begin{cases} \Delta h = 0 & \text{in } 4\widetilde{Q} \subset Q_1 \\ h = u & \text{on } \partial(4\widetilde{Q}). \end{cases}$$

Then, by (7.12), (7.10) and (7.14) (note that $4\widetilde{Q}$ contains x_0),

$$\|\nabla h\|^2_{L^\infty(2\widetilde{Q})} \leq C_n \frac{1}{|4\widetilde{Q}|}\int_{4\widetilde{Q}} |\nabla h|^2\, dx \leq C_n \frac{1}{|4\widetilde{Q}|}\int_{4\widetilde{Q}} |\nabla u|^2\, dx \leq C_n \eta. \tag{7.15}$$

In addition, by (7.11) and (7.14),

$$\frac{1}{|2\widetilde{Q}|}\int_{2\widetilde{Q}} |\nabla (u-h)|^2\, dx \leq \varepsilon^2 \frac{2^n}{|4\widetilde{Q}|}\int_{4\widetilde{Q}} |\nabla u|^2\, dx \leq 2^n \varepsilon^2 \eta. \tag{7.16}$$

In order to have a localized estimate, we need to introduce the following *restricted* maximal operator

$$M^*(|\nabla u|^2)(x) = \sup_{Q \ni x, Q \subset 2\widetilde{Q}} \frac{1}{|Q|}\int_Q |\nabla u|^2\, dz$$

where the supremum is taken over cubes contained in $2\widetilde{Q}$. By taking $\Omega = 2\widetilde{Q}$ in Example 4.17 of Chapter 4, we have that M^* satisfies a weak-type $(1,1)$ inequality: there exists a constant $C_* > 0$ depending only on n such that, for any $t > 0$,

$$|\{x \in 2\widetilde{Q} : M^*(|\nabla u|^2)(x) > t\}| \leq \frac{C_*}{t}\int_{2\widetilde{Q}} |\nabla u|^2\, dx. \tag{7.17}$$

We compare M^* with M. We claim that if $x \in Q$ then

$$M(|\nabla u|^2)(x) \leq \max\{M^*(|\nabla u|^2)(x), 6^n \eta\}. \tag{7.18}$$

Indeed, if the supremum in the definition of $M(|\nabla u|^2)(x)$ is attained at a cube contained in $2\widetilde{Q}$, then $M(|\nabla u|^2)(x) = M^*(|\nabla u|^2)(x)$. Otherwise, let Q' be a cube containing x and such that $Q' \not\subset 2\widetilde{Q}$. Then $Q_0 := 6Q' \supset \widetilde{Q} \ni x_0$ and, by (7.14),

$$\frac{1}{|Q'|}\int_{Q'} |\nabla u|^2\, dx \leq 6^n \frac{1}{|Q_0|}\int_{Q_0} |\nabla u|^2\, dx \leq 6^n \eta.$$

Thus, (7.18) follows.

We are ready to put all the estimates together. Recall that $N = \max\{6^n, 5C_n\}$, where C_n is as in (7.12) and (7.15). Then, by hypothesis, (7.18), the fact that, because

of (7.15), $M^*(|\nabla h|^2)(x) \le \|\nabla h\|^2_{L^\infty(2\widetilde{Q})} < N\eta/4$ for all $x \in Q$, the weak-type $(1,1)$ estimate for M^* in (7.17) with $t = N\eta/4$, and (7.16),

$$\begin{aligned}
\delta|Q| &< |\{M(|\nabla u|^2) > N\eta\} \cap Q| \\
&\le |\{M^*(|\nabla u|^2) > N\eta\} \cap Q| \\
&\le |\{M^*(|\nabla (u-h)|^2) + M^*(|\nabla h|^2) > N\eta/2\} \cap Q| \\
&\le |\{M^*(|\nabla (u-h)|^2) \ge N\eta/4\} \cap Q| + |\{M^*(|\nabla h|^2) \ge N\eta/4\} \cap Q| \\
&\le \frac{4C_*}{N\eta} \int_{2\widetilde{Q}} |\nabla(u-h)|^2 \, dx \\
&\le C\varepsilon^2 |2\widetilde{Q}| = C\varepsilon^2|Q|
\end{aligned}$$

where $C > 0$ depends only on n. If $C\varepsilon^2 < \delta$, we reach a contradiction. □

The main lemma and the Calderón–Zygmund AB lemma finally imply Theorem 7.11.

Proof of Theorem 7.11. We extend ∇u by defining $\nabla u = 0$ outside of Q_1. Let $N > 1$ be as in Lemma 7.16 and call $C_M > 0$ the constant depending on n in the weak-type $(1,1)$ estimate for the maximal function M:

$$|\{x \in \mathbb{R}^n : M(|\nabla u|^2)(x) > t\}| \le \frac{C_M}{t} \int_{Q_1} |\nabla u|^2 \, dx. \tag{7.19}$$

Let $\delta > 0$ to be fixed and set $\varepsilon = \varepsilon(\delta) > 0$ as in Lemma 7.16. By considering $v = (N\delta|Q_1|/C_M)^{1/2} u / \|\nabla u\|_{L^2(Q_1)}$, we can assume the normalization

$$\frac{1}{|Q_1|} \int_{Q_1} |\nabla u|^2 \, dx = \frac{N}{C_M} \delta. \tag{7.20}$$

Fix any $\eta > 1$. In the Calderón–Zygmund AB lemma (Lemma 7.15), we select

$$A = \{M(|\nabla u|^2) > N\eta\} \cap Q_{1/4} \qquad \text{and} \qquad B = \{M(|\nabla u|^2) > \eta\} \cap Q_{1/4}.$$

Then $A \subset B \subset Q_{1/4}$. By the weak-type $(1,1)$ estimate for the maximal function M in (7.19) with $t = N\eta > N$ and (7.20),

$$|A| \le \frac{C_M}{N\eta} \int_{Q_1} |\nabla u|^2 \, dx < \delta |Q_1|.$$

If $Q \subset Q_{1/4}$ is a dyadic cube such that

$$\frac{|A \cap Q|}{|Q|} \ge \delta$$

then, by the main lemma (Lemma 7.16),

$$\widetilde{Q} \subset B.$$

Therefore, by Lemma 7.15 (with $Q_{1/4}$ in place of Q_1), $|A| \le \delta|B|$, namely,

$$|\{M(|\nabla u|^2) > N\eta\} \cap Q_{1/4}| \le \delta|\{M(|\nabla u|^2) > \eta\} \cap Q_{1/4}|.$$

Since this estimate holds for arbitrary $\eta > 1$, we can iterate it to get

$$\begin{aligned} |\{M(|\nabla u|^2) > N^k\eta\} \cap Q_{1/4}| &= |\{M(|\nabla u|^2) > N(N^{k-1}\eta)\} \cap Q_{1/4}| \\ &\le \delta|\{M(|\nabla u|^2) > N^{k-1}\eta\} \cap Q_{1/4}| \\ &\le \delta^2|\{M(|\nabla u|^2) > N^{k-2}\eta\} \cap Q_{1/4}| \\ &\le \delta^k|\{M(|\nabla u|^2) > \eta\} \cap Q_{1/4}| \\ &\le \delta^k|Q_{1/4}| \end{aligned}$$

for all $k \ge 1$. Therefore,

$$\sum_{k=1}^{\infty} N^{kp/2}|\{M(|\nabla u|^2) > N^k\eta\} \cap Q_{1/4}| \le |Q_{1/4}| \sum_{k=1}^{\infty} N^{kp/2}\delta^k.$$

If we choose $\delta > 0$ depending on n and p such that $N^{p/2}\delta = \frac{1}{2}$ then the geometric series above converges to 1. Thus, by Lemma 7.12, $M(|\nabla u|^2) \in L^{p/2}(Q_{1/4})$ and

$$\begin{aligned} \int_{Q_{1/4}} |\nabla u|^p \, dx &\le \int_{Q_{1/4}} (M(|\nabla u|^2))^{p/2} \, dx \\ &\le C|Q_{1/4}| \\ &= C|Q_{1/4}| \frac{C_M}{N\delta|Q_1|} \|\nabla u\|_{L^2(Q_1)}^2 = C\|\nabla u\|_{L^2(Q_1)}^2 \end{aligned}$$

where $C > 0$ depends only on n and p. □

NOTES ON CHAPTER 7

The Schauder estimates we presented here are an appropriate modification of Caffarelli's technique [3,4]. The Calderón–Zygmund perturbation method was developed by Caffarelli–Peral in [8], where even more general equations are considered. These geometric techniques give regularity of weak solutions (that is, they are not *a priori* estimates methodologies) and are robust enough to apply, with nontrivial modifications, to nonlinear equations.

There are other ways of proving Schauder and Calderón–Zygmund estimates for divergence form equations. For example, [43] uses a "freezing the coefficient" method together with interpolation inequalities to obtain *a priori* Schauder estimates for divergence form equations with lower-order terms, and the solvability of the Dirichlet problem to extend the estimates to weak solutions. We will show how this classical *a priori* estimate technique works for nondivergence form equations in Chapter 11. On the other hand, [44] implements the "freezing the coefficient" technique in combination with approximations by harmonic functions to control the growth of local integrals of ∇u and deduce the Schauder estimates. As for Sobolev estimates, Di Fazio in [27] applies the "freezing the coefficient" technique and the boundedness of singular integrals and commutators to obtain *a priori* $W^{1,p}$ estimates for weak solutions.

III

Nondivergence form equations

CHAPTER 8

Viscosity solutions and the ABP estimate

In this chapter we introduce the basics on viscosity solutions to nondivergence form elliptic equations, including their relation with the Pucci extremal operators, and show that they satisfy the Alexandroff–Bakelman–Pucci (ABP) estimate.

8.1 NONDIVERGENCE FORM EQUATIONS

We consider nondivergence form equations

$$Lu = \operatorname{trace}(a(x)D^2u) = a^{ij}(x)\partial_{ij}u = f \qquad \text{in } \Omega$$

where $\Omega \subset \mathbb{R}^n$, $n \geq 2$, is a bounded domain. The matrix of coefficients $a(x) = (a^{ij}(x))_{i,j=1}^n$ is defined in Ω, symmetric $a^{ij}(x) = a^{ji}(x)$, and uniformly elliptic: there exist ellipticity constants $0 < \lambda \leq \Lambda$ such that

$$\lambda|\xi|^2 \leq a^{ij}(x)\xi_i\xi_j \leq \Lambda|\xi|^2$$

for any $\xi \in \mathbb{R}^n$, for all $x \in \Omega$. The right hand side f is a function defined on Ω.

For this and the next two chapters, no assumption is made on the smoothness of the coefficients $a(x)$. Then, unlike divergence form equations, the equation $Lu = f$ is totally infinitesimal in nature. For divergence form equations we always have some "global" information available from the beginning (namely, $u \in H^1(\Omega)$), but that is not the case for nondivergence form equations.

To understand what the equation $Lu = 0$ means, suppose that $u \in C^2(\Omega)$ and let $x_0 \in \Omega$. Since $D^2u(x_0)$ is a symmetric matrix, we can orthogonally diagonalize it as

$$D^2u(x_0) = PDP^{-1} = PDP^T$$

where $D = \operatorname{diag}(\lambda_1, \ldots, \lambda_n)$ is a diagonal matrix formed by the eigenvalues of $D^2u(x_0)$ and the columns of P are the eigenvectors of $D^2u(x_0)$. Therefore

$$Lu(x_0) = \operatorname{trace}(a(x_0)PDP^T) = \operatorname{trace}(P^Ta(x_0)PD) = \operatorname{trace}(b(x_0)D).$$

DOI: 10.1201/9781032679853-8

If b^{kk} denote the diagonal entries of $b(x_0)$, then $Lu(x_0) = 0$ can be written as

$$\sum_{k=1}^{n} b^{kk}\lambda_k = 0. \tag{8.1}$$

Observe that $b(x_0) = (b^{ij}(x_0))_{i,j=1}^n$ is also uniformly elliptic with the same ellipticity constants than $a(x)$ so that, if $e_1, \ldots, e_n$ are the standard basis vectors of $\mathbb{R}^n$,

$$0 < \lambda \leq \langle b(x_0)e_k, e_k\rangle = b^{kk} \leq \Lambda$$

Thus, unless $\lambda_k = 0$ for all k, the equation (8.1) says that some eigenvalues λ_k are positive and some others are negative. We can rewrite (8.1) as

$$\sum_{\lambda_k>0} b^{kk}\lambda_k = \sum_{\lambda_k<0} b^{kk}(-\lambda_k).$$

In particular, in dimension $n = 2$ this says that at x_0 we have

$$b^{11}\lambda_1 = b^{22}(-\lambda_2).$$

In other words, if u at x_0 *bends upward* in the direction of the eigenvector corresponding to the eigenvalue λ_1, then it has to *bend downward* in the direction of the eigenvector corresponding to the eigenvalue λ_2 accordingly, in a proportional way, depending only on ellipticity. As a matter of fact, if we let

$$\lambda_{\max} = \max\{\lambda_k : \lambda_k > 0\} \qquad \text{and} \qquad \lambda_{\min} = \min\{\lambda_k : \lambda_k < 0\}$$

then the equation $Lu = 0$ at x_0, that is, (8.1) says that

$$\lambda_{\max} \sim -\lambda_{\min}$$

with proportionality constants depending on n, λ and Λ. Indeed,

$$\lambda(-\lambda_{\min}) \leq \sum_{\lambda_k<0} b^{kk}(-\lambda_k) = \sum_{\lambda_k>0} b^{kk}\lambda_k \leq (n-1)\Lambda\lambda_{\max}$$

and

$$\lambda\lambda_{\max} \leq \sum_{\lambda_k>0} b^{kk}\lambda_k = \sum_{\lambda_k<0} b^{kk}(-\lambda_k) \leq (n-1)\Lambda(-\lambda_{\min}).$$

In conclusion, since we have no regularity on the coefficients $a(x)$, the only information we can deduce from $Lu = 0$ is that at every point the maximum eigenvalue of D^2u is positive, the minimum one is negative and they are proportional $\lambda_{\max} \sim -\lambda_{\min}$ with proportionality constants depending only on dimension and ellipticity. However, we have no control at all on how this property is changing from point to point. Thus, $Lu = 0$ is a completely infinitesimal equation.

The Pucci extremal operators mentioned in the introduction, and that we define next, appear naturally from considering (8.1).

For a symmetric matrix M of size $n \times n$, we denote by $\lambda_k = \lambda_k(M)$, $k = 1, \ldots, n$, its real eigenvalues. The **maximal Pucci extremal operator** $\mathcal{M}^+$ acting on M is

$$\mathcal{M}^+(M) = \Lambda \sum_{\lambda_k > 0} \lambda_k + \lambda \sum_{\lambda_k < 0} \lambda_k$$

and the **minimal Pucci extremal operator** $\mathcal{M}^-$ acting on M is

$$\mathcal{M}^-(M) = \lambda \sum_{\lambda_k > 0} \lambda_k + \Lambda \sum_{\lambda_k < 0} \lambda_k.$$

Now, if u is a C^2 solution to $Lu = 0$ and, as before, λ_k are the eigenvalues of the symmetric matrix $D^2u(x_0)$ then, since by (8.1) we have

$$\sum_{\lambda_k > 0} b^{kk}\lambda_k + \sum_{\lambda_k < 0} b^{kk}\lambda_k = 0$$

and $\lambda \leq b^{kk} \leq \Lambda$, we immediately deduce that

$$\mathcal{M}^+(D^2u(x_0)) \geq 0 \qquad \text{and} \qquad \mathcal{M}^-(D^2u(x_0)) \leq 0. \tag{8.2}$$

Thus, any C^2 solution u to $Lu = 0$ is a **subsolution** to the maximal Pucci equation $\mathcal{M}^+(D^2u) \geq 0$ and a **supersolution** to the minimal Pucci equation $\mathcal{M}^-(D^2u) \leq 0$.

Conversely, if $u \in C^2(\Omega)$ satisfies both inequalities in (8.2) at every point in Ω then there is a uniformly elliptic matrix $a(x)$ such that $Lu = a^{ij}(x)\partial_{ij}u = 0$, with $a(x)$ not necessarily continuous. To see this, let us observe the following characterization.

Lemma 8.1. *Let $\mathcal{A}$ be the family of all symmetric matrices A of size $n \times n$ such that $\lambda|\xi|^2 \leq A^{ij}\xi_i\xi_j \leq \Lambda|\xi|^2$ for all $\xi \in \mathbb{R}^n$. Then, for every symmetric matrix M,*

$$\mathcal{M}^+(M) = \sup_{A \in \mathcal{A}} \operatorname{trace}(AM)$$

and

$$\mathcal{M}^-(M) = \inf_{A \in \mathcal{A}} \operatorname{trace}(AM)$$

Proof. Let P be an orthogonal matrix such that $M = PDP^T$, where $D = \operatorname{diag}(\lambda_1, \ldots, \lambda_n)$ is the diagonal matrix of eigenvalues of M. Since $B = P^TAP \in \mathcal{A}$, its diagonal entries satisfy $\lambda \leq B^{kk} \leq \Lambda$ and, by definition of $\mathcal{M}^+$,

$$\operatorname{trace}(AM) = \operatorname{trace}(P^TAPD) = \operatorname{trace}(BD) \leq \mathcal{M}^+(M).$$

If $\lambda_1 \leq \cdots \leq \lambda_m \leq 0 \leq \lambda_{m+1} \leq \cdots \leq \lambda_n$ then the supremum is achieved at the diagonal matrix B whose first m diagonal entries are equal to λ and the rest are equal to Λ. The proof for $\mathcal{M}^-$ is analogous. □

Therefore, if $u \in C^2(\Omega)$ satisfies (8.2) then, by Lemma 8.1, the continuous function of $A \in \mathcal{A}$ defined as $F(A) = \operatorname{trace}(AD^2u(x_0))$ satisfies

$$\inf_{A \in \mathcal{A}} F(A) \leq 0 \leq \sup_{A \in \mathcal{A}} F(A).$$

Since $\mathcal{A}$ is a compact set, F attains the intermediate value 0 at some matrix $A = A(x_0) \in \mathcal{A}$, that is,

$$\operatorname{trace}(A(x_0)D^2u(x_0)) = 0.$$

But notice that the matrix attaining this intermediate value changes from point to point in a nonnecessarily continuous way.

8.2 VISCOSITY SOLUTIONS

We want to define a proper concept of weak solution to the nondivergence form equation

$$Lu = \text{trace}(a(x)D^2u) = a^{ij}(x)\partial_{ij}u = 0.$$

We cannot integrate by parts as with divergence form equations. In turn, we need to base the notion on the geometry of the maximum principle.

A first intuition to understand the idea of viscosity solution is the following. Suppose we want to say that a continuous function u is subharmonic at a point x_0:

$$\Delta u(x_0) \geq 0.$$

We cannot take two derivatives of u. The equation is essentially saying that "u is bending upwards at x_0." If this is our intuition, then the following should happen: if φ is any C^2 function that **touches** u **from above at** x_0, that is, if

$$\varphi(x_0) = u(x_0) \qquad \text{and} \qquad \varphi(x) \geq u(x) \text{ for nearby } x \neq x_0$$

then φ should also be "bending upwards at x_0." In other words, we should have

$$\Delta\varphi(x_0) \geq 0.$$

Another way of understanding the idea is by using the classical maximum principle for nondivergence form elliptic equations. Let $u \in C^2(\Omega)$ be a subsolution:

$$Lu \geq 0 \qquad \text{in } \Omega.$$

Let $\varphi \in C^2(\Omega)$ be a strict supersolution, namely,

$$L\varphi < 0 \qquad \text{in } \Omega.$$

Obviously,

$$L(u - \varphi) > 0 \qquad \text{in } \Omega.$$

If $u - \varphi$ has a local maximum at a point $x_0 \in \Omega$, then $D^2(u - \varphi)(x_0) \leq 0$ and an orthogonal change of variables shows that this is a contradiction with the equation $L(u - \varphi)(x_0) > 0$. In other words, if $u - \varphi$ has a local maximum at $x_0 \in \Omega$ then $L(u - \varphi)(x_0) \leq 0$, from which there holds

$$L\varphi(x_0) \geq 0.$$

By lowering φ by a constant, the fact that $u - \varphi$ has a local maximum of 0 at x_0 means that $(u - \varphi)(x) \leq 0$ for x near x_0, so φ touches u from above at x_0. We conclude that u cannot be touched from above at x_0 by any strict supersolution φ.

Since, as we have seen, nondivergence form equations are pointwise infinitesimal equations, we continue to assume that $a(x)$ and f are defined at every point $x \in \Omega$. Let $u \in C(\Omega)$. We have the following definitions.

(1) We say that u is a **viscosity subsolution** to $Lu = f$ in Ω, written

$$Lu \geq f \qquad \text{in the viscosity sense in } \Omega$$

if for any $x_0 \in \Omega$ and any function φ that is C^2 in an open neighborhood of x_0 and such that $\varphi(x_0) = u(x_0)$ and $\varphi(x) \geq u(x)$ for all x near x_0 (that is, φ **touches u from above at** x_0) there holds

$$L\varphi(x_0) \geq f(x_0).$$

(2) We say that u is a **viscosity supersolution** to $Lu = f$ in Ω, written

$$Lu \leq f \qquad \text{in the viscosity sense in } \Omega$$

if for any $x_0 \in \Omega$ and any function φ that is C^2 in an open neighborhood of x_0 and such that $\varphi(x_0) = u(x_0)$ and $\varphi(x) \leq u(x)$ for all x near x_0 (that is, φ **touches u from below at** x_0) there holds

$$L\varphi(x_0) \leq f(x_0).$$

Note that u is a viscosity supersolution if and only if $-u$ is a viscosity subsolution.

(3) We say that u is a **viscosity solution** to $Lu = f$ in Ω, or that is satisfies $Lu = f$ in the viscosity sense in Ω, if it is both a viscosity subsolution and a viscosity supersolution to $Lu = f$ in Ω.

Sometimes we will just refer to u as a viscosity (sub/super)solution without being explicit about the equation and the domain.

Remark 8.2. The touching functions φ in the definitions above are the natural replacements of the usual test functions for weak solutions to divergence form equations. Weak solutions have a natural functional setting, namely, the Sobolev space $H^1(\Omega)$, in which divergence form operators $\operatorname{div}(a(x)\nabla u)$ can be seen as unbounded linear operators. However, this is not the case for viscosity solutions: the nondivergence form operator $\operatorname{trace}(a(x)D^2u)$ has no functional analytic meaning as an unbounded linear operator on the space of continuous viscosity solutions.

We collect various simple properties of viscosity solutions.

Proposition 8.3 (Basic properties of viscosity solutions). *Let $a(x)$ and f be as above.*

(1) *In the definition of viscosity (sub/super)solution one can replace the C^2 touching functions φ by quadratic polynomials P.*

(2) *If u is a C^2 pointwise classical (sub/super)solution, then it is also a viscosity (sub/super)solution. Conversely, any C^2 viscosity (sub/super)solution u is also a pointwise classical (sub/super)solution.*

(3) *If u and v are viscosity subsolutions, then $w = \sup(u, v)$ is also a viscosity subsolution. Similarly, if u and v are viscosity supersolutions, so is $w = \inf(u, v)$.*

(4) *Suppose that* $a(x), f \in C(\Omega)$. *Let* $\{u_m\}_{m=1}^{\infty} \subset C(\Omega)$ *be a sequence such that* $Lu_m \geq f$ *in the viscosity sense in* Ω. *Assume that* u_m *converges to a continuous function* u *uniformly on compact subsets of* Ω. *Then* $Lu \geq f$ *in the viscosity sense in* Ω. *Analogous results hold for viscosity supersolutions and solutions.*

Proof. We consider only subsolutions. The proofs for supersolutions and solutions follow similar reasonings.

For (1), if u is a viscosity subsolution, then any quadratic polynomial P touching u from above at x_0 is a C^2 function, so $LP(x_0) \geq f(x_0)$ trivially. For the converse, suppose that we know that for any quadratic polynomial P that touches u from above at x_0 we have $LP(x_0) \geq f(x_0)$. To prove that u is a viscosity subsolution according to our definition, let φ be a function that is C^2 is a neighborhood of x_0 and that touches u from above at x_0. Fix any $\varepsilon > 0$. Let

$$P(x) = u(x_0) + \nabla\varphi(x_0)(x - x_0) + \frac{1}{2}(x - x_0)^T D^2\varphi(x_0)(x - x_0) + \frac{\varepsilon}{2}|x - x_0|^2.$$

Then, since $u(x_0) = \varphi(x_0)$ and φ is second-order differentiable in a neighborhood of x_0, it follows that there is a (possibly smaller) neighborhood of x_0 in which

$$\varphi(x) \leq P(x).$$

In other words, P is a quadratic polynomial that touches u from above at x_0. Therefore, since $D^2P = D^2\varphi(x_0) + \varepsilon I$,

$$LP(x_0) = \mathrm{trace}(a(x_0)D^2P(x_0)) = L\varphi(x_0) + \varepsilon\sum_{k=1}^{n} a^{kk}(x_0) \geq f(x_0).$$

By letting $\varepsilon \to 0$ in the last inequality, we obtain $L\varphi(x_0) \geq f(x_0)$ as desired.

To prove (2), let $u \in C^2(\Omega)$ satisfy $Lu \geq f$ at every point in Ω and let $x_0 \in \Omega$. Let P be a quadratic polynomial touching u by above at x_0. As $P - u$ has a local minimum at x_0, $D^2P(x_0) - D^2u(x_0) \geq 0$. Since $a(x_0)$ is uniformly elliptic and $D^2P(x_0) - D^2u(x_0)$ is symmetric and nonnegative definite,

$$LP(x_0) = \mathrm{trace}(a(x_0)D^2P(x_0)) \geq \mathrm{trace}(a(x_0)D^2u(x_0)) \geq f(x_0).$$

Therefore, by the statement in (1) that we just proved, u is a viscosity subsolution. Conversely, let u be a C^2 viscosity subsolution. Then, for any $\varepsilon > 0$, the C^2 function

$$\varphi(x) = u(x) + \frac{\varepsilon}{2}|x - x_0|^2$$

touches u from above at every point $x_0 \in \Omega$. Thus, since $a^{kk}(x) \leq \Lambda$ for every x,

$$Lu(x_0) = L\varphi(x_0) - \varepsilon\sum_{k=1}^{n} a^{kk}(x_0) \geq f(x_0) - \varepsilon n\Lambda.$$

Since $\varepsilon > 0$ was arbitrary, $Lu(x_0) \geq f(x_0)$, and u is a classical subsolution in Ω.

Let us consider (3). Let u and v be viscosity subsolutions and define $w = \sup(u, v)$. If φ touches w from above at x_0 then it touches u or v from above at x_0. Say it touches u. Then, since u is a viscosity subsolution, $L\varphi(x_0) \geq f(x_0)$.

For (4), let P be a quadratic polynomial touching u from above at $x_0 \in \Omega$. Fix an arbitrary $\varepsilon > 0$. Due to the uniform convergence of u_m to u over compact subsets, it is easy to see that for any $m_0 \geq 1$ and any ball $B_{1/m_0}(x_0) \subset \Omega$ there exist $m \geq m_0$ and a constant c_m such that $P(x) + \varepsilon|x - x_0|^2/2 + c_m$ touches u_m from above at some x_m in $B_{1/m_0}(x_0)$. Since u_m is a subsolution,

$$\operatorname{trace}(a(x_m)D^2P(x_m)) + \varepsilon \sum_{k=1}^{n} a^{kk}(x_m) \geq f(x_m).$$

Since $a(x), f \in C(\Omega)$, we can pass to the limit as $m_0 \to \infty$ to get

$$\operatorname{trace}(a(x_0)D^2P(x_0)) + \varepsilon \sum_{k=1}^{n} a^{kk}(x_0) \geq f(x_0).$$

We conclude by letting $\varepsilon \to 0$. □

The notion of viscosity solution can be extended to fully nonlinear equations like those involving the Pucci extremal operators $\mathcal{M}^{\pm}$. We say that $u \in C(\Omega)$ is a **subsolution to the Pucci extremal equation** $\mathcal{M}^{\pm}(D^2u) = f$ in Ω, written

$$\mathcal{M}^{\pm}(D^2u) \geq f \qquad \text{in the viscosity sense in } \Omega$$

if for any $x_0 \in \Omega$ and any function φ that is C^2 in an open neighborhood of x_0 and that touches u from above at x_0 there holds

$$\mathcal{M}^{\pm}(D^2\varphi(x_0)) \geq f(x_0).$$

Similarly, we can define the notions of viscosity supersolution and viscosity solution to $\mathcal{M}^{\pm}(D^2u) = f$ in Ω.

The following is a simple, yet very useful, property. It establishes that all viscosity solutions u to all equations of the form $Lu = f$ with coefficients $a(x)$ having ellipticity constants λ and Λ satisfy $\mathcal{M}^{-}(D^2u) \leq f \leq \mathcal{M}^{+}(D^2u)$. One of the main advantages of the Pucci extremal operators is that they do not depend on the point x explicitly as the equation $Lu = f$ does. We will come back to this in Chapter 10.

Lemma 8.4. *If $Lu \geq f$ in the viscosity sense then $\mathcal{M}^{+}(D^2u) \geq f$ in the viscosity sense. Similarly, if $Lu \leq f$ in the viscosity sense, then $\mathcal{M}^{-}(D^2u) \leq f$ in the viscosity sense. In particular, if u is a viscosity solution to $Lu = f$ in Ω then*

$$\mathcal{M}^{-}(D^2u) \leq f \leq \mathcal{M}^{+}(D^2u) \qquad \textit{in the viscosity sense in } \Omega.$$

Proof. We only consider subsolutions, the other cases are analogous. Suppose that u satisfies $Lu \geq f$ in the viscosity sense in Ω. If φ is a C^2 function in a neighborhood of x_0 that touches u from above at x_0 then, by Lemma 8.1, since $a(x_0) \in \mathcal{A}$,

$$f(x_0) \leq \operatorname{trace}(a(x_0)D^2\varphi(x_0)) \leq \sup_{A \in \mathcal{A}} \operatorname{trace}(AD^2\varphi(x_0)) = \mathcal{M}^{+}(D^2\varphi(x_0))$$

and $\mathcal{M}^{+}(D^2u) \geq f$ in the viscosity sense in Ω. □

For the rest of this third part of the book we consider viscosity solutions to $Lu = f$ and move on to building up the regularity theory.

8.3 THE ALEXANDROFF–BAKELMAN–PUCCI ESTIMATE

In this section we prove the Alexandroff–Bakelman–Pucci (ABP) estimate for viscosity solutions.

Theorem 8.5 (ABP estimate). *Let $u \in C(\overline{B_1})$ satisfy*

$$Lu \leq f \qquad \text{in the viscosity sense in } B_1$$

where $f \in L^\infty(B_1) \cap C(B_1)$. Suppose that $u \geq 0$ on ∂B_1. Consider $v = -u^- = \min(u, 0)$ and extend v to B_2 by setting $v = 0$ in $B_2 \setminus \overline{B_1}$. Let Γ_u be the convex envelope of v in B_2. Then

$$\left(\sup_{B_1} u^-\right)^n \leq C \int_{B_1 \cap \{u = \Gamma_u\}} (f^+)^n \, dx$$

where $C > 0$ is a constant depending only on n, λ and Λ.

From Theorem 8.5 we can easily deduce maximum principles.

Corollary 8.6 (Maximum principle). *Let $u \in C(\overline{B_1})$.*

(1) *If $Lu \leq 0$ in the viscosity sense in B_1 and $u \geq 0$ on ∂B_1, then $u \geq 0$ in B_1.*

(2) *If $Lu \geq 0$ in the viscosity sense in B_1 and $u \leq 0$ on ∂B_1, then $u \leq 0$ in B_1.*

(3) *If u is a viscosity solution to $Lu = f$ in B_1, then*

$$\|u\|_{L^\infty(B_1)} \leq \|u\|_{L^\infty(\partial B_1)} + C\|f\|_{L^n(B_1)}$$

where $C > 0$ is a constant depending only on n, λ and Λ.

Proof. For (1), we can directly apply Theorem 8.5 with $f = 0$ to find that $u^- = 0$ in B_1, that is, $u \geq 0$ in B_1. The statement in (2) follows directly from (1) by considering $-u$. For (3), let $x_0 \in \overline{B_1}$ such that $\|u\|_{L^\infty(B_1)} = |u(x_0)|$. If $x_0 \in \partial B_1$ then there is nothing to prove. Suppose then that $x_0 \in B_1$ and consider first the case when $u(x_0) < 0$. Since $L(u - \inf_{\partial B_1} u) = f$ and $u - \inf_{\partial B_1} u \geq 0$ on ∂B_1, by Theorem 8.5,

$$|u(x_0)| = \sup_{B_1}(u - \inf_{\partial B_1} u)^- + \inf_{\partial B_1} u \leq C\|f\|_{L^n(B_1)} + \|u\|_{L^\infty(\partial B_1)}.$$

If $u(x_0) > 0$ then we apply the same reasoning to $-u$. Thus, (3) follows. □

We devote the rest of this section to the proof of Theorem 8.5. The key object in the ABP estimate is the convex envelope of a function, that we define next.

Recall first that the supremum of a family of convex functions is again convex. Also, a convex function in an open convex set is continuous and, in addition, locally Lipschitz in the interior.

A function ℓ defined on $\mathbb{R}^n$ is an **affine function** (sometimes called a **hyperplane**) if $\ell(x) = l_0 + l(x)$, where $l_0 \in \mathbb{R}$ and $l(x)$ is a linear function. Let w be a function defined on a set A and let $x_0 \in A$. A **supporting hyperplane** for w at x_0 is an affine function ℓ that touches w from below at x_0 in A. In other words, $\ell(x_0) = w(x_0)$ and $\ell(x) \leq w(x)$ for any $x \in A$. The set of slopes of all supporting hyperplanes of w at x_0 is called the **subdifferential** $\partial w(x_0)$ of w at x_0 in A:

$$\partial w(x_0) = \{\nabla \ell : \ell \text{ is a supporting hyperplane of } w \text{ at } x_0 \text{ in } A\}.$$

It is known that if w is convex in an open convex set A, then there exists a supporting hyperplane for w at any $x_0 \in A$. Moreover, $\partial w(x_0)$ is a singleton if and only if w is differentiable at x_0 and, in this case, $\partial w(x_0) = \{\nabla w(x_0)\}$. Furthermore, $\partial w(x) = \{\nabla w(x)\}$ for almost every $x \in A$.

Let v be a continuous function defined on an open convex set A. The **convex envelope** Γ_v of v in A is defined as the largest convex function in A such that $\Gamma_v \leq v$ in A, namely, for any $x \in A$,

$$\begin{aligned}\Gamma_v(x) &= \sup\{w(x) : w \leq v \text{ in } A, w \text{ convex in } A\}\\ &= \sup\{\ell(x) : \ell \leq v \text{ in } A, \ell \text{ is affine}\}.\end{aligned}$$

The second equality in this definition follows because any convex function has a supporting hyperplane at every point. The set

$$\{v = \Gamma_v\} = \{x \in A : v(x) = \Gamma_v(x)\}$$

is the **contact set** of v, and its points are called **contact points**.

We need the change of variables formula (or area formula) for Lipschitz maps.

Lemma 8.7. *If $T = T(x) : \overline{B_1} \to \mathbb{R}^n$ is a Lipschitz map then T is differentiable a.e. in B_1 and*

$$|T(B_1)| \leq \int_{B_1} |\det \nabla T|\, dx.$$

Let $u \in C(\overline{B_1})$ such that $u \geq 0$ on ∂B_1. As in the statement of Theorem 8.5, let

$$u^-(x) = -\min(u, 0).$$

Since $u \geq 0$ on ∂B_1, we have that $-u^-(x) = \min(u,0) = 0$ on ∂B_1. Hence, the extension by zero outside of B_1 of $-u^-$ is a continuous function in B_2. Moreover, B_2 is an open convex set. From now on, we denote by Γ_u the convex envelope of $v = -u^-$ in $A = B_2$ and assume that $u^- \not\equiv 0$. Observe that the contact set is contained in B_1 (otherwise, it would contradict the convexity of Γ_u):

$$\{u = \Gamma_u\} = \{x \in B_2 : -u^-(x) = \Gamma_u(x)\} \subset B_1.$$

Theorem 8.8 (ABP estimate for functions)**.** *Let $u \in C(\overline{B_1})$ such that $u \geq 0$ on ∂B_1 and let Γ_u be defined as above. If $\Gamma_u \in C^{1,1}(\overline{B_1})$ then*

$$\big(\sup_{B_1} u^-\big)^n \leq C \int_{B_1} \det(D^2\Gamma_u)\, dx$$

where $C > 0$ is a constant depending only on n.

Proof. Consider the gradient map of the convex function Γ_u

$$x \longmapsto T(x) = \nabla\Gamma_u(x).$$

The main quantity to estimate from above and from below is the volume of the image:

$$|\nabla\Gamma_u(B_1)|.$$

Estimate from above. This follows directly by the change of variables for Lipschitz maps. Since, by assumption, $\Gamma_u \in C^{1,1}(\overline{B_1})$, we have that the gradient map $T = \nabla\Gamma_u : \overline{B_1} \to \mathbb{R}^n$ is a Lipschitz map. Therefore, by Lemma 8.7 applied to $T = \nabla\Gamma_u$ with $\nabla T = D^2\Gamma_u$,

$$|\nabla\Gamma_u(B_1)| \leq \int_{B_1} |\det(D^2\Gamma_u)|\,dx = \int_{B_1} \det(D^2\Gamma_u)\,dx. \tag{8.3}$$

The second equality in (8.3) holds because $\Gamma_u \in C^{1,1}(\overline{B_1})$ is convex, so that $D^2\Gamma_u(x)$ is a nonnegative definite matrix for a.e. $x \in B_1$. Observe that $\Gamma_u \in C^{1,1}(\overline{B_1})$ is a crucial hypothesis to obtain (8.3).

Estimate from below. For this part of the proof we do not actually need the assumption that $\Gamma_u \in C^{1,1}(\overline{B_1})$. Let ℓ be an affine function (hyperplane) with slope

$$|\omega| = |\nabla\ell| \leq \frac{\sup_{B_1} u^-}{4}.$$

We raise this hyperplane from $-\infty$ upward until it touches Γ_u for the first time. Since the slope of ℓ is smaller than $(\sup_{B_1} u^-)/4$, this vertical translation of ℓ will touch the graph of Γ_u before touching ∂B_2. In fact, it will touch Γ_u inside of B_1 because the contact set $\{u = \Gamma_u\}$ is inside of B_1. This means that ℓ is a supporting hyperplane for Γ_u at some point in B_1. Thus ω belongs to the image of B_1 by the subdifferential, that is, $\omega \in \partial\Gamma_u(B_1)$. We have just proved that

$$B_{(\sup_{B_1} u^-)/4}(0) \subset \partial\Gamma_u(B_1). \tag{8.4}$$

But, since Γ_u is convex, it is locally Lipschitz in B_1, and thus $\partial\Gamma_u(B_1) = \nabla\Gamma_u(B_1)$ except for a set of measure zero. By taking measure in (8.4),

$$\omega_n\left(\frac{\sup_{B_1} u^-}{4}\right)^n \leq |\nabla\Gamma_u(B_1)|. \tag{8.5}$$

Therefore, by plugging together (8.3) and (8.5), we find that there exists a constant $C > 0$ that depends only on n such that

$$\left(\sup_{B_1} u^-\right)^n \leq C \int_{B_1} \det(D^2\Gamma_u)\,dx.$$

□

We stress once again that the estimate from below

$$\big(\sup_{B_1} u^-\big)^n \leq C_n |\nabla \Gamma_u(B_1)|$$

does not need any additional regularity assumptions on $\nabla \Gamma_u$. Indeed, Γ_u is convex, therefore differentiable a.e. and the proof of (8.5) follows without any problems.

Theorem 8.8 is valid for *any* continuous function u such that Γ_u is $C^{1,1}$. However, $\Gamma_u \in C^{1,1}$ is not true in general. But if u is a viscosity supersolution, then we can show this regularity of Γ_u and establish the ABP estimate of Theorem 8.5.

Proving the following nontrivial regularity properties of the convex envelope Γ_u is the most important step in the proof of the ABP estimate for supersolutions.

Lemma 8.9. *Let $u \in C(\overline{B_1})$ satisfy*

$$Lu \leq f \qquad \text{in the viscosity sense in } B_1$$

where $f \in L^\infty(B_1) \cap C(B_1)$. Suppose that $u \geq 0$ on ∂B_1 and let Γ_u be the convex envelope of $v = -u^-$ in B_2 as defined above. The following statements hold.

(1) (Regularity of Γ_u at contact points). *Let $x_0 \in \{u = \Gamma_u\} \subset B_1$ be a contact point and consider a ball $B_\delta(x_0) \subset B_1$, for some $\delta > 0$. Then*

$$f(x_0) \geq 0$$

and there exists an affine function ℓ such that

$$\ell(x) \leq \Gamma_u(x) \leq \ell(x) + C\big(\sup_{B_\delta} f^+\big)|x - x_0|^2$$

for every $x \in B_{\nu\delta}(x_0) \subset B_1$, where $C > 0$ and $0 < \nu < 1$ depend only on n, λ and Λ. As a consequence,

$$\det(D^2\Gamma_u(x_0)) \leq C f^+(x_0)^n \tag{8.6}$$

for a.e. $x_0 \in \{u = \Gamma_u\}$.

(2) $\Gamma_u \in C^{1,1}(\overline{B_1})$.

(3) *We have*

$$\det(D^2\Gamma_u(x)) = 0$$

for a.e. $x \in B_1 \setminus \{u = \Gamma_u\}$.

Clearly, the ABP estimate for supersolutions, Theorem 8.5, is a direct consequence of Theorem 8.8 and Lemma 8.9. Therefore, we are left to show Lemma 8.9. Before we do that, let us understand the intuition behind each statement.

To see why the last conclusion of Lemma 8.9(1) should hold, suppose for the moment that u is a C^2 classical supersolution, that is, $Lu \leq f$ pointwise in B_1. Since $u - \Gamma_u$ has a local minimum (of zero) at every contact point, we will have

$$0 \leq D^2\Gamma_u \leq D^2 u \qquad \text{on } \{u = \Gamma_u\}.$$

Then
$$0 \le \det(D^2\Gamma_u) \le \det(D^2 u).$$
Notice that the Monge–Ampère operator $\det(D^2u)$ is involved. Now we have two different ways to conclude.

Using the meaning of the equation. As in Section 8.1, the pointwise equation $Lu \le f$ means that, at every point,
$$\lambda_{\max} \le C(-\lambda_{\min} + f)$$
where $\lambda_{\max}$ and $\lambda_{\min}$ denote the maximum and minimum eigenvalues of D^2u, respectively, and $C > 0$ depends only on n, λ and Λ. Since all the eigenvalues λ_k of D^2u at contact points are nonnegative, we actually have
$$\lambda_{\max} \le C f^+ \qquad \text{on } \{u = \Gamma_u\}.$$
Then, on the contact set,
$$0 \le \det(D^2u) = \prod_{k=1}^{n} \lambda_k \le (\lambda_{\max})^n \le C(f^+)^n.$$

Using the arithmetic mean-geometric mean inequality. If A and B are nonnegative definite symmetric matrices of size $n \times n$ then
$$\det(AB) \le \left(\frac{\operatorname{trace}(AB)}{n}\right)^n.$$
Let $A = a^{ij}(x)$ and $B = D^2u$. Then, on $\{u = \Gamma_u\}$,
$$\begin{aligned} 0 \le \det(D^2u) &= \frac{1}{\det(a^{ij}(x))} \det(a^{ij}(x)\partial_{ij}u) \\ &\le \frac{1}{\lambda^n}\left(\frac{a^{ij}(x)\partial_{ij}u}{n}\right)^n \\ &\le C(f^+)^n. \end{aligned}$$

Now (2) in Lemma 8.9 only says that $\Gamma_u \in C^{1,1}$. If $\Gamma_u \in C^2$, then Γ_u would be a C^2 function touching the viscosity supersolution u from below at every contact point $x \in \{u = \Gamma_u\}$. Therefore, we could take $\varphi = \Gamma_u$ in the definition of viscosity supersolution and get
$$0 \le \det(D^2\Gamma_u) \le C(L\Gamma_u)^n \le C(f^+)^n \qquad \text{on } \{u = \Gamma_u\}.$$
The problem is that we cannot ensure that $\Gamma_u \in C^2$.

Finally, one could guess that (3) is going to be true directly from the picture of the convex envelope Γ_u. For instance, in dimension one, the convex envelope is flat outside of the contact set.

Proof of Lemma 8.9. For (1), suppose that the contact point x_0 is the origin, that is $u(0) = \Gamma_u(0)$. By subtracting a supporting hyperplane of Γ_u at $x_0 = 0$, we can assume that $0 \leq \Gamma_u \leq u$ in B_δ and $u(0) = 0$.

Let us first show that $f(0) \geq 0$. The function $\varphi(x) = -\varepsilon|x|^2/2$ is C^2 and touches the supersolution u from below at 0. Hence

$$L\varphi(0) = \operatorname{trace}(a(0)D^2\varphi(0)) = -\varepsilon\sum_{k=1}^{n} a^{kk}(0) \leq f(0).$$

But $a^{kk}(0) \leq \Lambda$, so we deduce that

$$-n\Lambda\varepsilon \leq L\varphi(0) \leq f(0)$$

and conclude by taking $\varepsilon \to 0$.

Next we prove that

$$0 \leq \Gamma_u(x) \leq C(\sup_{B_\delta} f^+)|x|^2$$

for $x \in B_{\nu\delta}$, where $C > 0$ and $0 < \nu < 1$ depend only on n, λ and Λ. Let us call

$$w = \Gamma_u.$$

We need to estimate, for any small $r > 0$,

$$C_r = \frac{1}{r^2}\max_{B_r} w.$$

Let us fix any $0 < r \leq \delta/4$. By convexity, w attains its maximum in $\overline{B_r}$ at some point on the boundary:

$$\max_{\overline{B_r}} w = \max_{\partial B_r} w.$$

After a rotation, we can assume that

$$\max_{\overline{B_r}} w = w(0, \ldots, 0, r).$$

We have

$$w(x', r) \geq w(0, \ldots, 0, r)$$

for any $x = (x', r) \in B_\delta$, where we use the notation x' for points in $\mathbb{R}^{n-1}$. Indeed, w at any point in B_r is smaller than $w(0, \ldots, 0, r)$, so, by convexity w is bigger than $w(0, \ldots, 0, r)$ at every point of the form $x = (x', r + \rho) \in B_\delta$, for any $\rho > 0$, and the inequality is obtained by taking $\rho \to 0$. Consider now the set

$$A = B_{\delta/2} \cap \{-r < x_n < r\}.$$

We construct a quadratic polynomial P that touches u from below in A and curves upward very much in the direction of x_n. Set

$$\widetilde{P}(x) = (x_n + r)^2 - b|x'|^2$$

for some $b > 0$ that we will determine next. We estimate $\widetilde{P}$ on the boundary of A.

(*i*) If $x_n = -r$ then $\widetilde{P} \leq 0$.

(*ii*) If $x \in \partial A$ is such that $|x| = \delta/2$ and $-r < x_n < r$ then, as $r \leq \delta/4$, we can estimate $\delta^2/4 \leq |x'|^2 + r^2 \leq |x'|^2 + \delta^2/16$. In particular, $|x'|^2 \geq \delta^2/8$ and

$$\widetilde{P} \leq 4r^2 - b\delta^2/8 = 0$$

as soon as we take $b = 32r^2/\delta^2$.

(*iii*) If $x_n = r$ then

$$\widetilde{P} = 4r^2 - \frac{32r^2}{\delta^2}|x'|^2 \leq 4r^2.$$

Hence, if we let

$$P(x) = \frac{C_r}{4}\widetilde{P}(x) = \frac{\max_{B_r} w}{4r^2}\left((x_n + r)^2 - \frac{32r^2}{\delta^2}|x'|^2\right)$$

then, by the estimates on $\widetilde{P}$ above and the fact that $0 \leq w \leq u$,

$$P \leq w \leq u \qquad \text{on } \partial A.$$

Moreover,

$$P(0) > 0 = w(0) = u(0).$$

Hence, by continuity, an appropriate vertical translation of P will touch the supersolution u from below somewhere in $A \subset B_{\delta/2}$. At that touching point $z \in A$, since D^2P is a constant diagonal matrix D,

$$LP(z) = \operatorname{trace}(a(z)D^2P(z)) = \sum_{k=1}^{n} a^{kk}(z)D^{kk} \leq f(z) \leq \sup_{B_\delta} f^+.$$

We compute

$$D^2P = \frac{C_r}{4}\operatorname{diag}(-64r^2/\delta^2, \ldots, -64r^2/\delta^2, 2).$$

Then, by using that $\lambda \leq a^{kk}(z) \leq \Lambda$,

$$\left(-\frac{16r^2}{\delta^2}\Lambda(n-1) + \frac{\lambda}{2}\right)C_r \leq \sup_{B_\delta} f^+.$$

Now we choose r sufficiently small, depending only on n, λ, Λ and δ, so that

$$-\frac{16r^2}{\delta^2}\Lambda(n-1) \geq -\frac{\lambda}{4}$$

that is, $r \leq (\frac{\lambda}{64\Lambda(n-1)})^{1/2}\delta =: \nu\delta$, and conclude that

$$C_r \leq \frac{4}{\lambda}\sup_{B_\delta} f^+.$$

This gives that, for any $|x| = r < \nu\delta$, $w(x) \leq \frac{4}{\lambda}(\sup_{B_\delta} f^+)|x|^2$, as desired.

To prove the last statement in (1), notice that since Γ_u is convex in B_2, by Aleksandrov's theorem, Γ_u is pointwise second-order differentiable a.e. in B_2. Since at every contact point $x_0 \in \{u = \Gamma_u\}$ and ball $B_\delta(x_0) \subset B_1$, Γ_u has a tangent plane $\ell(x)$ from below and a tangent paraboloid of the form $\ell(x) + C(\sup_{B_\delta} f^+)|x - x_0|^2/2$ from above, we find that at a.e. contact point x_0,

$$0 \leq D^2\Gamma_u(x_0) \leq C(\sup_{B_\delta} f^+)I \tag{8.7}$$

where $C > 0$ depends only on n, λ and Λ. Taking $\delta \to 0$,

$$0 \leq D^2\Gamma_u(x_0) \leq Cf^+(x_0)I$$

and (8.6) holds for a.e. contact point x_0.

For (2), observe that (1), see (8.7), says that $D^2\Gamma_u$ is bounded a.e. in $\{u = \Gamma_u\} \subset B_1$. Next, let $x_0 \in \overline{B_1} \setminus \{u = \Gamma_u\}$ and let ℓ be a supporting hyperplane for Γ_u at x_0 in B_2. We may assume that $\Gamma_u = 0$ on ∂B_2. We claim that

$$x_0 \in \text{Convex hull of } \{x_1, \ldots, x_{n+1}\}$$

where the points $x_1, \ldots, x_{n+1}$ (not necessarily all distinct) are contact points, except for a possible one $x_{n+1} \in \partial B_2$ and, on this convex hull, $\ell = \Gamma_u$. With this, we conclude that $D^2\Gamma_u(x_0) = 0$ for a.e. $x_0 \in \overline{B_1} \setminus \{u = \Gamma_u\}$ and, therefore, Γ_u is $C^{1,1}$ in $\overline{B_1}$. Now the idea of the proof of the claim is the following. Let x_0 be a noncontact point and let ℓ be an affine function that realizes the supremum in the definition of $\Gamma_u(x_0)$. By the maximality of Γ_u, ℓ must touch u at one point (otherwise, ℓ could be raised a bit more). Even more, ℓ touches u in at least 2 points. If not, we slightly tilt ℓ around $(x_0, \Gamma_u(x_0))$ to get a plane ℓ_t such that $\ell_t < u$. We can raise ℓ_t by some quantity $\alpha > 0$ until it touches u. But then

$$\ell_t(x_0) + \alpha = \ell(x_0) + \alpha > \Gamma_u(x_0)$$

contradicting the maximality of $\Gamma_u(x_0)$. So there are two points $(a, \ell(a)) = (a, u(a)) = (a, \Gamma_u(a))$ and $(b, \Gamma_u(b))$ that are in the graph of ℓ, as well as $(x_0, \Gamma_u(x_0))$. Since Γ_u is convex, $\Gamma_u \equiv \ell$ in the convex hull of those 3 points. In the general case, let ℓ be as before, that is, $\ell(x_0) = \Gamma_u(x_0)$ and having at least one contact point with u. Then the set $C =$ (closed convex hull of $\{x \in \overline{B_2} : \ell(x) = -u^-(x)\}$) is nonempty. By the argument we just did, if $x_0 \notin C$, we can tilt the plane around x_0 and move it upward to obtain a contradiction. Hence $x_0 \in C$ and Caratheodory's theorem says that x_0 is a convex combination of $n+1$ points living in the set $\{x \in \overline{B_2} : \ell(x) = -u^-(x)\}$. Notice that there is at most one x_i with $|x_i| = 2$, since if this is not true then $\Gamma_u \equiv 0$. Notice as well that if one $x_i \in B_2 \setminus B_1$ then also $\Gamma_u \equiv 0$. By convexity, $\Gamma_u = \ell$ in C. The claim is proved.

The proof of (2) shows that for any $x_0 \in B_1 \setminus \{u = \Gamma_u\}$ there exists an open neighborhood of x_0 on which Γ_u is affine. As a consequence we get (3). □

NOTES ON CHAPTER 8

Viscosity solutions, initially developed by Crandall–Lions [23], are primarily useful for nonlinear PDEs as the notion gives a flexible, powerful setting for proving existence and uniqueness of solutions. Standard references are the *user's guide* by Crandall–Ishii–Lions [22], where the Perron's method for existence of viscosity solutions can be found, and Caffarelli–Cabré [7], that focuses on the regularity theory.

The ABP estimate for the case of viscosity solutions was proved by Caffarelli [4], see also [3]. For Section 8.3, we followed Caffarelli–Cabré [7]. The classical proof of the ABP estimate for strong solutions can be found in Gilbarg–Trudinger [43].

Results about convex functions, including Aleksandrov's theorem, as well as Lemma 8.7, can be seen in Evans–Gariepi [33]. For the statement of Caratheodory's theorem used in the proof of Lemma 8.9(2), see Theorem 3.25 in Rudin [55] and its following lemma.

CHAPTER 9

The Krylov–Safonov Harnack inequality

In this chapter we prove the Krylov–Safonov Harnack inequality for nonnegative viscosity solutions to nondivergence form elliptic equations, which in particular implies that viscosity solutions are locally Hölder continuous.

9.1 THE KRYLOV–SAFONOV HARNACK INEQUALITY

We consider viscosity (sub/super)solutions to equations in nondivergence form

$$Lu = \text{trace}(a(x)D^2u) = a^{ij}(x)\partial_{ij}u = 0$$

in a cube $Q_r = (-r,r)^n \subset \mathbb{R}^n$, $n \geq 2$, of side length $2r > 0$. The matrix of coefficients $a(x) = (a^{ij}(x))_{i,j=1}^n$ is symmetric $a^{ij}(x) = a^{ji}(x)$ and uniformly elliptic: there are constants $0 < \lambda \leq \Lambda$ such that $\lambda|\xi|^2 \leq a^{ij}(x)\xi_i\xi_j \leq \Lambda|\xi|^2$, for all $\xi \in \mathbb{R}^n$ and every $x \in Q_r$. No assumption on the regularity of the coefficients is made.

The main result of this chapter is the following Krylov–Safonov Harnack inequality for nonnegative viscosity solutions.

Theorem 9.1 (Harnack inequality). *There exists a constant $C_H > 0$ depending only on n, λ and Λ such that for any viscosity solution $u \in C(\overline{Q_{4\sqrt{n}}})$ to*

$$\begin{cases} Lu = 0 & \text{in } Q_{4\sqrt{n}} \\ u \geq 0 & \text{in } Q_{4\sqrt{n}} \end{cases}$$

we have

$$\sup_{Q_{1/4}} u \leq C_H \inf_{Q_{1/4}} u.$$

As a consequence, viscosity solutions u to $Lu = 0$ are locally Hölder continuous.

The choice of the domain $Q_{4\sqrt{n}}$ is only for convenience, as we will need to place two concentric balls in between Q_3 and the cube where the equation holds. In fact, the following chain of inclusions holds:

$$B_1 \subset Q_1 \subset Q_3 \subset B_{3\sqrt{n}} \subset B_{4\sqrt{n}} \subset Q_{4\sqrt{n}}.$$

DOI: 10.1201/9781032679853-9

By the usual rescaling and covering arguments (see, for example, the proofs of Theorem 2.16 or Corollary 5.3), Theorem 9.1 can be established with a bounded domain Ω in place of $Q_{4\sqrt{n}}$ and a subdomain $\Omega' \subset\subset \Omega$ in place of $Q_{1/4}$, and the constant $C_H > 0$ will also depend on $\operatorname{dist}(\Omega', \partial\Omega)$.

The local Hölder continuity of u follows by showing that the Harnack inequality implies an oscillation decay exactly as we did in Section 6.4.

For simplicity in the exposition, the right hand side of the equation $Lu = f$ is taken to be 0. The techniques apply when there is a continuous and bounded nonzero right hand side as well, giving the Harnack inequality in this case, see Remark 9.8.

Theorem 9.1 will be a direct consequence of the following normalized result.

Theorem 9.2 (Normalized Harnack inequality). *There exists a constant $C_H > 0$ depending only on n, λ and Λ such that for any viscosity solution $u \in C(\overline{Q_{4\sqrt{n}}})$ to*

$$\begin{cases} Lu = 0 & \text{in } Q_{4\sqrt{n}} \\ u \geq 0 & \text{in } Q_{4\sqrt{n}} \end{cases}$$

if

$$\inf_{Q_{1/4}} u \leq 1,$$

then

$$\sup_{Q_{1/4}} u \leq C_H.$$

To see why Theorem 9.2 is enough, let u be as in Theorem 9.1 and define

$$u_\delta = \frac{u}{\inf_{Q_{1/4}} u + \delta} \qquad \text{for } \delta > 0.$$

Then $u_\delta \geq 0$, $Lu_\delta = 0$ in the viscosity sense in $Q_{4\sqrt{n}}$ and $\inf_{Q_{1/4}} u_\delta \leq 1$. Hence, by Theorem 9.2,

$$\sup_{Q_{1/4}} u_\delta \leq C_H$$

which gives

$$\sup_{Q_{1/4}} u \leq C_H\Big(\inf_{Q_{1/4}} u + \delta\Big)$$

and Theorem 9.1 is obtained by letting $\delta \to 0$.

The normalized Harnack inequality in Theorem 9.2 says that there exists $C_H > 0$ such that if u is below 1 somewhere in $Q_{1/4}$ then

$$\mu_u(t) = |\{u > t\} \cap Q_{1/4}| = 0 \qquad \text{for all } t \geq C_H.$$

The idea is to first prove that if u is a supersolution that is controlled at a point in $Q_{1/4}$ then its distribution function satisfies the power decay $\mu_u(t) \leq dt^{-\varepsilon}$ for some $\varepsilon, d > 0$, namely, u is in **weak-L^ε**. The second step is to show that if a subsolution u is in weak-L^ε then it must be bounded. The latter will be achieved by applying the weak-L^ε estimate to both u and the supersolution $C_1 - C_2 u$, where $C_1, C_2 > 0$. In particular, if u is a solution then $|\{u > t\} \cap Q_{1/4}| = 0$ for all large t.

Let us be more specific, comparing with the case of harmonic functions. The following two steps immediately imply Theorem 9.2.

Step 1. Pointwise control gives control in measure. Let $u \geq 0$ be a viscosity supersolution to $Lu = 0$ in $Q_{4\sqrt{n}}$ such that

$$\inf_{Q_{1/4}} u \leq 1.$$

Then $u \in$ weak-$L^\varepsilon(Q_1)$: there exist $\varepsilon, d > 0$ depending only on n, λ and Λ such that

$$|\{u > t\} \cap Q_1| \leq dt^{-\varepsilon} \qquad \text{for every } t > 0. \tag{9.1}$$

This property is obvious for superharmonic functions: if $u \geq 0$ is superharmonic in B_4 and $u(0) = 1$, then, by the super-mean value property,

$$1 = u(0) \geq \frac{1}{|B_1|} \int_{B_1} u \, dz.$$

In other words, $u \in L^1(B_1) \subset$ weak-L^1 with $\|u\|_{L^1(B_1)} \leq |B_1| = d$.

Step 2. Control in measure gives pointwise control. Let $u \geq 0$ be a viscosity subsolution to $Lu = 0$ in $Q_{4\sqrt{n}}$ such that $u \in$ weak-$L^\varepsilon(Q_1)$ as in (9.1). Then, in $Q_{1/4}$, u is bounded by a constant $C_H > 0$ that depends only on n, λ and Λ:

$$\sup_{Q_{1/4}} u \leq C_H.$$

This is also clear for subharmonic functions. If $u \geq 0$ is subharmonic in B_4 and $\|u\|_{L^1(B_1)} \leq |B_1| = d$ then, for any $x \in B_{1/4}$, by the sub-mean value property,

$$u(x) \leq \frac{1}{|B_{1/4}(x)|} \int_{B_{1/4}(x)} u \, dz \leq 4^n \frac{1}{|B_1|} \int_{B_1} u \, dz \leq 4^n.$$

9.2 THE WEAK-L^ε ESTIMATE FOR SUPERSOLUTIONS

In this section we cover Step 1: pointwise control gives control in measure.

Lemma 9.3 (Supersolutions are in weak-L^ε). *There exist constants $\varepsilon, d > 0$ depending only on n, λ and Λ such that if $u \in C(\overline{Q_{4\sqrt{n}}})$, $u \geq 0$, satisfies $Lu \leq 0$ in the viscosity sense in $Q_{4\sqrt{n}}$ and*

$$\inf_{Q_3} u \leq 1$$

then

$$|\{u \geq t\} \cap Q_1| \leq dt^{-\varepsilon} \qquad \textit{for every } t > 0.$$

The proof of Lemma 9.3 for nonnegative viscosity supersolutions is done in two stages. First we show that if $\inf_{Q_3} u \leq 1$, then there are constants $0 < \mu < 1$ and $M > 1$ depending only on n, λ and Λ such that

$$|\{u \leq M\} \cap Q_1| > \mu |Q_1|.$$

In other words, if u is controlled at a point in Q_3 then it cannot be mostly infinity everywhere in Q_1, it has to be universally bounded in a nontrivial portion of Q_1. Notice how well localized this estimate is. Indeed, u could be equal to 1 only at a point that is very close to ∂Q_3 and, as such, far from Q_1. In the second stage, we iterate this inequality to get the power decay

$$|\{u > M^k\} \cap Q_1| \leq (1-\mu)^k |Q_1| \tag{9.2}$$

for all $k \geq 1$. Then Lemma 9.3 readily follows. Indeed, let $t > 0$ and take k such that $M^k < t \leq M^{k+1}$. Then, by setting $d = (1-\mu)^{-1}|Q_1|$ and choosing ε such that $1 - \mu = M^{-\varepsilon}$,

$$\begin{aligned} |\{u \geq t\} \cap Q_1| &\leq |\{u > M^k\} \cap Q_1| \\ &\leq (1-\mu)^k |Q_1| = d(M^{k+1})^{-\varepsilon} \leq dt^{-\varepsilon}. \end{aligned} \tag{9.3}$$

We begin with the case $k = 1$, which uses the ABP estimate of Chapter 8 (see Theorem 8.5) in a crucial way.

Lemma 9.4. *There exist constants $0 < \mu < 1$ and $M > 1$ depending only on n, λ and Λ such that if $u \in C(\overline{Q_{4\sqrt{n}}})$, $u \geq 0$, satisfies $Lu \leq 0$ in the viscosity sense in $Q_{4\sqrt{n}}$ and*

$$\inf_{Q_3} u \leq 1$$

then

$$|\{u \leq M\} \cap Q_1| > \mu |Q_1|.$$

Proof. In order to localize the estimate in Q_1, we need an appropriate barrier function γ that will be added to u so as to pull u down very much in Q_1, make $L(u+\gamma) < 0$ outside of Q_1 and still have $u + \gamma \geq 0$ on $\partial B_{4\sqrt{n}} \subset \overline{Q_{4\sqrt{n}}}$. In this way, the contact set for the convex envelope of $-(u+\gamma)^-$ in $B_{8\sqrt{n}}$ will be inside Q_1 and, on that set, we can control how much u goes down.

We construct γ in terms of n, λ and Λ and satisfying the necessary properties by using the typical radially symmetric barrier function $|x|^{-\sigma}$. Let $\sigma = n\Lambda/\lambda > 0$. This choice of σ will be apparent soon.

- We need $\gamma \leq -2$ in Q_3 and $\gamma \geq 0$ in $\mathbb{R}^n \setminus B_{4\sqrt{n}}$. We can do this by first letting $\gamma(x) = c_2 - c_1|x|^{-\sigma}$ for $|x| \geq 1/4$, imposing $\gamma = -2$ on $\partial B_{3\sqrt{n}}$ (note that $B_{3\sqrt{n}} \supset Q_3$) and $\gamma = 0$ on $\partial B_{4\sqrt{n}}$, and solving a linear system to find $c_1, c_2 > 0$. Then we extend γ to $|x| < 1/4$ in a smooth way so as to keep $\gamma \leq -2$ in $B_{3\sqrt{n}}$.

- We fixed $\sigma = n\Lambda/\lambda$ large in order to have γ very concave outside of Q_1 so that $L\gamma < 0$ outside of Q_1 and $(L\gamma)^+$ is supported in Q_1. Indeed, for $|x| > 1/4$,

  $$\partial_{ij}\gamma(x) = c_1\sigma|x|^{-\sigma-2}(\delta_{ij} - (\sigma+2)|x|^{-2}x_i x_j)$$

  and, by ellipticity,

  $$\begin{aligned} L\gamma &= c_1\sigma|x|^{-\sigma-2}(\operatorname{trace}(a(x)) - (\sigma+2)|x|^{-2}a^{ij}(x)x_i x_j) \\ &\leq c_1\sigma|x|^{-\sigma-2}(n\Lambda - (\sigma+2)\lambda) < 0. \end{aligned}$$

  On the other hand, since γ is smooth, $D^2\gamma$ is bounded in B_1 and then $L\gamma \leq C_0$ for all $|x| \leq 1/4$, where $C_0 > 0$ depends only on n, λ and Λ.

Therefore, we can take a continuous cutoff function $\eta \in C_c(Q_1)$, $0 \leq \eta \leq 1$, with $\eta = 1$ in $\overline{B_{1/4}}$ so that

$$L\gamma \leq C_0\eta \qquad \text{in } \mathbb{R}^n \tag{9.4}$$

Let

$$-M = \min_{\mathbb{R}^n} \gamma. \tag{9.5}$$

Then $M > 1$ depends only on n, λ and Λ.

Consider now

$$\tilde{u} = u + \gamma.$$

By using that $\inf_{Q_3} u \leq 1$,

$$\inf_{Q_3} \tilde{u} \leq \inf_{Q_3} u - 2 \leq -1.$$

It is clear that

$$\tilde{u} \geq 0 \qquad \text{on } \partial B_{4\sqrt{n}}.$$

Moreover, since γ is smooth,

$$L\tilde{u} \leq C_0\eta \qquad \text{in the viscosity sense in } B_{4\sqrt{n}}. \tag{9.6}$$

Indeed, if a C^2 function φ touches $\tilde{u} = u + \gamma$ from below at x_0, then $\varphi - \gamma \in C^2$ touches the supersolution u from below at x_0, so that, by (9.4),

$$L\varphi(x_0) = L(\varphi - \gamma)(x_0) + L\gamma(x_0) \leq L\gamma(x_0) \leq C_0\eta(x_0).$$

We apply the ABP estimate in Theorem 8.5 to $\tilde{u} \in C(\overline{B_{4\sqrt{n}}})$ with $f = C_0\eta \in C(\overline{B_{4\sqrt{n}}})$ (notice that changing $B_{4\sqrt{n}}$ by B_1 only affects the constant C in the ABP estimate by a factor depending only on n) to get

$$\begin{aligned} 1 &\leq \big(\sup_{Q_3} \tilde{u}^-\big)^n \\ &\leq \big(\sup_{B_{4\sqrt{n}}} \tilde{u}^-\big)^n \\ &\leq C \int_{B_{4\sqrt{n}} \cap \{\tilde{u} = \Gamma_{\tilde{u}}\}} (C_0\eta)^n \, dx \\ &\leq C|\{\tilde{u} = \Gamma_{\tilde{u}}\} \cap Q_1| \\ &\leq C|\{\tilde{u} \leq 0\} \cap Q_1| \\ &= C|\{u \leq -\gamma\} \cap Q_1| \\ &\leq C|\{u \leq M\} \cap Q_1| \end{aligned} \tag{9.7}$$

where we used that $0 \leq \eta \leq 1$ is supported in $Q_1 \subset B_{4\sqrt{n}}$, $\tilde{u} \leq 0$ on the contact set $\{\tilde{u} = \Gamma_{\tilde{u}}\}$ and (9.5). The lemma follows by taking $0 < \mu < 1/(2^n C)$. □

For the second stage, that is, to iterate the estimate in Lemma 9.4 and get the power decay (9.2) for all $k \geq 1$, we need the Calderón–Zygmund AB lemma that we proved and used in Chapter 7, see Lemma 7.15. As it happened there, this is not surprising, because these covering lemmas are effective tools to estimate distribution functions, see also Sections 4.3 and 4.5.

We state the result here again for convenience.

Lemma 9.5 (Calderón–Zygmund AB lemma). *Let $A \subset B \subset Q_1$ be measurable sets and $0 < \delta < 1$ such that*

(a) $|A| \leq \delta |Q_1|$ and

(b) if $Q \subset Q_1$ is a dyadic cube such that

$$\frac{|A \cap Q|}{|Q|} > \delta,$$

then the predecessor $\widetilde{Q} \subset Q_1$ of Q satisfies

$$\widetilde{Q} \subset B.$$

Then

$$|A| \leq \delta |B|.$$

We are ready to iterate Lemma 9.4. By (9.3), the following result (that the reader should compare with Lemma 7.16 and the proof of Theorem 7.11) gives the weak-L^ε estimate in Lemma 9.3.

Lemma 9.6. *Let $0 < \mu < 1$, $M > 1$ and u be as in Lemma 9.4. Then, for any $k \geq 1$,*

$$|\{u > M^k\} \cap Q_1| \leq (1-\mu)^k |Q_1|.$$

Proof. The proof is by induction in k. The case $k = 1$ is exactly Lemma 9.4. Indeed,

$$|\{u \leq M\} \cap Q_1| \geq \mu |Q_1|$$

if and only if

$$|\{u > M\} \cap Q_1| \leq (1-\mu)|Q_1|.$$

For the induction step, suppose that

$$|\{u > M^{k-1}\} \cap Q_1| \leq (1-\mu)^{k-1} |Q_1|.$$

Let

$$A = \{u > M^k\} \cap Q_1$$

and

$$B = \{u > M^{k-1}\} \cap Q_1.$$

Notice that $A \subset B \subset Q_1$. If the Calderón–Zygmund AB lemma (Lemma 9.5) applies to these sets with $\delta = 1 - \mu$, then

$$|A| \leq (1-\mu)|B|$$

and this finishes the induction step:

$$|\{u > M^k\} \cap Q_1| = |A| \leq \delta |B| = (1-\mu)|\{u > M^{k-1}\} \cap Q_1| \leq (1-\mu)^k |Q_1|.$$

Thus, we must check that A and B verify hypotheses (a) and (b) of Lemma 9.5 with $\delta = 1 - \mu$. Since $A = \{u > M^k\} \subset \{u > M\}$, by the base step,

$$|A| \leq |\{u > M\} \cap Q_1| \leq (1-\mu)|Q_1| = \delta |Q_1|$$

and (a) is satisfied. For (b), let $Q \subset Q_1$ be a dyadic cube such that

$$\frac{|A \cap Q|}{|Q|} > \delta.$$

This is the same as saying that

$$|\{u > M^k\} \cap Q| > (1-\mu)|Q|$$

or, equivalently,

$$|\{u/M^{k-1} \leq M\} \cap Q| < \mu |Q|. \tag{9.8}$$

We have to show that the predecessor $\widetilde{Q}$ of Q is contained in $B = \{u > M^{k-1}\} \cap Q_1$. Suppose that this is not the case, that is, that there is a point $\widetilde{x} \in \widetilde{Q}$ such that

$$u(\widetilde{x}) \leq M^{k-1}. \tag{9.9}$$

Now we perform a translation and rescaling from Q to Q_1 to fall back into the hypotheses of Lemma 9.4. We translate and expand the cube $Q = Q_r(x_0)$ to center it at the origin and make it of side length 2. Namely, we apply the transformation

$$x = x_0 + rz \qquad \text{for } z \in Q_1 \text{ and } x \in Q = Q_r(x_0).$$

Then the predecessor $\widetilde{Q}$ of Q is transformed into a cube $\widetilde{Q_1}$ of side length 4 that is a predecessor of Q_1 and, in particular, $\widetilde{Q_1} \subset Q_3$. We look at the corresponding translated, rescaled and normalized function

$$u^*(z) = \frac{u(x)}{M^{k-1}}.$$

Then $u^* \geq 0$ is a viscosity supersolution in $Q_{4\sqrt{n}}$ to a translated and rescaled equation having the same ellipticity constants as the original equation. Let $\widetilde{z}$ be the point in $\widetilde{Q_1} \subset Q_3$ corresponding to $\widetilde{x}$. By (9.9),

$$\inf_{Q_3} u^* \leq u^*(\widetilde{z}) = \frac{u(\widetilde{x})}{M^{k-1}} \leq 1.$$

Therefore, Lemma 9.4 applies to u^* and we get

$$|\{u^* \leq M\} \cap Q_1| > \mu |Q_1|.$$

In other words, rescaling back to Q,

$$\frac{|\{u^* \leq M\} \cap Q_1|}{|Q_1|} = \frac{|\{u/M^{k-1} \leq M\} \cap Q|}{|Q|} > \mu$$

which is a contradiction to (9.8). The contradiction came from assuming that $\widetilde{Q}$ was not contained in B. Hence, $\widetilde{Q} \subset B$ and the proof is completed. □

9.3 SUBSOLUTIONS IN WEAK-L^ε ARE BOUNDED AND CONCLUSION

In this section we perform Step 2 (control in measure gives pointwise control) and conclude with the proof of the normalized Harnack inequality, Theorem 9.2.

Lemma 9.7 (Subsolutions in weak-L^ε are bounded). *Let $d, \varepsilon > 0$ be the constants of Lemma 9.3. There exists a constant $C_H > 0$ depending only on n, λ and Λ such that if $u \in C(\overline{Q_{4\sqrt{n}}})$, $u \geq 0$, satisfies $Lu \geq 0$ in the viscosity sense in $Q_{4\sqrt{n}}$ and*

$$|\{u \geq t\} \cap Q_1| \leq dt^{-\varepsilon} \qquad \text{for all } t > 0$$

then

$$\sup_{Q_{1/4}} u \leq C_H.$$

Proof. We show that we can find constants $C_H >> 1$ and $0 < \delta < 1$ depending only on n, λ and Λ such that if

$$u(x_0) = M > C_H \qquad \text{for some } x_0 \in Q_{1/4}$$

then there exists a sequence $\{x_k\}_{k=1}^{\infty} \subset Q_{1/2}$ such that

$$u(x_k) \geq (1+\delta)^k M \qquad \text{for all } k \geq 1.$$

This contradicts the boundedness of $u \in C(\overline{Q_{4\sqrt{n}}})$ and therefore we must have

$$u \leq C_H \qquad \text{in } Q_{1/4}.$$

To proceed with this argument, suppose that $u(x_0) = M > C_H$ for some $x_0 \in Q_{1/4}$. We are going to determine C_H and δ in the process. Consider a cube $Q_r(x_0)$ centered at x_0 and with side length $2r > 0$ sufficiently small to be determined. We want to find a point $x_1 \in Q_{4\sqrt{n}r}(x_0)$ such that

$$u(x_1) \geq (1+\delta)M.$$

By contradiction, suppose that is not the case, namely, assume that

$$u \leq (1+\delta)M \qquad \text{in } Q_{4\sqrt{n}r}(x_0) \text{ for any } \delta > 0.$$

Let

$$A = \{u > M/2\} \cap Q_r(x_0)$$

and

$$B = \{u \leq M/2\} \cap Q_r(x_0).$$

Certainly,

$$|A| + |B| = |Q_r(x_0)|.$$

The idea is to show that, because u stays below $(1+\delta)M$ in $Q_{4\sqrt{n}r}(x_0)$ for any $\delta > 0$, then we can make both

$$|A|, |B| < \frac{1}{2}|Q_r(x_0)|$$

which is a contradiction.

First, the measure of A can be estimated by using the hypothesis that u satisfies the weak-L^ε estimate of Lemma 9.3. We have

$$|A| \leq |\{u > M/2\} \cap Q_1| \leq d(M/2)^{-\varepsilon}.$$

From here we can choose the size of the cube $Q_r(x_0)$: let $r > 0$ such that

$$d(M/2)^{-\varepsilon} = \frac{1}{3}|Q_r(x_0)|. \tag{9.10}$$

Notice that for $M > C_H$ very large, r is very small. In order to have everything happening well inside of $Q_{4\sqrt{n}}$, the constant C_H has to be initially chosen large enough, at least bounded from below by some constant depending on n, d and ε:

$$M > C_H \geq C(n, d, \varepsilon) > 0. \tag{9.11}$$

Then

$$|\{u > M/2\} \cap Q_r(x_0)| < \frac{1}{2}|Q_r(x_0)|. \tag{9.12}$$

Second, to estimate the measure of B, we renormalize and flip the picture to apply Lemma 9.3 again. As in the proof of Lemma 9.6, consider the transformation

$$x = x_0 + rz \qquad \text{for } z \in Q_{4\sqrt{n}} \text{ and } x \in Q_{4\sqrt{n}r}(x_0).$$

In particular, this sends the cube $Q_r(x_0)$ into Q_1. We look at the translated, rescaled, flipped and normalized picture

$$\tilde{u}(z) = \frac{(1+\delta)M - u(x)}{\delta M}.$$

Then $\tilde{u}$ satisfies the following properties in $Q_{4\sqrt{n}}$: $\tilde{u} \geq 0$ because $u \leq (1+\delta)M$ everywhere in $Q_{4\sqrt{n}r}(x_0)$ by assumption; $\tilde{u}$ is a supersolution to a translated and rescaled equation having the same ellipticity constants as the original equation because u is a subsolution in $Q_{4\sqrt{n}r}(x_0)$; and, since $u(x_0) = M$, we have $\tilde{u}(0) = 1$ and thus

$$\inf_{Q_3} \tilde{u} \leq 1.$$

Hence, $\tilde{u}$ satisfies the weak-L^ε estimate of Lemma 9.3:

$$|\{\tilde{u} \geq t\} \cap Q_1| \leq dt^{-\varepsilon} \qquad \text{for any } t > 0.$$

By rescaling back to u and $Q_r(x_0)$,

$$|\{u \leq (1 + \delta - t\delta)M\} \cap Q_r(x_0)| \leq dt^{-\varepsilon}|Q_r(x_0)|.$$

Choose $t > 0$ such that

$$(1 + \delta - t\delta)M = \frac{M}{2}$$

that is,

$$t = \frac{1/2 + \delta}{\delta}$$

to get

$$|\{u \leq M/2\} \cap Q_r(x_0)| \leq d\Big(\frac{\delta}{1/2+\delta}\Big)^{\varepsilon} |Q_r(x_0)|.$$

Now we make δ sufficiently small depending only on d and ε so that

$$d\Big(\frac{\delta}{1/2+\delta}\Big)^{\varepsilon} < \frac{1}{2}$$

and finally conclude that

$$|B| = |\{u \leq M/2\} \cap Q_r(x_0)| < \frac{1}{2}|Q_r(x_0)|.$$

This is in contradiction with (9.12).

To recapitulate, we have proved that there exist $C_H > 0$ large and $\delta > 0$ small depending only on n, λ and Λ such that if

$$u(x_0) > C_H \qquad \text{at some } x_0 \in Q_r(x_0) \subset Q_{1/4}$$

then

$$u(x_1) > (1+\delta)C_H \qquad \text{at some } x_1 \in Q_{4\sqrt{n}r}(x_0)$$

provided (see (9.10) and (9.11))

$$r = C_H^{-\varepsilon/n}.$$

An iteration of this claim gives a sequence of points $\{x_k\}_{k=1}^{\infty}$ such that, for any $k \geq 1$,

$$u(x_k) > (1+\delta)^k C_H \qquad \text{at some } x_k \in Q_{4\sqrt{n}r_k}(x_{k-1})$$

where

$$r_k = ((1+\delta)^{k-1}C_H)^{-\varepsilon/n}.$$

We want the sequence $\{x_k\}_{k=1}^{\infty}$ to stay within $Q_{1/2}$: for any $j \geq 1$,

$$|x_j| \leq \sum_{k=j+1}^{\infty} |x_{k-1} - x_k| \leq C_n C_H^{-\varepsilon/n} \sum_{k=1}^{\infty} \frac{1}{[(1+\delta)^{\varepsilon/n}]^k} = DC_H^{-\varepsilon/n}$$

where $D > 0$ depends only on n, λ and Λ. Now, if necessary, we increase $C_H > 0$ in terms of n, λ and Λ so that $DC_H^{-\varepsilon/n} < 1/2$. As we said at the beginning of the proof, this contradicts the boundedness of u and the lemma is proved. □

Proof of Theorem 9.2. Let $u \in C(\overline{Q_{4\sqrt{n}}})$, $u \geq 0$, be a viscosity solution to $Lu = 0$ in $Q_{4\sqrt{n}}$ such that $\inf_{Q_{1/4}} u \leq 1$. In particular, u is a viscosity supersolution with $\inf_{Q_3} u \leq \inf_{Q_{1/4}} u \leq 1$. Then u satisfies the weak-L^{ε} estimate of Lemma 9.3. Therefore, since u is also a viscosity subsolution, Lemma 9.7 gives that $\sup_{Q_{1/4}} u \leq C_H$. □

Remark 9.8 (Harnack inequality with nonzero right hand side). The Harnack inequality also holds for nonnegative viscosity solutions u to $Lu = f$ in $Q_{4\sqrt{n}}$ whenever f is continuous and bounded and, in this case, it takes the form

$$\sup_{Q_{1/4}} u \le C_H\Big(\inf_{Q_{1/4}} u + \|f\|_{L^n(Q_{4\sqrt{n}})}\Big).$$

The proof follows the same lines as the proof of Theorem 9.1 with adequate modifications. By considering $\varepsilon_0(u/\|f\|_{L^n(Q_{4\sqrt{n}})})$ for some $\varepsilon_0 > 0$, we can always assume that $\|f\|_{L^n(Q_{4\sqrt{n}})} \le \varepsilon_0$. The first difference comes in the proof of the weak-L^ε estimate, where instead of (9.6) we will have that

$$L(u+\gamma) \le f + C_0\eta.$$

Then the extra term involving $\|f\|_{L^n(Q_{4\sqrt{n}})}$ that will appear in (9.7) can be reabsorbed in the left hand side by choosing ε_0 sufficiently small depending only on n, λ and Λ. The rest of the proofs carry through without difficulties because this smallness condition on the L^n norm of f is preserved under the necessary rescalings and translations. The Harnack inequality in this form also implies the local Hölder continuity of viscosity solutions to $Lu = f$, with the Hölder norm of u bounded by $C(\|u\|_{L^\infty} + \|f\|_{L^n})$.

NOTES ON CHAPTER 9

The Harnack inequality for nonnegative strong solutions to uniformly elliptic equations in nondivergence form was first proved by Krylov–Safonov [48, 49], see also Gilbarg–Trudinger [43].

Caffarelli gave the proof of the Harnack inequality for viscosity solutions to fully nonlinear elliptic equations in [4], see also [3] for the case we consider here. Our presentation follows the fully nonlinear proof in Caffarelli–Cabré [7], where the case of a nonzero right hand side is considered as well; see also Han–Lin [44].

CHAPTER 10

Savin's method of sliding paraboloids

In this chapter we prove the Harnack inequality for nonnegative viscosity solutions to nondivergence form elliptic equations that we saw in Chapter 9 with a different technique: Savin's method of sliding paraboloids. In particular, Savin's strategy avoids the consideration of the convex envelope and the ABP estimate is replaced by a very geometric point-to-measure estimate.

10.1 SAVIN'S SLIDING PARABOLOIDS FOR HARNACK INEQUALITY

We consider viscosity solutions to nondivergence form equations

$$Lu = \mathrm{trace}(a(x)D^2u) = a^{ij}(x)\partial_{ij}u = 0 \qquad \text{in } B_1 \subset \mathbb{R}^n$$

for $n \geq 2$. The matrix of coefficients $a(x) = (a^{ij}(x))_{i,j=1}^n$ is defined in B_1, symmetric $a^{ij}(x) = a^{ji}(x)$ and uniformly elliptic: there are constants $0 < \lambda \leq \Lambda$ such that $\lambda|\xi|^2 \leq a^{ij}(x)\xi_i\xi_j \leq \Lambda|\xi|^2$, for all $\xi \in \mathbb{R}^n$ and every $x \in B_1$. No extra assumption on the regularity of the coefficients is made.

The aim of this chapter is to implement Savin's method of sliding paraboloids to prove the following (normalized) Harnack inequality for nonnegative viscosity solutions. The reader should review Chapter 8 for basics about viscosity solutions and Section 9.1 for more about the Harnack inequality and its consequences.

Theorem 10.1 (Harnack inequality). *There exists a constant $C_H > 0$ depending only on n, λ and Λ such that for any viscosity solution $u \in C(\overline{B_1})$ to*

$$\begin{cases} Lu = 0 & \text{in } B_1 \\ u \geq 0 & \text{in } B_1 \end{cases}$$

we have

$$u \leq C_H u(0) \qquad \text{in } B_{1/2}.$$

The proof of Theorem 10.1 by Savin's method of sliding paraboloids can be split into two steps.

DOI: 10.1201/9781032679853-10

Step 1. We first slide paraboloids of constant opening from below until they touch the graph of u for the first time. At these touching points we use the equation to show that we can estimate the measure of the set of touching points by the set of vertices of the corresponding sliding paraboloids. We call this the *point-to-measure estimate.* In particular, this is an initial estimate on the measure of the set where u can be large. Then, by increasing the opening of the paraboloids (that is, by making them *narrower*) and using a localization result and a covering lemma, we show that the set of touching points almost covers $B_{1/2}$ in measure. Making the paraboloids narrower means that u can be touched at more points and at larger values. This implies that u is comparable to $u(0)$ in $B_{1/2}$ except for a set of very small measure. The reader should compare these ideas with the proof of the weak-L^ε estimate for supersolutions, Lemma 9.3.

Step 2. Next, if u is much larger that $u(0)$ somewhere in $B_{1/2}$ then, by the same methods as in Step 1 (this time, sliding paraboloids from above or, equivalently, by flipping the picture) we find that u is much bigger than $u(0)$ in a set of positive measure, which contradicts the fact that u was comparable to $u(0)$ in $B_{1/2}$ except for a very small set. Therefore, u must be of the order of $u(0)$ in $B_{1/2}$. This second step should be compared with the reasoning used for the proof of the boundedness of subsolutions that satisfy the weak-L^ε estimate, Lemma 9.7.

10.2 THE POINT-TO-MEASURE ESTIMATE FOR SUPERSOLUTIONS

In some sense, in Savin's method, the point-to-measure estimate plays the role of the ABP estimate, Theorem 8.5, in the proof of the Harnack inequality of Chapter 9. In particular, Theorem 10.2 avoids the consideration of the convex envelope of u. Before presenting the estimate, we fix some notation.

Let $a > 0$ and $z \in \mathbb{R}^n$. A **(concave) paraboloid** P of opening a and vertex z is a quadratic polynomial of the form

$$P(x) = -\frac{a}{2}|x - z|^2 + c \qquad x \in \mathbb{R}^n$$

where $c \in \mathbb{R}$. If the paraboloid P coincides with a function u at a point, namely, if $P(x_0) = u(x_0)$ at some x_0, then we can compute c and write

$$P(x) = -\frac{a}{2}|x - z|^2 + \frac{a}{2}|x_0 - z|^2 + u(x_0).$$

Theorem 10.2 (Point-to-measure estimate for supersolutions). *Let $u \in C(\overline{B_1})$ satisfy $Lu \leq 0$ in the viscosity sense in B_1. Let $B \subset \overline{B_1}$ be a closed set and fix $a > 0$. For each vertex $z \in B$ we slide paraboloids P of opening a and vertex z from below until they touch the graph of u for the first time. Let A denote the set of contact points and assume that $A \subset B_1$. Then A is compact and there exists $0 < c < 1$ depending only on n, λ and Λ such that*

$$|A| \geq c|B|.$$

In other words, the measure of the set of contact points is controlled from below by the measure of the set of vertices.

Remark 10.3 (Point-to-measure estimate for subsolutions). If u satisfies $Lu \geq 0$ in the viscosity sense in B_1 and $A \subset B_1$ denotes the set of points where u can be touched from above by convex paraboloids P of opening $a > 0$ and vertex $z \in B$ of the form $P(x) = \frac{a}{2}|x-z|^2 + c$ for some $c \in \mathbb{R}$ then, by flipping the picture, that is, by considering $-u$, the same conclusion of Theorem 10.2 holds, namely, $|A| \geq c|B|$.

The fact that the set of contact points A in Theorem 10.2 is compact has nothing to do with viscosity solutions. Indeed, to see that $A \subset B_1$ is closed and, therefore, compact, let $x_k \in A$ such that $x_k \to x_0 \in \overline{B_1}$ as $k \to \infty$. By definition of A, there exist corresponding paraboloids P_k with vertices $z_k \in B$ such that P_k touches u from below at x_k in B_1. By assumption, $B \subset \overline{B_1}$ is closed, so B is compact. Thus, up to a subsequence, $z_k \to z_0 \in B$. The continuity of u gives that, for each $x \in B_1$,

$$\begin{aligned} P_k(x) &= -\frac{a}{2}|x-z_k|^2 + \frac{a}{2}|x_k - z_k|^2 + u(x_k) \\ &\to -\frac{a}{2}|x-z_0|^2 + \frac{a}{2}|x_0 - z_0|^2 + u(x_0) =: P(x). \end{aligned}$$

Since $P_k \leq u$ for all k, it must be that $P \leq u$ in B_1. Moreover $P(x_0) = u(x_0)$. Therefore, P is a paraboloid of opening a and vertex $z_0 \in B$ that touches u from below at x_0, and $x_0 \in A$.

The proof of Theorem 10.2 is first done for the case when u has some mild regularity, namely, when u can be touched from above at every point in B_1 by a convex paraboloid of fixed opening. The general case follows by passing to the minimal Pucci extremal equation and a technical, yet powerful regularization procedure suitable for viscosity solutions. In order to give a clear presentation of the ideas, we will prove these two cases separately.

Proof of Theorem 10.2 for u regular. Suppose that at every point $x_0 \in B_1$ there is a convex quadratic polynomial of opening $b > 0$ touching u from above in B_1, namely,

$$u(x) \leq \frac{b}{2}|x-x_0|^2 + \ell(x) \qquad \text{for all } x \in B_1 \tag{10.1}$$

where $\ell(x) = l_0 + l(x)$ is an affine function, namely, $l_0 \in \mathbb{R}$ and $l(x)$ is linear. This implies that u is **(uniformly) semiconcave** in B_1, that is, the function

$$w(x) = u(x) - \frac{b}{2}|x|^2$$

is concave in B_1. Indeed, if $x+h$ and $x-h$ are in B_1 then, by (10.1),

$$\frac{w(x+h)+w(x-h)-2w(x)}{|h|^2} = \frac{u(x+h)+u(x-h)-2u(x)}{|h|^2} - b \leq 0.$$

In particular, by choosing $x = \frac{x_1+x_2}{2}$ and $h = \frac{x_1-x_2}{2}$, $w(\frac{x_1+x_2}{2}) \geq \frac{1}{2}(w(x_1)+w(x_2))$ for all $x_1, x_2 \in B_1$ and, since w is continuous, w is concave. By properties of concave functions, we get that u is locally Lipschitz and pointwise second-order differentiable a.e. in B_1.

Under this regularity assumption on u and the hypotheses of Theorem 10.2, we have that we can touch u from below at every $x_0 \in A$ by a concave paraboloid P of vertex z_0 and opening a and from above by a convex quadratic polynomial of opening b. In particular, u is differentiable at every $x_0 \in A$.

With this, we can uniquely determine the vertex z_0 of P in the following way. Since $u - P$ attains a minimum at x_0,

$$\nabla u(x_0) = \nabla P(x_0) = -a(x_0 - z_0)$$

which implies that

$$z_0 = x_0 + \frac{1}{a}\nabla u(x_0).$$

We can thus consider the map T from the set of touching points A onto the set of vertices B given by

$$T(x_0) = z_0 = x_0 + \frac{1}{a}\nabla u(x_0) \qquad x_0 \in A.$$

Since u admits touching quadratic polynomials of fixed opening from above and below in A, it follows that D^2u is bounded a.e. in A or, equivalently, ∇u is Lipschitz in A. Hence, T is a Lipschitz map from A to B. By the change of variables formula for Lipschitz maps (see Lemma 8.7),

$$|B| = |T(A)| \le \int_A |\det \nabla T|\, dx = \int_A \left|\det\left(I + \frac{1}{a}D^2u\right)\right| dx. \tag{10.2}$$

We need to estimate the right hand side of (10.2) by $|A|$. Here is where the equation is used. Recall that u is second-order differentiable a.e. in B_1, that is, for a.e. $z \in B_1$ we have a quadratic polynomial approximation of u:

$$\begin{aligned} u(x) &= Q(z,x) + o(|x-z|^2) \\ &= u(z) + \nabla u(z)\cdot(x-z) + \frac{1}{2}(x-z)^T D^2u(z)(x-z) + o(|x-z|^2) \end{aligned}$$

for all $x \in B_1$. Fix any $z \in A$ for which this quadratic approximation of u holds. We claim that there is $C > 0$ depending only on n, λ and Λ such that

$$-aI \le D^2u(z) \le CaI. \tag{10.3}$$

The first inequality in (10.3) is straightforward because u can always be touched from below at $z \in A$ by a concave paraboloid of opening a. For the second inequality, notice that for any $\varepsilon > 0$ the quadratic polynomial

$$Q(z,x) - \frac{\varepsilon}{2}|x-z|^2$$

touches u from below at z in a neighborhood of z. Since u is a viscosity supersolution,

$$LQ = \operatorname{trace}(a(z)(D^2u(z) - \varepsilon I)) \le 0. \tag{10.4}$$

By letting $\varepsilon \to 0$,

$$\operatorname{trace}(a(z)D^2u(z)) \leq 0. \tag{10.5}$$

Next suppose, by way of contradiction, that

$$D^2u(z) > CaI.$$

Then, for any $1 \leq k \leq n$,

$$D^2u(z) > Ca(e_k \otimes e_k) > Ca(e_k \otimes e_k) - aI \tag{10.6}$$

where $\{e_k\}_{k=1}^n$ denotes the standard basis of vectors in $\mathbb{R}^n$ and $e_k \otimes e_k = e_k e_k^T$. Using (10.6) and the fact that the matrix of coefficients $a(z) = (a^{ij}(z))_{i,j=1}^n$ is uniformly elliptic,

$$\operatorname{trace}(a(z)D^2u(z)) \geq (Ca)a^{kk}(z) - a\operatorname{trace}(a(z)) \geq Ca\lambda - an\Lambda.$$

If $C > n\Lambda/\lambda$ then we contradict the equation in (10.5). Hence, (10.3) holds.

Now we apply (10.3) into (10.2) to conclude that

$$|B| \leq \int_A \det((C+1)I)\,dx = (C+1)^n|A|$$

as desired. □

To prove Theorem 10.2 for the general case, let us begin by recalling that if u is a viscosity supersolution to $Lu = 0$ in B_1, then u is also a viscosity supersolution to the minimal Pucci extremal equation:

$$\mathcal{M}^-(D^2u) \leq 0 \qquad \text{in the viscosity sense in } B_1$$

see Lemma 8.4. The introduction of the minimal Pucci extremal operator is convenient for various reasons, one of them being that the equation $\mathcal{M}^-(D^2u) \leq 0$ is translation invariant, not depending explicitly on the points $x \in B_1$.

Now we must observe that Theorem 10.2 continues to hold for the case when u is a regular supersolution to the more general Pucci extremal equation $\mathcal{M}^-(D^2u) = 0$. Indeed, everything in the proof we just did above is true in this case until we have to prove (10.3). The equation in (10.4) must now be replaced by

$$\mathcal{M}^-(D^2Q) = \mathcal{M}^-(D^2u(z) - \varepsilon I) \leq 0.$$

Lemma 8.1 says that the function $\mathcal{M}^-(M)$ of symmetric matrices M is an infimum of linear functions of the form $\operatorname{trace}(AM)$. Therefore, $\mathcal{M}^-(M)$ is a concave function of M and, in particular, continuous in M. Hence, by taking $\varepsilon \to 0$ above,

$$\mathcal{M}^-(D^2u(z)) \leq 0$$

which replaces (10.5). Using again the characterization of Lemma 8.1, it is clear that if M_1 and M_2 are symmetric matrices such that $M_1 \leq M_2$ then

$$\mathcal{M}^-(M_1) \leq \mathcal{M}^-(M_2)$$

namely, $\mathcal{M}^-(M)$ is monotone increasing. Therefore, if (10.6) holds,

$$0 \geq \mathcal{M}^-(D^2u(z)) \geq \mathcal{M}^-(Ca(e_k \otimes e_k) - aI) = [\lambda(C-1) - \Lambda(n-1)]a$$

and a contraction is reached by choosing $C > \Lambda(n-1)/\lambda + 1$. Hence, (10.3) is valid and we have proved Theorem 10.2 for regular supersolutions u to the minimal Pucci extremal equation $\mathcal{M}^-(D^2u) = 0$.

To remove the regularity assumption, we regularize u by using a suitable method of regularization of viscosity solutions called inf/sup-convolution. The first two statements in the following lemma hold for any continuous function u in a bounded domain Ω. The third one is specific for viscosity supersolutions and uses in a particular way the fact that the Pucci extremal equation $\mathcal{M}^-(D^2u) = 0$ is translation invariant.

Lemma 10.4 (inf-convolutions and supersolutions). *Let Ω be a bounded domain and fix a subdomain $\Omega' \subset\subset \Omega$. Let $u \in C(\Omega)$. For $\varepsilon > 0$, define the inf-convolution u_ε of u in Ω' as*

$$u_\varepsilon(x) = \min_{z \in \overline{\Omega'}} \left\{ u(z) + \frac{1}{\varepsilon}|z - x|^2 \right\} \qquad \text{for } x \in \Omega'.$$

The following statements hold.

(1) *$u_\varepsilon \in \operatorname{Lip}(\Omega')$ and $u_\varepsilon \nearrow u$ uniformly in Ω' as $\varepsilon \to 0$.*

(2) *For every point $x_0 \in \Omega'$, u_ε can be touched from above at x_0 in Ω' by a convex quadratic polynomial of opening $2/\varepsilon$: there is an affine function $\ell(x)$ such that $\ell(x_0) = u_\varepsilon(x_0)$ and*

$$u_\varepsilon(x) \leq \frac{1}{\varepsilon}|x - x_0|^2 + \ell(x) \qquad \text{for all } x \in \Omega'.$$

(3) *If u satisfies*

$$\mathcal{M}^-(D^2u) \leq 0 \qquad \text{in the viscosity sense in } \Omega$$

and $\Omega'' \subset\subset \Omega'$ then there exists $\varepsilon_0 > 0$ depending only on u, Ω' and Ω'' such that for every $0 < \varepsilon < \varepsilon_0$ the inf-convolution u_ε satisfies

$$\mathcal{M}^-(D^2u_\varepsilon) \leq 0 \qquad \text{in the viscosity sense in } \Omega''.$$

Proof. For (1), let $x_0, x_1 \in \Omega'$. For any $x \in \overline{\Omega'}$, we have

$$\begin{aligned} u_\varepsilon(x_0) &\leq u(x) + \frac{1}{\varepsilon}|x - x_0|^2 \\ &\leq u(x) + \frac{1}{\varepsilon}|x - x_1|^2 + \frac{1}{\varepsilon}|x_1 - x_0|^2 + \frac{2}{\varepsilon}|x - x_1||x_1 - x_0| \\ &\leq u(x) + \frac{1}{\varepsilon}|x - x_1|^2 + \frac{3}{\varepsilon}\operatorname{diam}(\Omega')|x_1 - x_0|. \end{aligned}$$

By taking the infimum over $x \in \overline{\Omega'}$ on the right hand side,

$$u_\varepsilon(x_0) - u_\varepsilon(x_1) \leq \frac{3}{\varepsilon}\operatorname{diam}(\Omega')|x_1 - x_0|$$

and $u \in \operatorname{Lip}(\Omega')$.

To prove that $u_\varepsilon \nearrow u$ uniformly in Ω', note first that if $\varepsilon_1 \le \varepsilon_2$ then $1/\varepsilon_2 \le 1/\varepsilon_1$ and, by definition, $u_{\varepsilon_2} \le u_{\varepsilon_1}$. By choosing $x = x_0 = z$ in the definition of u_ε,

$$u_\varepsilon(x_0) \le u(x_0) \qquad \text{for any } x_0 \in \Omega'.$$

Second, since $u \in C(\Omega)$, for any $x_0 \in \Omega'$ there is $x_0^* \in \overline{\Omega'}$ such that the minimum in the definition of u_ε is attained:

$$u_\varepsilon(x_0) = u(x_0^*) + \frac{1}{\varepsilon}|x_0^* - x_0|^2 \ge u(x_0^*). \tag{10.7}$$

From these two facts,

$$|x_0 - x_0^*|^2 = \varepsilon\big(u_\varepsilon(x_0) - u(x_0^*)\big) \le \varepsilon\big(u(x_0) - u(x_0^*)\big) \le \varepsilon \operatorname*{osc}_{\Omega'} u \tag{10.8}$$

which shows that $x_0^* \to x_0$ as $\varepsilon \to 0$. Therefore, as $\varepsilon \to 0$,

$$0 \le \sup_{x_0 \in \Omega'} \big(u(x_0) - u_\varepsilon(x_0)\big) \le \sup_{x_0 \in \Omega'} \big(u(x_0) - u(x_0^*)\big) \to 0$$

because of (10.8) and the fact that u is uniformly continuous on $\overline{\Omega'}$.

Consider next (2). Given $x_0 \in \Omega'$, let $x_0^* \in \overline{\Omega'}$ be the point of minimum in the definition of $u_\varepsilon(x_0)$ as in (10.7). By letting $z = x_0^*$ in the definition of u_ε,

$$u_\varepsilon(x) \le \frac{1}{\varepsilon}|x - x_0^*|^2 + u(x_0^*) \qquad \text{for all } x \in \Omega'$$

with equality at $x = x_0$. It is easy to verify that $\frac{1}{\varepsilon}|x - x_0^*|^2 + u(x_0^*) = \frac{1}{\varepsilon}|x - x_0|^2 + \ell(x)$ for some affine function $\ell(x)$ such that $\ell(x_0) = u_\varepsilon(x_0)$, and (2) is proved.

Let us prove the supersolution property in (3). Let $x_0 \in \Omega''$ and let $P(x)$ be a quadratic polynomial that touches u_ε from below at x_0 in Ω''. As in (10.7), let x_0^* be the point where the minimum in the definition of $u_\varepsilon(x_0)$ is attained. We translate the polynomial P to $x_0^* - x_0$ by considering

$$Q(x) = P(x + x_0 - x_0^*) - \frac{1}{\varepsilon}|x_0 - x_0^*|^2.$$

By (10.8), we can choose $\varepsilon_0 > 0$ such that, for any $0 < \varepsilon < \varepsilon_0$, if $x_0 \in \Omega''$ then $x_0^* \in \Omega'$. Let $x \in \Omega'$ be sufficiently close to x_0^* so that $x + x_0 - x_0^* \in \Omega'$. By definition,

$$u(x) \ge u^\varepsilon(x + x_0 - x_0^*) - \frac{1}{\varepsilon}|x_0 - x_0^*|^2 \ge Q(x)$$

and

$$u(x_0^*) = u^\varepsilon(x_0) - \frac{1}{\varepsilon}|x_0 - x_0^*|^2 = P(x_0) - \frac{1}{\varepsilon}|x_0 - x_0^*|^2 = Q(x_0^*).$$

Hence, Q touches u from below at x_0^*. Since u is a supersolution,

$$\mathcal{M}^-(D^2Q(x_0^*)) = \mathcal{M}^-(D^2P(x_0)) \le 0$$

and u_ε satisfies $\mathcal{M}^-(D^2u_\varepsilon) \le 0$ in the viscosity sense in Ω''. □

Remark 10.5 (sup-convolutions and subsolutions). Geometrically speaking, u_ε is the (lower) envelope of the family of convex paraboloids P of opening $2/\varepsilon$ with vertex z such that $P(z) = u(z)$. A parallel result to Lemma 10.4 can be established for subsolutions by defining the sup-convolution

$$u^\varepsilon(x) = \max_{z\in\overline{\Omega'}} \left\{ u(z) - \frac{1}{\varepsilon}|z-x|^2 \right\} \qquad \text{for } x \in \Omega'.$$

In this case, if u satisfies $\mathcal{M}^+(D^2u) \geq 0$ in the viscosity sense in Ω then also $\mathcal{M}^+(D^2u^\varepsilon) \geq 0$ in the viscosity sense in Ω'', for all $\varepsilon > 0$ sufficiently small.

With Lemma 10.4 we conclude the proof of Theorem 10.2 for the general case.

Proof of Theorem 10.2 for general u. If u is a viscosity supersolution to $Lu = 0$ in B_1 then, by Lemma 8.4, $\mathcal{M}^-(D^2u) \leq 0$ in the viscosity sense in B_1. Let u_ε be the inf-convolution of u as in Lemma 10.4. Then u_ε can be touched from above by a convex quadratic polynomial of fixed opening at every point. Since $\mathcal{M}(D^2u_\varepsilon) \leq 0$ in the viscosity sense, by the proof for the regular case we established before the statement of Lemma 10.4,

$$|A_\varepsilon| \geq c|B|$$

where A_ε is the corresponding set of touching points for u_ε in B_1. Now,

$$\limsup_{k\to\infty} A_{1/k} = \bigcap_{m=1}^{\infty}\bigcup_{k=m}^{\infty} A_{1/k} \subset A.$$

Indeed, if $x_0 \notin A$ then there exists $\delta > 0$ such that, for any paraboloid P with vertex in the closed set B sliding from below, we have the strict separation $u(x_0) - \delta \geq P(x_0)$. Since $u_\varepsilon(x_0) \to u(x_0)$ as $\varepsilon \to 0$, we find that there is $m \geq 1$ such that $u_{1/k}(x_0) - \delta/2 \geq P(x_0)$ for all $k \geq m$ and for all such P. This means that none of the parabolids P sliding from below can touch $u_{1/k}$ at x_0 for all $k \geq m$ and, therefore, $x_0 \notin \bigcap_{m=1}^{\infty}\bigcup_{k=m}^{\infty} A_{1/k}$. Hence, $|A| \geq c|B|$. □

10.3 THE LOCALIZATION LEMMA

In this section we prove an important localization lemma. We show that if a viscosity supersolution u can be touched from below by a concave paraboloid P of opening a in a ball $B_r(x_0)$ at height at most a, then the set in which u can be touched from below by paraboloids of increased opening Ca at larger height Ca in any smaller ball, say, $B_{r/8}(x_0)$, makes up a universal proportion of $B_r(x_0)$, see Lemma 10.8.

For this, we first need a barrier to control how much u detaches from a touching paraboloid P, see Lemma 10.6.

The reader should compare these arguments with Lemma 9.4, where a barrier involving the function $|x|^{-\sigma}$ is constructed to localize the estimate.

Lemma 10.6 (Detachment from a touching paraboloid). *Fix any $0 < \eta < 1$. Let $u \in C(B_1)$ satisfy $Lu \leq 0$ in the viscosity sense in B_1. Let $B_r(x_0) \subset B_1$. Suppose that u can be touched from below in B_1 at some point $x_1 \in B_r(x_0)$ by a concave paraboloid*

P of opening $a > 0$ with vertex z_1. Then there are a constant $C_0 > 0$ depending only on n, λ, Λ and η and a point $x_2 \in \overline{B_{\eta r}(x_0)}$ such that

$$u(x_2) - P(x_2) \leq C_0 a r^2.$$

Proof. Fix any $0 < \eta < 1$. For simplicity, we let $B_r(x_0) = B_r$. The proof for any other point $x_0 \neq 0$ follows the same lines.

If the touching point x_1 is in $\overline{B_{\eta r}}$, then we can choose $x_2 = x_1$ and we trivially have $u(x_2) - P(x_2) = 0 \leq C_0 a r^2$. Therefore, we assume that $x_1 \in B_r \setminus \overline{B_{\eta r}}$.

Recall that we can write

$$P(x) = -\frac{a}{2}|x - z_1|^2 + \frac{a}{2}|x_1 - z_1|^2 + u(x_1). \tag{10.9}$$

Consider $w = u - P \geq 0$ in B_1. By ellipticity,

$$LP = a^{ij}(x)\partial_{ij}P = -a\,\mathrm{trace}(a(x)) \geq -an\Lambda.$$

Therefore, since P is smooth and u is a viscosity supersolution to $Lu = 0$, as we did in the proof of (9.6) we find that

$$Lw \leq an\Lambda \qquad \text{in the viscosity sense in } B_1. \tag{10.10}$$

We next construct a subsolution ϕ that will be used as a barrier to make $w - \phi$ a supersolution in the open ring $\eta r < |x| < r$. For this, we modify the power function $|x|^{-\sigma}$ appropriately. Since

$$\partial_{ij}|x|^{-\sigma} = \sigma|x|^{-\sigma-2}\big[(\sigma+2)|x|^{-2}x_i x_j - \delta_{ij}\big]$$

then, for any $|x| \leq r$, $x \neq 0$, by ellipticity,

$$\begin{aligned} a^{ij}(x)\partial_{ij}|x|^{-\sigma} &= \sigma|x|^{-\sigma-2}\big[(\sigma+2)|x|^{-2}a^{ij}(x)x_i x_j - \mathrm{trace}(a(x))\big] \\ &\geq \sigma r^{-\sigma-2}\big[(\sigma+2)\lambda - n\Lambda\big]. \end{aligned}$$

We can then choose $\sigma = (2n\Lambda + 1)/\lambda$ so that, for $|x| \leq r$, $x \neq 0$,

$$L(a\sigma^{-1}r^{\sigma+2}|x|^{-\sigma}) \geq a\big[(\sigma+2)\lambda - n\Lambda\big] > a(n\Lambda + 1). \tag{10.11}$$

We also make a vertical translation so that $\phi = 0$ for $|x| = r$. Hence, we define

$$\phi(x) = a\sigma^{-1}r^{\sigma+2}\big(|x|^{-\sigma} - r^{-\sigma}\big) \qquad \text{for } \eta r \leq |x| \leq r.$$

Since ϕ is smooth, by (10.10) and (10.11),

$$L(w - \phi) \leq -a < 0 \qquad \text{in the viscosity sense in } \eta r < |x| < r. \tag{10.12}$$

We look for the point x_2 in the closed ring $\eta r \leq |x| \leq r$ where the continuous function $w - \phi$ attains its minimum. Observe that the minimum is in fact negative because x_1 is in the ring, $w(x_1) = u(x_1) - P(x_1) = 0$ and $\phi(x_1) > 0$. Then the point of minimum cannot be on ∂B_r as $w - \phi = w \geq 0$ there. If x_2 is an interior point in the ring then

the constant $c = (w - \phi)(x_2)$ touches $w - \phi$ from below at x_2 and, by using that $w - \phi$ is a supersolution, see (10.12),

$$0 = Lc \le -a < 0.$$

Hence, it must be that x_2 is on the inner sphere $\partial B_{\eta r} \subset \overline{B_{\eta r}}$. Moreover,

$$0 \ge w(x_2) - \phi(x_2) = u(x_2) - P(x_2) - a\sigma^{-1}r^2(\eta^{-\sigma} - 1)..$$

The conclusion follows with $C_0 = \sigma^{-1}(\eta^{-\sigma} - 1) > 0$. □

Next we introduce a convenient notation for the set of points where not only u can be touched from below by a paraboloid P with vertex in B_1 but also the value of u is controlled in terms of the opening of P. These are in fact the sets we need to measure, see our observations after the statement of Theorem 10.1.

Definition 10.7. Let $u \in C(\overline{B_1})$ and fix $a > 0$. We denote by A_a the set of points $x \in B_1$ such that u can be touched from below at x by a (concave) paraboloid P of opening a and vertex $z \in \overline{B_1}$ and, in addition,

$$u(x) \le a.$$

As we saw after Remark 10.3, the set A_a is closed.

Lemma 10.8 (Localization lemma). *Fix $a > 0$. Let $u \in C(B_1)$ satisfy $Lu \le 0$ in the viscosity sense in B_1. Let $B_r(x_0) \subset B_1$ such that*

$$\overline{B_r(x_0)} \subset B_1 \qquad \text{and} \qquad \overline{B_r(x_0)} \cap A_a \neq \varnothing.$$

Then there exist constants $C > 1$ and $0 < c < 1$ depending only on n, λ and Λ such that

$$|A_{Ca} \cap B_{r/8}(x_0)| \ge c|B_r(x_0)|.$$

Proof. Without loss of generality, let

$$x_1 \in B_r(x_0) \cap A_a \neq \varnothing.$$

Otherwise, replace r by $r + \varepsilon$ and the result follows by letting $\varepsilon \to 0$.

Let $P(x)$ be the concave paraboloid of opening $a > 0$ and vertex $z_1 \in \overline{B_1}$ that touches u from below in B_1 at $x_1 \in B_r(x_0)$. By definition of A_a, $u(x_1) \le a$. Recall that $P(x)$ is given by (10.9). Choose $\eta = 1/16$ in the detachment from a touching paraboloid result, Lemma 10.6. Then there are a point $x_2 \in \overline{B_{r/16}(x_0)}$ and a constant $C_0 > 0$ depending only on n, λ and Λ such that

$$u(x_2) - P(x_2) \le C_0 a r^2.$$

Now, for $C' > 0$ to be determined, we slide from below the family of paraboloids

$$Q(x) = P(x) - C'\frac{a}{2}|x - z|^2 + c$$

for vertices z very close to x_2, say, $|z - x_2| < r/64$, until they touch the graph of u. Using the formula for P in (10.9), it is not difficult to see that we can write

$$Q(x) = -\frac{(C'+1)a}{2}\left|x - \frac{z_1 + C'z}{C'+1}\right|^2 + \tilde{c}$$

for some new $\tilde{c} \in \mathbb{R}$. In other words, the opening of these new touching paraboloids Q is $(C'+1)a > 0$ and the vertex is a convex combination of z_1 and z:

$$\frac{1}{C'+1}z_1 + \frac{C'}{C'+1}z \qquad \text{for } |z - x_2| < \frac{r}{64}.$$

Let B be the set of these vertices and let A denote the set of corresponding touching points. Observe that $|B|$ is of the order of r^n. We next show that the touching set A is localized around x_0 and that $u \leq Ca$ there. We then apply Theorem 10.2 with these sets A and B to conclude.

For that, we first estimate the independent term c in Q, which is a measure of how much we need to slide Q in order to touch u. Since Q is below u and $|z - x_2| < r/64$, at the point $x_2 \in \overline{B_{r/16}(x_0)}$ we have

$$\begin{aligned} c &= Q(x_2) - P(x_2) + C'\frac{a}{2}|x_2 - z|^2 \\ &\leq u(x_2) - P(x_2) + C'\frac{a}{2}|x_2 - z|^2 \\ &\leq C_0 ar^2 + C'\frac{a}{2}\left(\frac{r}{64}\right)^2. \end{aligned}$$

If $|x - x_2| \geq r/16$ and $|z - x_2| < r/64$, then $|x - z| \geq 3r/64 > r/32$. Therefore, with this and the estimate of the independent term c above, we can choose $C' > 0$ large depending only on C_0 such that, for any x satisfying $|x - x_2| \geq r/16$,

$$\begin{aligned} Q(x) &= P(x) - C'\frac{a}{2}|x - z|^2 + c \\ &\leq P(x) - C'\frac{a}{2}\left(\frac{r}{32}\right)^2 + C_0 ar^2 + C'\frac{a}{2}\left(\frac{r}{64}\right)^2 \\ &\leq P(x) - C'\frac{a}{2}\left(\frac{r}{64}\right)^2 + C_0 ar^2 \\ &< P(x) \leq u(x). \end{aligned}$$

In other words, Q is strictly below u outside of the ball $B_{r/16}(x_2)$. This shows that the set of touching points A is localized around x_0, that is, $A \subset B_{r/16}(x_2) \subset B_{r/8}(x_0)$.

Next, we need to show that $A \subset A_{(C'+1)a}$, namely, that at any $x \in A$ we have $u(x) \leq (C'+1)a$. For this, it is enough to give a uniform estimate of Q in B_1. Since P attains in maximum at $x = x_1$, see (10.9), $x_1, z_1 \in B_1$ and $r < 1$, by the estimate

on the independent term c above,

$$\begin{aligned}
Q(x) &\leq P(x) + c \\
&\leq \frac{a}{2}|x_1 - z_1|^2 + u(x_1) + C_0 a r^2 + C'\frac{a}{2}\left(\frac{r}{64}\right)^2 \\
&\leq 2a + a + C_0 a r^2 + C'\frac{a}{2}\left(\frac{r}{64}\right)^2 \\
&\leq (C'+1)a
\end{aligned}$$

for C' large enough depending only on C_0. This and the previous paragraph prove that $A \subset A_{Ca} \cap B_{r/8}(x_0)$ with $C = C' + 1 > 1$.

Thus, by Theorem 10.2,

$$\begin{aligned}
|A_{Ca} \cap B_{r/8}(x_0)| \geq |A| \geq c|B| \\
&= c\left|\left\{\frac{1}{C'+1}z_1 + \frac{C'}{C'+1}z : |z - x_2| < \frac{r}{64}\right\}\right| \\
&= c\left(\frac{C'}{C'+1}\right)^n |B_{r/64}(x_2)| \\
&= c\left(\frac{C'}{C'+1}\right)^n \left(\frac{1}{64}\right)^n |B_r(x_0)|
\end{aligned}$$

and the lemma is proved. □

Remark 10.9 (The sets A_a are increasing in a). Once the existence of $C > 1$ in Lemma 10.8 has been established, one can always take C larger. In fact, if $\tilde{a} > a$ then $A_a \subset A_{\tilde{a}}$. To see this, suppose that a paraboloid P of opening a and vertex $z \in \overline{B_1}$ touches u from below at $x_0 \in B_1$ and $u(x_0) \leq a$. We have

$$\begin{aligned}
P(x) &= -\frac{a}{2}|x - z|^2 + \frac{a}{2}|x_0 - z|^2 + u(x_0) \\
&= -\frac{a}{2}|x - x_0|^2 + a(x - x_0)\cdot(z - x_0) + u(x_0).
\end{aligned}$$

Consider the concave quadratic polynomial

$$\tilde{P}(x) = -\frac{\tilde{a}}{2}|x - x_0|^2 + a(x - x_0)\cdot(z - x_0) + u(x_0).$$

Clearly $\tilde{P}(x) \leq P(x) \leq u(x)$ and $\tilde{P}(x_0) = u(x_0)$, namely, $\tilde{P}$ touches u from below at x_0. Furthermore, there is a constant c such that

$$\tilde{P}(x) = -\frac{\tilde{a}}{2}|x - \tilde{z}|^2 + c$$

where $\tilde{z} = (a/\tilde{a})z + (1 - a/\tilde{a})x_0$, a convex combination of $z \in \overline{B_1}$ and $x_0 \in B_1$, so that $\tilde{z} \in \overline{B_1}$. As $u(x_0) \leq a \leq \tilde{a}$, we conclude that $x_0 \in A_{\tilde{a}}$.

10.4 THE COVERING LEMMA

Recall that the proof of Lemma 9.6 required a consequence of the Calderón–Zygmund cube decomposition, see Lemma 9.5. Here we also need a covering lemma.

Lemma 10.10. *Let $\{D_k\}_{k=0}^{\infty}$ be a family of closed sets and $c > 0$ such that the following properties are satisfied:*

(1) $D_0 \subset D_1 \subset D_2 \subset \cdots \subset D_k \subset \cdots \subset \overline{B_{1/3}}$, *with* $D_0 \neq \varnothing$*;*

(2) *for any $x \in B_1$ and $r > 0$ such that*

$$B_r(x) \subset B_1 \qquad B_{r/8}(x) \subset B_{1/3} \qquad \overline{B_r(x)} \cap D_k \neq \varnothing$$

we have

$$|B_{r/8}(x) \cap D_{k+1}| \geq c|B_r(x)|.$$

Then there exists $0 < \mu < 1$ depending only on c and n such that

$$|B_{1/3} \setminus D_k| \leq (1-\mu)^k |B_{1/3}|.$$

Lemma 10.10 is the parallel for balls to the Calderón–Zygmund AB lemma for cubes we used in Chapter 9, Lemma 9.5, with $\delta = 1 - \mu > 0$. To see this, let

$$A = B_{1/3} \setminus D_{k+1} \qquad \text{and} \qquad B = B_{1/3} \setminus D_k.$$

Then $A \subset B \subset B_{1/3}$ and $|A| = |B_{1/3}| - |D_{k+1}| \leq (1-\mu)|B_{1/3}|$. By contrapositive, hypothesis (2) in Lemma 10.10 says that whenever $|B_{r/8}(x) \cap D_{k+1}| < \mu |B_{r/8}(x)|$ we have $\overline{B_r(x)} \cap D_k = \varnothing$ or, equivalently, if

$$\frac{|A \cap B_{r/8}(x)|}{|B_{r/8}(x)|} \geq 1 - \mu,$$

then

$$B_r(x) \cap B_{1/3} \subset B_{1/3} \setminus D_k = B.$$

In other words, $B_{r/8}(x)$ and $B_r(x)$ play the roles of the dyadic cube Q and its predecessor $\widetilde{Q}$ in hypothesis (*b*) of Lemma 9.5, respectively. The estimate $|A| \leq (1-\mu)|B|$ is here iterated to obtain the conclusion of Lemma 10.10, see (10.14).

Proof of Lemma 10.10. We estimate $|B_{1/3} \setminus D_k|$ by means of a covering argument. For each $x \in B_{1/3} \setminus D_k$, set $r = r_x = \operatorname{dist}(x, D_k) < 2/3$ and consider the open cover of $B_{1/3} \setminus D_k$ by the family of balls $B_r(x)$, $r = r_x$. Then the Vitali covering lemma says that we can subtract an at most countably infinite family of balls $\{B_{r_i}(x_i)\}_{i=1}^{\infty}$ that cover $B_{1/3} \setminus D_k$ and such that the family $\{B_{r_i/3}(x_i)\}_{i=1}^{\infty}$ is disjoint. We have

$$|B_{1/3} \setminus D_k| \leq \left| \bigcup_{i=1}^{\infty} B_{r_i}(x_i) \cap B_{1/3} \right| \leq \sum_{i=1}^{\infty} |B_{r_i}(x_i) \cap B_{1/3}|.$$

We relate the measure of $B_{r_i}(x_i) \cap B_{1/3}$ back to the measure of $B_{r_i/3}(x_i) \cap D_{k+1}$ using the hypothesis. We claim that if $x_0 \in B_{1/3}$ and $r = r_{x_0} = \operatorname{dist}(x_0, D_k)$ then

$$|B_r(x_0) \cap B_{1/3}| \leq \frac{1}{\mu}|B_{r/3}(x_0) \cap D_{k+1}| \tag{10.13}$$

where $\mu > 0$ depends on c and n. If this claim is true, we can estimate

$$\begin{aligned}
|B_{1/3} \setminus D_k| &\leq \sum_{i=1}^{\infty} |B_{r_i}(x_i) \cap B_{1/3}| \\
&\leq \frac{1}{\mu} \sum_{i=1}^{\infty} |B_{r_i/3}(x_i) \cap D_{k+1}| \\
&= \frac{1}{\mu} \sum_{i=1}^{\infty} |B_{r_i/3}(x_i) \cap (D_{k+1} \setminus D_k)| \\
&= \frac{1}{\mu} \left| \bigcup_{i=1}^{\infty} B_{r_i/3}(x_i) \cap (D_{k+1} \setminus D_k) \right| \\
&\leq \frac{1}{\mu} |D_{k+1} \setminus D_k|.
\end{aligned}$$

With this,

$$|B_{1/3} \setminus D_{k+1}| \leq |B_{1/3} \setminus D_k| - |D_{k+1} \setminus D_k| \leq (1 - \mu)|B_{1/3} \setminus D_k|$$

and we iterate to obtain the conclusion:

$$\begin{aligned}
|B_{1/3} \setminus D_k| &\leq (1 - \mu)|B_{1/3} \setminus D_{k-1}| \\
&\leq (1 - \mu)^2 |B_{1/3} \setminus D_{k-2}| \\
&\leq (1 - \mu)^k |B_{1/3} \setminus D_0| \\
&\leq (1 - \mu)^k |B_{1/3}|.
\end{aligned} \tag{10.14}$$

We are left to prove the claim (10.13). Observe that we cannot directly apply hypothesis (2) to $B_r(x_0)$ because we cannot ensure that $B_r(x_0) \subset B_1$ nor $B_{r/8}(x_0) \subset B_{1/3}$. Therefore, we consider a related, shifted ball centered at a point closer to the origin. For this, define

$$x_1 = x_0 - \frac{r}{6} \frac{x_0}{|x_0|} = \left(1 - \frac{r}{6|x_0|}\right) x_0.$$

We first check that

$$B_{r/6}(x_1) \subset B_{r/3}(x_0) \cap B_{1/3}. \tag{10.15}$$

Indeed, for any $z \in B_{r/6}(x_1)$, we have $|z - x_0| \leq |z - x_1| + |x_1 - x_0| \leq r/6 + r/6 = r/3$ and $|z| \leq |z - x_1| + |x_1| \leq r/6 + (1 - r/(6|x_0|))|x_0| = |x_0| < 1/3$. Next, note that

$$\operatorname{dist}(x_1, D_k) \leq \operatorname{dist}(x_0, D_k) + |x_1 - x_0| \leq r + \frac{r}{6} = \frac{7}{6} r =: \tilde{r}.$$

We thus consider the ball $B_{\tilde{r}}(x_1)$ and verify that it satisfies the hypotheses of (2).

- $B_{\tilde{r}}(x_1) \subset B_1$. If $z \in B_{\tilde{r}}(x_1)$ then $|z| \leq |z - x_1| + |x_1| < \tilde{r} + (1 - r/(6|x_0|))|x_0| = |x_0| + r < 1/3 + 2/3 = 1$.
- $B_{\tilde{r}/8}(x_1) \subset B_{1/3}$. If $z \in B_{\tilde{r}/8}(x_1)$ then, as above, $|z| < \tilde{r}/8 + (1 - r/(6|x_0|))|x_0| = 7r/48 + |x_0| - r/6 < |x_0| < 1/3$.
- $\overline{B_{\tilde{r}}(x_1)} \cap D_k \neq \varnothing$. This follows because $\operatorname{dist}(x_1, D_k) \leq \tilde{r}$ and D_k is closed.

Hence, since (10.15) holds and $r/6 > \frac{7}{8}(r/6) = \tilde{r}/8$, by the conclusion in (2),

$$\begin{aligned}
|B_{r/3}(x_0) \cap D_{k+1}| &\geq |B_{r/6}(x_1) \cap D_{k+1}| \\
&\geq |B_{\tilde{r}/8}(x_1) \cap D_{k+1}| \\
&\geq c|B_{\tilde{r}}(x_1)| \\
&= cc_n r^n \\
&\geq \mu |B_{1/3} \cap B_r(x_0)|
\end{aligned}$$

and (10.13) is proved. □

10.5 CONCLUSION: PROOF OF THE HARNACK INEQUALITY

We proceed with the proof of Theorem 10.1. Let $u \in C(\overline{B_1})$ be a nonnegative viscosity solution to $Lu = 0$ in B_1. We prove that

$$u \leq C_H u(0) \qquad \text{in } B_{1/6}.$$

The claim in Theorem 10.1 follows by the usual translation and rescaling argument. Assume that

$$u(0) = \frac{a}{20} > 0.$$

Then we must prove that

$$u(x) \leq aC_1 \qquad \text{for all } x \in B_{1/6}.$$

We begin by sliding a concave paraboloid P of opening a and vertex 0 from below until it touches the graph of u for the first time at $x_0 \in B_1$. Then the touching point x_0 is not far from the origin and we have control on $u(x_0)$. Indeed, recall that

$$P(x) = -\frac{a}{2}|x|^2 + \frac{a}{2}|x_0|^2 + u(x_0).$$

If $|x_0| > 1/3$ then, as $u \geq 0$,

$$\frac{a}{20} = u(0) \geq P(0) = \frac{a}{2}|x_0|^2 + u(x_0) \geq \frac{a}{18}$$

which is impossible. Therefore, the contact point x_0 must be in $\overline{B_{1/3}}$. Furthermore, since the vertex of P is 0,

$$u(x_0) = P(x_0) \leq P(0) \leq u(0) \leq a.$$

By recalling Definition 10.7, we have proved that

$$D_0 = A_a \cap \overline{B_{1/3}} \neq \varnothing.$$

Next, we increase the opening of the touching paraboloids, that is, we consider the closed sets

$$D_k = A_{aC^k} \cap \overline{B_{1/3}} \qquad \text{for } k \geq 0 \tag{10.16}$$

where $C > 0$ is the constant from the localization lemma, Lemma 10.8. Observe that, by Remark 10.9, we can always make C larger, so we can assume that

$$C > 4.$$

To estimate the measure of (the complement) of the sets D_k, we apply the covering lemma, Lemma 10.10. We already know that $D_0 \neq \varnothing$. By Remark 10.9,

$$D_0 \subset D_1 \subset D_2 \subset \cdots \subset D_k \subset \cdots \subset \overline{B_{1/3}}$$

and hypothesis (1) of Lemma 10.10 is satisfied. To check that hypothesis (2) holds, consider a ball $B_r(x) \subset B_1$ such that $B_{r/8}(x) \subset B_{1/3}$ and $\overline{B_r(x)} \cap D_k \neq \varnothing$. Then, by the localization lemma, Lemma 10.8, with aC^k in place of a,

$$|D_{k+1} \cap B_{r/8}(x)| = |A_{aC^{k+1}} \cap B_{r/8}(x)| \geq c|B_r(x)|$$

for some $c > 0$ depending only on n, λ and Λ. By Lemma 10.10, there exists $0 < \mu < 1$ depending only on c and n such that

$$|B_{1/3} \setminus D_k| \leq (1-\mu)^k |B_{1/3}|. \tag{10.17}$$

Also, from the definition of the set A_{aC^k},

$$u(x) \leq aC^k \qquad \text{for } x \in D_k.$$

At this point, Step 1 described in Section 10.1 is completed: we have shown that the set D_k where u is comparable to $u(0) \sim a$, which here means $u \leq aC^k$, almost covers $B_{1/3}$ in measure, namely, its measure is bounded below by $(1-(1-\mu)^k)|B_{1/3}|$.

The next task for Step 2 is to prove that if u is very large at some point in $B_{1/6}$ as compared to $u(0) \sim a$, that is, $u \geq aC^{k+1}$ for some $k \geq k_0$, where $k_0 \geq 1$ depends on n, λ and Λ, then u becomes even larger at a nearby point: $u \geq aC^{k+2}$. Iterating this claim, a sequence of points in $\overline{B_{1/4}}$ is found where u tends to infinity, contradicting the boundedness of u. Again, the reader should compare this strategy with the one implemented in the proof of Lemma 9.7.

We proceed to assume that for some $k \geq k_0$ there exists a point $x_k \in B_{1/6}$ such that

$$u(x_k) \geq aC^{k+1}$$

and show that there is a point $x_{k+1} \in \partial B_{\rho_k}(x_k)$ such that

$$u(x_{k+1}) \geq aC^{k+2}. \tag{10.18}$$

The integer $k_0 \geq 1$ and the radius $\rho_k < 1$ will depend on n, λ and Λ and will be found in the process.

Suppose, by way of contradiction, that (10.18) does not hold, namely, that

$$u < aC^{k+2} \qquad \text{on } \partial B_{\rho_k}(x_k).$$

Over the ball $B_k = \{x : |x - x_k| < \rho_k\}$, we slide convex paraboloids with large opening and vertex close to x_k of the form

$$P(x) = \frac{2aC^{k+2}}{\rho_k^2}|x - z|^2 + c \qquad \text{for } |z - x_k| \leq \rho_k/C$$

from above until they touch the graph of u for the first time in B_k. Let A denote the set of contact points. We now show that the contact points $x_0 \in A$ are in the interior of B_k and that $u(x_0) \geq 2aC^k$.

For this, fix a point $x_0 \in A$ and a corresponding paraboloid P that touches u from above in B_k at x_0. If necessary, slide P further down until it intersects u at x_k and let us denote this paraboloid by $\tilde{P}$. We can write $\tilde{P}$ and estimate it from below in B_k by using that its vertex is close to x_k, $|z - x_k| \leq \rho_k/C$, the hypothesis that $u(x_k) \geq aC^{k+1}$ and the fact that $C > 4$:

$$\begin{aligned}
\tilde{P}(x) &= \frac{2aC^{k+2}}{\rho_k^2}|x - z|^2 - \frac{2aC^{k+2}}{\rho_k^2}|x_k - z|^2 + u(x_k) \\
&\geq \frac{2aC^{k+2}}{\rho_k^2}|x - z|^2 - 2aC^k + aC^{k+1} \\
&\geq \frac{2aC^{k+2}}{\rho_k^2}|x - z|^2 + 2aC^k.
\end{aligned}$$

Therefore,

$$u(x_0) = P(x_0) \geq \tilde{P}(x_0) \geq 2aC^k$$

which shows that

$$A \subset \{x_0 \in \overline{B_k} : u(x_0) \geq 2aC^k\}.$$

Let us see that, actually, x_0 is in the interior of B_k. Indeed, if $|x_0 - x_k| = \rho_k$ then, by the estimate on $\tilde{P}$ above,

$$\begin{aligned}
u(x_0) = P(x_0) &\geq \tilde{P}(x_0) \\
&\geq \frac{2aC^{k+2}}{\rho_k^2}|x_0 - z|^2 + 2aC^k \\
&\geq \frac{2aC^{k+2}}{\rho_k^2}(|x_0 - x_k| - |x_k - z|)^2 + 2aC^k \\
&\geq \frac{2aC^{k+2}}{\rho_k^2}(\rho_k - \rho_k/C)^2 + 2aC^k \\
&= 2aC^k(C - 1)^2 + 2aC^k \\
&> aC^{k+2}
\end{aligned}$$

contradicting our assumption that $u < aC^{k+2}$ on $\partial B_{\rho_k}(x_k)$. Consequently,

$$A \subset \{x_0 \in B_k : u(x_0) \geq 2aC^k\} \subset B_k \tag{10.19}$$

By the point-to-measure estimate for subsolutions, see Remark 10.3,

$$|A| \geq c|\{z : |z - x_k| \leq \rho_k/C\}| = c\omega_n \frac{1}{C^n}\rho_k^n \geq c_0\rho_k^n$$

where $c_0 > 0$ is small and depends only on n, λ and Λ. Now we choose ρ_k as

$$\rho_k = \left(\frac{2(1-\mu)^k}{c_0}|B_{1/3}|\right)^{1/n}$$

so that we can rewrite the estimate above as

$$|A| \geq 2(1-\mu)^k|B_{1/3}|. \tag{10.20}$$

Next we fix k_0 to make sure that all the contact points in A are not far from the origin. For any contact point $x_0 \in A$,

$$|x_0| \leq |x_0 - x_k| + |x_k| \leq \rho_k + 1/6.$$

Therefore, if k_0 is sufficiently large depending only on n, λ and Λ and such that

$$\sum_{k=k_0}^{\infty} \rho_k < 1/4 - 1/6, \tag{10.21}$$

then all the contact points x_0 satisfy $|x_0| < 1/4$ and we get

$$A = A \cap B_{1/3}. \tag{10.22}$$

We are ready to deduce the contradiction. By the definition of D_k in (10.16),

$$\{u > aC^k\} \cap B_{1/3} \subset B_{1/3} \setminus D_k.$$

With this, (10.17), (10.20), (10.22) and (10.19),

$$\begin{aligned}
|\{u > aC^k\} \cap B_{1/3}| &\leq |B_{1/3} \setminus D_k| \\
&\leq (1-\mu)^k|B_{1/3}| \\
&\leq \frac{1}{2}|A| \\
&= \frac{1}{2}|A \cap B_{1/3}| \\
&\leq \frac{1}{2}|\{u \geq 2aC^k\} \cap B_{1/3}| \\
&\leq \frac{1}{2}|\{u > aC^k\} \cap B_{1/3}|
\end{aligned}$$

which is a contradiction. This completes the proof of (10.18).

We finally use (10.18) and (10.21) to finish the proof of Theorem 10.1. Suppose that there is a point $x_{k_0} \in B_{1/6}$ such that

$$u(x_{k_0}) > aC^{k_0+1}.$$

Then there is a point $x_{k_0+1} \in \partial B_{\rho_{k_0}}(x_{k_0})$ such that

$$u(x_{k_0+1}) > aC^{k_0+2}.$$

Repeating this process, we can find a sequence of points $x_{k+1} \in \partial B_{\rho_k}(x_k)$ such that

$$u(x_{k+1}) > aC^{k+2} \qquad \text{for } k > k_0.$$

By the choice of k_0 in (10.21), the points x_{k+1} do not escape out of $B_{1/4}$:

$$|x_{k+1}| \leq |x_{k_0}| + \sum_{k=k_0}^{\infty} |x_k - x_{k+1}| \leq 1/6 + \sum_{k=k_0}^{\infty} \rho_k < 1/4.$$

In particular, u is unbounded in $\overline{B_{1/4}}$, which is a contradiction to the continuity of u in B_1. Therefore, it must be that

$$u \leq aC^{k_0+1} \qquad \text{in } B_{1/6}$$

and this completes the proof of Theorem 10.1.

Remark 10.11 (Method of sliding paraboloids with nonzero right hand side). Savin's method of sliding paraboloids can be easily modified to prove the Harnack inequality for nonnegative viscosity solutions u to $Lu = f$ in B_1 whenever f is continuous and bounded and, in this case, it takes the form

$$\sup_{B_{1/2}} u \leq C_H\Big(u(0) + \|f\|_{L^\infty(B_1)}\Big).$$

Indeed, by considering $u/(\|f\|_{L^\infty(B_1)}/a)$ for some $a > 0$, we can always assume that $\|f\|_{L^\infty(B_1)} \leq a$. In the proof of the point-to-measure estimate, Theorem 10.2, when u is regular, equations (10.4) and (10.5) must be replaced by $LQ \leq a$ and $\operatorname{trace}(a(z)D^2u(z)) \leq a$, respectively. The contradiction there follows by choosing C even larger, independent of a. To continue with the general case, the minimal Pucci extremal equation now becomes $\mathcal{M}^-(D^2u) \leq a$. It is easy to see that, in this case, the conclusion of Lemma 10.4(3) is that $\mathcal{M}^-(D^2u_\varepsilon) \leq a$ in the viscosity sense in Ω''. Similar simple modifications must be done in the proof of the detachment from a paraboloid lemma, Lemma 10.6. Observe that the proof of the localization lemma, Lemma 10.8, only uses the statement of Lemma 10.6. Finally, the proof we presented in this section remains unchanged under the assumption that $\|f\|_{L^\infty(B_1)} \leq a$.

NOTES ON CHAPTER 10

The method of sliding paraboloids for viscosity solutions to fully nonlinear elliptic equations was developed by Savin in [58], where other applications are also shown.

Our presentation follows Savin [58] and Stinga–Vaughan [71].

For properties of concave functions see, for instance, Evans–Gariepi [33]. The useful method of inf- and sup- convolutions for viscosity solutions was first introduced by Jensen in [45], see also Caffarelli–Cabré [7] for other important applications to regularity estimates. The Vitali covering lemma can be found in Evans–Gariepi [33] or Stein [63].

CHAPTER 11

Perturbation theory for nondivergence form equations

In this chapter we prove the Schauder estimates for viscosity solutions to nondivergence form elliptic equations with continuous coefficients. For this, we implement Caffarelli's method. In addition, we show the Calderón–Zygmund estimates for strong solutions, emphasizing the role that *a priori* estimates play in the theory. The Calderón–Zygmund estimates will be proven using the classical technique of *freezing the coefficients*.

11.1 SCHAUDER ESTIMATES

We consider viscosity solutions u to nondivergence form equations

$$Lu = \operatorname{trace}(a(x)D^2u) = a^{ij}(x)\partial_{ij}u = f \qquad \text{in } B_1 \subset \mathbb{R}^n$$

for $n \geq 2$. The matrix of coefficients $a(x) = (a^{ij}(x))_{i,j=1}^n$ is defined in B_1, symmetric $a^{ij}(x) = a^{ji}(x)$ and uniformly elliptic: there are constants $0 < \lambda \leq \Lambda$ such that $\lambda|\xi|^2 \leq a^{ij}(x)\xi_i\xi_j \leq \Lambda|\xi|^2$, for all $\xi \in \mathbb{R}^n$ and every $x \in B_1$. We assume that $a(x) \in C(B_1)$ and that $f \in C(B_1) \cap L^\infty(B_1)$.

Our aim is to show the Schauder estimate that says that if $a(x), f \in C^{0,\alpha}(B_1)$ for some $0 < \alpha < 1$, then $u \in C^{2,\alpha}(B_{1/2})$.

It is obvious that if $g \in C^\alpha(0)$, then

$$[g]_{C_n^\alpha(0)} := \sup_{0<r<1} \frac{1}{r^\alpha}\left(\frac{1}{|B_r|}\int_{B_r} |g - g(0)|^n \, dx\right)^{1/n} \leq [g]_{C^\alpha(0)}.$$

We then say that a function $g : B_1 \to \mathbb{R}$ is **Hölder continuous of order α at 0 in the L^n sense** if $[g]_{C_n^\alpha(0)} < \infty$.

As we will see in a moment, it is enough to prove the following pointwise estimate at, say, the origin.

DOI: 10.1201/9781032679853-11

Theorem 11.1 (Schauder estimates for viscosity solutions). *Fix $0 < \alpha < 1$. Let $u \in C(B_1)$ be a viscosity solution to $Lu = f$ in B_1. Suppose that $a(x)$ and f are Hölder continuous of order α at 0 in the L^n sense. Then $u \in C^{2,\alpha}(0)$, that is, there exists a polynomial P_∞ of degree 2 such that*

$$\|u - P_\infty\|_{L^\infty(B_r)} \leq C_* r^{2+\alpha} \qquad \text{for any } 0 < r < 1$$

and

$$|P_\infty(0)| + |\nabla P_\infty(0)| + |D^2 P_\infty(0)| \leq C_*$$

where

$$C_* \leq C(\|u\|_{L^\infty(B_1)} + |f(0)| + [f]_{C_n^\alpha(0)})$$

and $C > 0$ is a constant depending only on n, λ, Λ, α and $[a(x)]_{C_n^\alpha(0)}$. In particular,

$$\|u\|_{C^{2,\alpha}(0)} \leq C(\|u\|_{L^\infty(B_1)} + |f(0)| + [f]_{C_n^\alpha(0)}).$$

The origin in Theorem 11.1 was chosen for simplicity. In fact, the proof can be done at any point $x \in B_{1/2}$ to conclude that if $a(x), f \in C^{0,\alpha}(B_1)$ then

$$\|u\|_{C^{2,\alpha}(B_{1/2})} \leq C(\|u\|_{L^\infty(B_1)} + \|f\|_{C^{0,\alpha}(B_1)})$$

where $C > 0$ depends only on n, λ, Λ, α and $[a(x)]_{C^\alpha(B_1)}$. The pointwise estimate of Theorem 11.1 is sufficient for our purposes because of the following characterization due to Campanato.

Theorem 11.2 (Campanato's theorem). *Let u be a measurable function defined on a Lipschitz domain Ω and fix $0 < \alpha < 1$. Then $u \in C^{2,\alpha}(\overline{\Omega})$ if and only if the following holds. There exists a constant $D > 0$ such that for every $z \in \overline{\Omega}$ there exists a polynomial $P_z(x)$ of degree 2 such that*

$$|u(x) - P_z(x)| \leq D|x - z|^{2+\alpha} \qquad \text{for every } x \in B_1(z) \cap \Omega.$$

In this case, the norm $\|u\|_{C^{2,\alpha}(\Omega)}$ is equivalent to

$$\sup\left\{|P_z| + \frac{|u(x) - P_z(x)|}{|x - z|^{2+\alpha}} : x \in B_1(z) \cap \Omega, x \neq z, z \in \overline{\Omega}\right\}$$

where $|P_z|$ denotes the sum of the modulus of the coefficients of the polynomial P_z.

The regularity of the coefficients and the right hand side in Theorem 11.1 are measured with respect to the L^n norm. This is because the Krylov–Safonov Harnack inequality involves the L^n norm of f, see Remark 9.8. In fact, if we were to use the Harnack inequality with nonzero right hand side proved with Savin's method of sliding paraboloids, we will have to measure the regularity of $a(x)$ and f at the origin with respect to the L^∞ norm, see Remark 10.11.

At this point, the reader should recall the proofs of the Schauder estimates for the Laplacian by the method of maximum principle (Section 3.7) and for divergence form equations by the method of compactness (Section 7.1). In particular, for the

case of this section, the maximum principle for the Poisson equation of Theorem 3.14 and the Caccippoli inequality of Lemma 7.4 are naturally replaced by the maximum principle of Corollary 8.6, which is a consequence of the ABP estimate, Theorem 8.5.

As in the above-mentioned other cases, the idea of the proof of Theorem 11.1 is to transfer the regularity from harmonic functions by reducing to the case when $a(x) \sim I$, the identity matrix, and $f \sim 0$. To implement this, we need a basic Hölder global regularity result for viscosity solutions to $Lu = f$. Let us begin by stating the following consequence of the Harnack inequality with nonzero right hand side, see Remark 9.8.

Theorem 11.3 (Interior Hölder regularity with right hand side). *Let $u \in C(B_1)$ be a viscosity solution to $Lu = f$ in B_1. Suppose that $f \in C(B_1) \cap L^\infty(B_1)$. Then there exist $C > 0$ and $0 < \bar{\alpha} < 1$ depending only on n, λ and Λ such that*

$$\|u\|_{C^{\bar{\alpha}}(\overline{B_{1/2}})} \leq C(\|u\|_{L^\infty(B_1)} + \|f\|_{L^n(B_1)})$$

and, by scaling, for any ball $B_r \subset B_1$,

$$\|u\|_{C^{\bar{\alpha}}(\overline{B_{r/2}})} \leq \frac{C}{r^{\bar{\alpha}}}(\|u\|_{L^\infty(B_r)} + r\|f\|_{L^n(B_r)}). \tag{11.1}$$

Combining Theorem 11.3 and a barrier argument we can obtain the following boundary regularity result. The proof is in two steps. First, with a barrier, the regularity of u up to the boundary is proved, see (11.2). Second, the interior and boundary estimates are *patched together*, see Cases I and II in the following proof.

Theorem 11.4 (Boundary regularity). *Let $u \in C(\overline{B_1})$ be a viscosity solution to*

$$\begin{cases} Lu = 0 & \text{in } B_1 \\ u = \varphi & \text{on } \partial B_1 \end{cases}$$

where

$$\varphi \in C^\beta(\partial B_1)$$

for some $0 < \beta < 1$. Let $\bar{\alpha} > 0$ be the Hölder exponent from (11.1). Then $u \in C^\tau(\overline{B_1})$ for $\tau = \min\{\beta/2, \bar{\alpha}\} \in (0,1)$ with

$$\|u\|_{L^\infty(B_1)} \leq \|\varphi\|_{L^\infty(\partial B_1)}$$

and

$$[u]_{C^\tau(\overline{B_1})} \leq C[\varphi]_{C^\beta(\partial B_1)}$$

where $C > 0$ is a constant depending only on n, λ, Λ and τ.

Proof. By the maximum principle, Corollary 8.6, $\|u\|_{L^\infty(B_1)} \leq \|\varphi\|_{L^\infty(\partial B_1)}$.

Let us prove that, for any $x_0 \in \partial B_1$, $u \in C^{\beta/2}(x_0)$ with

$$\sup_{x \in \overline{B_1}} \frac{|u(x) - u(x_0)|}{|x - x_0|^{\beta/2}} \leq 2^{\beta/2} \sup_{x \in \partial B_1} \frac{|\varphi(x) - \varphi(x_0)|}{|x - x_0|^\beta}. \tag{11.2}$$

After a translation, a rotation and by subtracting a constant, we can assume, without loss of generality, that $B_1 = B_1(e_n)$, $x_0 = 0$ and $\varphi(x_0) = \varphi(0) = u(0) = 0$. Let

$$K = \sup_{x \in \partial B_1} \frac{|\varphi(x) - \varphi(x_0)|}{|x - x_0|^\beta} = \sup_{x \in \partial B_1} \frac{|\varphi(x)|}{|x|^\beta}.$$

Notice that if $x \in \partial B_1(e_n)$ then

$$1 = |x - e_n|^2 = x_1^2 + \cdots + x_{n-1}^2 + (x_n - 1)^2 = |x|^2 - 2x_n + 1$$

so that $|x|^2 = 2x_n$. Therefore, if $x \in \partial B_1(e_n)$ then

$$u(x) = \varphi(x) \le K|x|^\beta = 2^{\beta/2} K x_n^{\beta/2}.$$

Let us define $h(x) = 2^{\beta/2} K x_n^{\beta/2}$ for $x \in B_1(e_n)$. Since $\beta/2 - 1 < 0$, by ellipticity,

$$\begin{aligned} Lh(x) &= \operatorname{trace}(a(x)D^2 h(x)) \\ &= 2^{\beta/2} K a_{nn}(x) \frac{\beta}{2}\left(\frac{\beta}{2} - 1\right) x_n^{\beta/2-2} \\ &\le 2^{\beta/2} K \lambda \frac{\beta}{2}\left(\frac{\beta}{2} - 1\right) x_n^{\beta/2-2} < 0 \end{aligned}$$

for any $x \in B_1(e_n)$. Then h is a classical strict supersolution to $Lh = 0$ in $B_1(e_n)$ such that $h \ge u$ on $\partial B_1(e_n)$. It follows that $h - u$ satisfies $L(h - u) < 0$ in the viscosity sense in $B_1(e_n)$ and $h - u \ge 0$ on $\partial B_1(e_n)$. Therefore, by the maximum principle for viscosity supersolutions in Corollary 8.6, $h - u \ge 0$ in $B_1(e_n)$ and so we get

$$u(x) \le h(x) = 2^{\beta/2} K x_n^{\beta/2} \le 2^{\beta/2} K |x|^{\beta/2} \qquad \text{for any } x \in B_1(e_n).$$

By applying this result to $-u$,

$$|u(x)| \le 2^{\beta/2} K |x|^{\beta/2} \qquad \text{for any } x \in B_1(e_n)$$

and this is (11.2).

To obtain the conclusion, we now estimate

$$\frac{|u(x) - u(z)|}{|x - z|^\tau} \qquad \text{for any } x, z \in B_1.$$

Fix any $x, z \in B_1$ and let $x_0, z_0 \in \partial B_1$ such that

$$d_x = \operatorname{dist}(x, \partial B_1) = |x - x_0|$$

and

$$d_z = \operatorname{dist}(z, \partial B_1) = |z - z_0|.$$

Without loss of generality, $d_z \le d_x$. There are two cases.

Case I. Suppose that $|x-z| > d_x/2 \geq d_z/2$. This means that x and z are far apart so we should use the boundary estimate (11.2). Observe that $|x_0-z_0| \leq d_x+|x-z|+d_z \leq 5|x-z|$. By using (11.2), since $|x-z|/2 \leq 1$ and $\tau \leq \beta/2$,

$$\begin{aligned}|u(x)-u(z)| &\leq |u(x)-u(x_0)| + |u(x_0)-u(z_0)| + |u(z_0)-u(z)| \\ &\leq C(d_x^{\beta/2} + |x_0-z_0|^{\beta/2} + d_z^{\beta/2})[\varphi]_{C^\beta(\partial B_1)} \\ &\leq C|x-z|^\tau [\varphi]_{C^\beta(\partial B_1)}.\end{aligned}$$

Case II. Suppose that $|x-z| \leq d_x/2$. In this case x and z are close to each other, so we will use the interior estimate (11.1). To correct the term $r^{-\bar{\alpha}}$ in (11.1) we need to use a boundary point x_0 and apply (11.2). We have $z \in \overline{B_{d_x/2}(x)} \subset B_{d_x}(x) \subset B_1$. We apply (11.1) to the solution $u-u(x_0)$ in $B_{d_x}(x)$, then (11.2) and finally use that $\tau \leq \bar{\alpha}, \beta/2$ and $d_x \leq 1$ to obtain

$$\begin{aligned}\frac{|u(x)-u(z)|}{|x-z|^\tau} &= \frac{|(u(x)-u(x_0))-(u(z)-u(x_0))|}{|x-z|^\tau} \\ &\leq \frac{C}{d_x^{\bar{\alpha}}}\|u-u(x_0)\|_{L^\infty(B_{d_x}(x))}|x-z|^{\bar{\alpha}-\tau} \\ &\leq \frac{C}{d_x^\tau}\|u-u(x_0)\|_{L^\infty(B_{d_x}(x))} \\ &\leq C d_x^{\beta/2-\tau}[\varphi]_{C^\beta(\partial B_1)} \\ &\leq C[\varphi]_{C^\beta(\partial B_1)}.\end{aligned}$$

□

The boundary regularity estimate of Theorem 11.4 is key in proving the following approximation lemma that establishes that if $|u| \leq 1$, the coefficients $a(x)$ are close to the identity matrix I and f is small in L^n then u is close to a harmonic function.

Lemma 11.5 (Approximation lemma). *Let $u \in C(B_1)$ be a viscosity solution to*

$$Lu = f \qquad \text{in } B_1$$

where $f \in C(B_1) \cap L^\infty(B_1)$, with

$$|u| \leq 1 \qquad \text{in } B_1.$$

Fix any $0 < \varepsilon < 1/16$. If

$$\|a(x) - I\|_{L^n(B_{3/4})} < \varepsilon$$

and $h \in C(\overline{B_{3/4}}) \cap C^\infty(B_{3/4})$ is the unique classical solution to

$$\begin{cases}\Delta h = 0 & \text{in } B_{3/4} \\ h = u & \text{on } \partial B_{3/4}\end{cases}$$

then

$$\|u-h\|_{L^\infty(B_{1/2})} \leq C(\varepsilon^\gamma + \|f\|_{L^n(B_1)})$$

where $C > 0$ and $0 < \gamma < 1$ are constants depending only on n, λ and Λ.

Proof. We already know that h exists, is unique and, by the maximum principle,

$$|h| \le 1 \qquad \text{in } B_{3/4}$$

see Chapter 2. We look at the equation for $u - h$. Since h is smooth,

$$L(u-h) = f - Lh = f - (L-\Delta)h = f - (a^{ij}(x) - \delta_{ij})\partial_{ij}h$$

in the viscosity sense in $B_{3/4}$. By the maximum principle in Corollary 8.6, for any $\delta > 0$, there exists a constant $C > 0$ depending on n, λ, Λ and δ such that

$$\begin{aligned} &\|u-h\|_{L^\infty(B_{3/4-\delta})} \\ &\quad \le \|u-h\|_{L^\infty(\partial B_{3/4-\delta})} + C\|f - (a^{ij}(x) - \delta_{ij})\partial_{ij}h\|_{L^n(B_{3/4-\delta})} \\ &\quad \le \|u-h\|_{L^\infty(\partial B_{3/4-\delta})} + C\|f\|_{L^n(B_1)} + C\varepsilon\|D^2h\|_{L^\infty(B_{3/4-\delta})} \end{aligned} \tag{11.3}$$

where in the last line the hypothesis on $a(x)$ was used.

We need to control the terms

$$\|u-h\|_{L^\infty(\partial B_{3/4-\delta})} \qquad \text{and} \qquad \|D^2h\|_{L^\infty(B_{3/4-\delta})}.$$

Before we continue, notice that we could have taken $\delta = 0$, so that the boundary term $\|u-h\|_{L^\infty(\partial B_{3/4})} = 0$. The problem now is that we would need to control $\|D^2h\|_{L^\infty(B_{3/4})}$, namely, the behavior of the second derivatives of h near the boundary of $B_{3/4}$. However, we only know that $h = u$ on $\partial B_{3/4}$ and that u is at most Hölder continuous, so D^2h blows up near $\partial B_{3/4}$.

Control on $\|u-h\|_{L^\infty(\partial B_{3/4-\delta})}$. From (11.1),

$$\|u\|_{C^{\bar\alpha}(\overline{B_{3/4}})} \le C(\|u\|_{L^\infty(B_1)} + \|f\|_{L^n(B_1)}) \le C(1 + \|f\|_{L^n(B_1)})$$

where $C > 0$ and $0 < \bar\alpha < 1$ depend only on n, λ and Λ. In particular, $u \in C^{\bar\alpha}(\partial B_{3/4})$. By Theorem 11.4 applied to the harmonic function h, we get that $h \in C^{\bar\alpha/2}(\overline{B_{3/4}})$ with the estimate

$$[h]_{C^{\bar\alpha/2}(\overline{B_{3/4}})} \le C[u]_{C^{\bar\alpha}(\partial B_{3/4})} \le C(1 + \|f\|_{L^n(B_1)}). \tag{11.4}$$

Now, both u and h are in $C^{\bar\alpha/2}(\overline{B_{3/4}})$ and $u - h = 0$ on $\partial B_{3/4}$. This implies that, for any $0 < \delta < 1/4$,

$$\|u-h\|_{L^\infty(\partial B_{3/4-\delta})} \le [u-h]_{C^{\bar\alpha/2}(\overline{B_{3/4}})}\delta^{\bar\alpha/2} \le C\delta^{\bar\alpha/2}(1 + \|f\|_{L^n(B_1)}).$$

Control on $\|D^2h\|_{L^\infty(B_{3/4-\delta})}$. We claim that

$$\|D^2h\|_{L^\infty(B_{3/4-\delta})} \le c\delta^{\bar\alpha/2-2}(1 + \|f\|_{L^n(B_1)})$$

where $c > 0$ depends only on n. Indeed, given any $x_0 \in B_{3/4-\delta}$, consider any point $x_1 \in \partial B_\delta(x_0) \subset \overline{B_{3/4}}$. The function $h - h(x_1)$ is harmonic in $B_{3/4}$, so we can apply

to it the interior derivative estimate of Theorem 2.11 in $B_\delta(x_0)$ together with the $C^{\bar{\alpha}/2}(\overline{B_{3/4}})$ estimate (11.4) to get

$$\begin{aligned}|D^2h(x_0)| &= |D^2(h-h(x_1))(x_0)| \\ &\leq \frac{c}{\delta^2}\sup_{B_\delta(x_0)}|h-h(x_1)| \\ &\leq \frac{c}{\delta^2}\delta^{\bar{\alpha}/2}[h]_{C^{\bar{\alpha}/2}(\overline{B_{3/4}})} \\ &\leq c\delta^{\bar{\alpha}/2-2}\big(1+\|f\|_{L^n(B_1)}\big).\end{aligned}$$

Now, going back to (11.3),

$$\|u-h\|_{L^\infty(B_{3/4-\delta})} \leq C\delta^{\bar{\alpha}/2}(1+\varepsilon\delta^{-2})\big(1+\|f\|_{L^n(B_1)}\big)+C\|f\|_{L^n(B_1)}.$$

The conclusion follows by taking $\delta=\varepsilon^{1/2}<1/4$ and then $\gamma=\bar{\alpha}/4$. □

Proof of Theorem 11.1. We begin with a series of reductions. By an orthogonal change of variables, we can assume that $a(0)=I$. After this, one can also assume that $f(0)=0$ by simply noticing that $L(u-f(0)\frac{|x|^2}{2n})=f-f(0)a^{ij}(x)\delta_{ij}/n$ in the viscosity sense in B_1. Let $\varepsilon_0>0$ small to be determined. By following similar scaling ideas as those in the proof of Theorem 7.1, it can be supposed that $[a]_{C_n^\alpha(0)}<\varepsilon_0$. Finally, by normalizing as $u/(\|u\|_{L^\infty(B_1)}+\frac{1}{\varepsilon_0}[f]_{C_n^\alpha(0)})$, it can be assumed that $|u|\leq 1$ and $[f]_{C_n^\alpha(0)}<\varepsilon_0$.

We prove that if

$$\left(\frac{1}{|B_r|}\int_{B_r}|a(x)-I|^n\,dx\right)^{1/n}+\left(\frac{1}{|B_r|}\int_{B_r}|f|^n\,dx\right)^{1/n}<\varepsilon_0 r^\alpha \qquad \text{for } 0<r<1$$

where $\varepsilon_0>0$ is sufficiently small depending only on n, λ, Λ and α, then there exists a polynomial P_∞ of degree 2 such that

$$\|u-P_\infty\|_{L^\infty(B_r(0))}\leq Cr^{2+\alpha} \qquad \text{for any } 0<r<1$$

and

$$|P_\infty(0)|+|\nabla P_\infty(0)|+|D^2P_\infty(0)|\leq C$$

where $C>0$ is a constant that depends only on n, λ, Λ and α.

We construct the polynomial P_∞ by induction. The claim is that there exist $0<\rho<1$, depending only on n, λ, Λ and α, and a sequence of harmonic polynomials

$$P_k(x)=\frac{1}{2}x^TA_kx+B_k\cdot x+C_k$$

such that for any $k\geq 0$

$$\|u-P_k\|_{L^\infty(B_{\rho^k})}<\rho^{k(2+\alpha)}$$

and

$$\rho^{2(k-1)}|A_k-A_{k-1}|+\rho^{k-1}|B_k-B_{k-1}|+|C_k-C_{k-1}|\leq c_0\rho^{(k-1)(2+\alpha)}$$

where $P_0 = P_{-1} = 0$ and $c_0 > 0$ is a constant depending only on n, λ, Λ and α. From this claim the conclusion follows, see the proofs of Theorems 3.19 and 7.3.

The claim clearly holds for $k = 0$. Let $k \geq 0$ and suppose that the claim holds for all $0 \leq \ell \leq k$. We want to prove that it is also true for $k+1$. Consider the function

$$\tilde{u}(z) = \frac{(u - P_k)(\rho^k z)}{\rho^{k(2+\alpha)}} \qquad \text{for } z \in B_1.$$

Then $\tilde{u} \in C(B_1)$ is a viscosity solution to

$$\tilde{a}^{ij}(z)\partial_{ij}\tilde{u} = \tilde{f}(z) \qquad \text{in } B_1$$

where

$$\tilde{a}(z) = \frac{a(\rho^k z)}{\rho^{k\alpha}}$$

and

$$\tilde{f}(z) = \frac{f(\rho^k z)}{\rho^{k\alpha}} + \frac{(\delta_{ij} - a^{ij}(\rho^k z))\partial_{ij} P_k}{\rho^{k\alpha}}$$

where we used that $\delta^{ij}\partial_{ij}P_k = \Delta P_k = 0$. We now choose ε_0 so that we can apply the approximation lemma, Lemma 11.5, to $\tilde{u}$. For any $z \in B_1$, by induction hypothesis,

$$|\tilde{u}(z)| = \frac{|u - P_k|(\rho^k z)}{\rho^{k(2+\alpha)}} \leq \frac{\|u - P_k\|_{L^\infty(B_{\rho^k})}}{\rho^{k(2+\alpha)}} < 1.$$

In addition,

$$\|\tilde{a} - I\|_{L^n(B_1)} = \frac{C_n}{\rho^{k\alpha}}\left(\frac{1}{|B_{\rho^k}|}\int_{B_{\rho^k}} |a(x) - I|^n \, dx\right)^{1/n} \leq C_n [a]_{C^\alpha_n(0)} < C_n\varepsilon_0$$

and

$$\begin{aligned}\|\tilde{f}\|_{L^n(B_1)} &\leq \frac{C_n}{\rho^{k\alpha}}\left(\frac{1}{|B_{\rho^k}|}\int_{B_{\rho^k}} |f(x)|^n \, dx\right)^{1/n} \\ &\quad + \frac{1}{\rho^{k\alpha}} \sup |D^2 P_k| \left(\frac{1}{|B_{\rho^k}|}\int_{B_{\rho^k}} |a(x) - I|^n \, dx\right)^{1/n} \\ &< C_n\varepsilon_0 + C_{\alpha,\rho}\varepsilon_0\end{aligned}$$

where in the last inequality we have estimated as follows:

$$|D^2 P_k| \leq \sum_{\ell=1}^{k} |D^2 P_\ell - D^2 P_{\ell-1}| \leq c_0 \sum_{\ell=1}^{k} \rho^{(\ell-1)\alpha} \leq C_{\alpha,\rho}.$$

Choose $\varepsilon_0 > 0$ depending on n, α and ρ such that $\varepsilon = C_n\varepsilon_0 + C_{\alpha,\rho}\varepsilon_0 < 1/16$. By Lemma 11.5, there is a harmonic function h in $B_{3/4}$ with $|h| \leq 1$ in $B_{3/4}$ such that

$$\|\tilde{u} - h\|_{L^\infty(B_{1/2})} \leq C(\varepsilon^\gamma + \varepsilon) \leq 2C\varepsilon^\gamma.$$

Let

$$\tilde{P}(z) = \frac{1}{2} z^T D^2 h(0) z + \nabla h(0) \cdot z + h(0).$$

Observe that, by interior derivative estimates for harmonic functions (Theorem 2.11), the coefficients of $\tilde{P}$ are bounded by a constant $c_n > 0$ that depends only on n. Therefore,

$$\|\tilde{u} - \tilde{P}\|_{L^\infty(B_\rho)} \leq \|\tilde{u} - h\|_{L^\infty(B_\rho)} + \|h - \tilde{P}\|_{L^\infty(B_\rho)} \leq 2C\varepsilon^\gamma + c_n\rho^3 \leq \rho^{2+\alpha}$$

where we first chose ρ small depending only on n and α so that $c_n\rho^3 \leq \frac{1}{2}\rho^{2+\alpha}$ and then decreased ε_0 so that $2C\varepsilon^\gamma \leq \frac{1}{2}\rho^{2+\alpha}$. After rescaling back,

$$|u(x) - P_k(x) - \rho^{k(2+\alpha)}\tilde{P}(x/\rho^k)| \leq \rho^{(k+1)(2+\alpha)}$$

for any $x \in B_{\rho^{k+1}}$. The conclusion follows by defining

$$P_{k+1}(x) = P_k(x) + \rho^{k(2+\alpha)}\tilde{P}(x/\rho^k).$$

□

11.2 CALDERÓN–ZYGMUND ESTIMATES

We consider nondivergence form equations

$$Lu = \operatorname{trace}(a(x)D^2u) = a^{ij}(x)\partial_{ij}u = f \qquad \text{in } B_1 \subset \mathbb{R}^n$$

for $n \geq 2$. The matrix of coefficients $a(x) = (a^{ij}(x))_{i,j=1}^n$ is defined in B_1, symmetric $a^{ij}(x) = a^{ji}(x)$, bounded $\|a\|_{L^\infty(B_1)} \leq \Lambda$ for some $\Lambda > 0$ and elliptic: there is a constant $\lambda > 0$ such that $a^{ij}(x)\xi_i\xi_j \geq \lambda|\xi|^2$, for all $\xi \in \mathbb{R}^n$ and every $x \in B_1$. We assume that $a(x) \in C(B_1)$ and that f is a measurable function on B_1.

We say that a second-order weakly differentiable function u in B_1 (that is, u is in $L^1_{\mathrm{loc}}(B_1)$ and the second-order distributional derivatives D^2u are also in $L^1_{\mathrm{loc}}(B_1)$) is a **strong solution** to $Lu = f$ in B_1 if

$$Lu(x) = a^{ij}(x)\partial_{ij}u(x) = f(x) \qquad \text{for a.e. } x \in B_1.$$

Our aim is to show the following *a priori* Calderón–Zygmund estimates.

Theorem 11.6 (Calderón–Zygmund estimates for strong solutions). *Fix $1 < p < \infty$ and $f \in L^p(B_1)$. Let $u \in W^{2,p}_{\mathrm{loc}}(B_1) \cap L^p(B_1)$ be a strong solution to $Lu = f$ in B_1. Then there exists a constant $C > 0$ depending only on n, λ, Λ, p and the modulus of continuity of $a(x)$ such that*

$$\|D^2u\|_{L^p(B_{1/2})} \leq C(\|u\|_{L^p(B_1)} + \|f\|_{L^p(B_1)}). \tag{11.5}$$

Theorem 11.6 is an *a priori* estimate in the sense that we are assuming that the solution u already satisfies $D^2u \in L^p_{\mathrm{loc}}(B_1)$. However, the important point is the estimate (11.5) for D^2u in terms of u itself and f and in which the constant C is independent of the solution u. As an application, consider the general **method of**

continuity. This is a classical technique that can be used to show the existence of a unique strong solution u, say, in the Banach space $X = W^{2,p}(B_1) \cap W_0^{1,p}(B_1)$ to

$$\begin{cases} Lu = f & \text{in } B_1 \\ u = 0 & \text{on } \partial B_1. \end{cases} \tag{11.6}$$

In our particular scenario, the method says that if two linear operators L_0 and L_1 are bounded from $X = W^{2,p}(B_1) \cap W_0^{1,p}(B_1)$ into the normed linear space $Y = L^p(B_1)$ and there exists $c > 0$ such that the convex combination (or intermediate) operator $L_t = (1-t)L_0 + tL_1$ satisfies the estimate

$$\|u\|_X \le c\|L_t u\|_Y$$

for all $u \in X$ and all $t \in [0,1]$, then $L_1 : X \to Y$ is onto if and only if $L_0 : X \to Y$ is onto. Notice that the estimate for L_t says, in particular, that L_t is one-to-one. In the application of the method to our PDE problem, we start with the Laplacian $L_0 = \Delta$ and we want to arrive to $L_1 = L$. Clearly, for any $u \in X = W^{2,p}(B_1) \cap W_0^{1,p}(B_1)$,

$$\|L_0 u\|_Y = \|\Delta u\|_{L^p(B_1)} \le \|D^2 u\|_{L^p(B_1)} \le \|u\|_X$$

and, since the coefficients $a(x)$ are bounded,

$$\|L_1 u\|_Y = \|a^{ij}(x)\partial_{ij} u\|_{L^p(B_1)} \le C\|D^2 u\|_{L^p(B_1)} \le \|u\|_X.$$

In other words, L_0 and L_1 are continuous from X to Y. The estimate for L_t now translates into the following *a priori* estimate: there is $c > 0$ such that, for every $u \in W^{2,p}(B_1) \cap W_0^{1,p}(B_1)$,

$$\|u\|_{W^{2,p}(B_1)} \le c\|((1-t)\Delta + tL)u\|_{L^p(B_1)} \qquad \text{for all } t \in [0,1]. \tag{11.7}$$

Note that the coefficients of $L_t = (1-t)\Delta + tL = a_t^{ij}(x)\partial_{ij}$ are also symmetric, continuous, bounded by $\Lambda + 1$ and elliptic with ellipticity constant $\min(1,\lambda) > 0$, uniformly in $t \in [0,1]$. If we are able to prove that the *a priori* estimate (11.7) holds, then L_0 is onto if and only if L_1 is onto. This means that for any $f \in L^p(B_1)$ there exists a unique solution $u \in W^{2,p}(B_1) \cap W_0^{1,p}(B_1)$ to (11.6) if and only if the problem

$$\begin{cases} \Delta u = f & \text{in } B_1 \\ u = 0 & \text{on } \partial B_1 \end{cases}$$

is uniquely solvable in $W^{2,p}(B_1) \cap W_0^{1,p}(B_1)$ for any $f \in L^p(B_1)$.

It is important to observe that, with the method of continuity, the regularity is proved as part of the construction of the solution.

Certainly, the required *a priori* estimate in (11.7) is a global $W^{2,p}(B_1)$ estimate. Here we only present the interior *a priori* estimate of Theorem 11.6. To obtain the global result one needs to prove the *a priori* estimates up to the boundary by techniques such as flattening the boundary and then patch them with the interior estimates. We will not pursue the global estimate here, but showcase the classical technique of *freezing the coefficients* that is used to prove Theorem 11.6.

An ingredient we need is the following interpolation inequality for Sobolev spaces.

Lemma 11.7 (Interpolation inequality for $W^{2,p}$). *There exists a constant $C > 0$ such that for any $u \in W^{2,p}(B_1)$, $1 < p < \infty$, and any $\varepsilon > 0$,*

$$\|\nabla u\|_{L^p(B_1)} \leq \varepsilon \|D^2 u\|_{L^p(B_1)} + \frac{C}{\varepsilon}\|u\|_{L^p(B_1)}.$$

Sketch of proof. By using the linear extension operator $E : W^{2,p}(B_1) \to W_0^{2,p}(B_2)$, the density of C_c^2 in $W_0^{2,p}$ and noticing that it is enough to prove the inequality in each variable separately, one is reduced to assume that $u \in C_c^2(\mathbb{R})$. It is sufficient to prove the inequality for any interval (a, b) of length $\varepsilon > 0$. Apply the mean value theorem for points $a < x_1 < a + \varepsilon/3$ and $b - \varepsilon/3 < x_2 < b$ to get an intermediate point $\bar{x}$ such that

$$|u'(\bar{x})| \leq \frac{3}{\varepsilon}(|u(x_1)| + |u(x_2)|).$$

Hence, for any $x \in (a, b)$,

$$|u'(x)| \leq \frac{3}{\varepsilon}(|u(x_1)| + |u(x_2)|) + \int_a^b |u''|\, dz.$$

Integrating over all those x_1 and x_2 and applying Hölder's inequality,

$$\begin{aligned}|u'(x)|^p &\leq \left(\int_a^b |u''|\, dz + \frac{C}{\varepsilon^2}\int_a^b |u|\, dz\right)^p \\ &\leq 2^{p-1}\left(\varepsilon^{p-1}\int_a^b |u''|^p\, dz + \frac{C^p}{\varepsilon^{p+1}}\int_a^b |u|^p\, dz\right).\end{aligned}$$

We conclude after integrating over all $x \in (a, b)$. □

Proof of Theorem 11.6. The idea is to find a radius $0 < R < 1/4$ such that

$$\|D^2 u\|_{L^p(B_{R/2}(x))} \leq C(\|Lu\|_{L^p(B_R(x))} + \|u\|_{L^p(B_R(x))}) \qquad \text{for all } x \in B_{1/2}$$

where $C > 0$ depends only on n, λ, Λ, p and the (uniform) modulus of continuity of the coefficients a in, say, $B_{3/4}$. The full estimate follows by covering $B_{1/2}$ by a finite number of these balls and adding up the resulting estimates. For simplicity, we let $x = 0$. The estimate for any other point $x \in B_{1/2}$ follows the same lines. Moreover, by an orthogonal change of variables, we can assume that $a(0) = I$.

Recall Corollary 4.9, which says that if $v \in W_0^{2,p}(B_R)$ then

$$\|D^2 v\|_{L^p(B_R)} \leq C\|\Delta v\|_{L^p(B_R)}$$

with $C > 0$ depending only on n and p. We want an estimate like this with Lv in place of Δv. To that end, we compare the Laplacian Δ, that corresponds to the operator L evaluated at the *frozen* coefficient $a(0)$, with L in the following way:

$$\Delta v = a^{ij}(0)\partial_{ij} v = (a^{ij}(0) - a^{ij}(x))\partial_{ij} v + Lv.$$

With this, Corollary 4.9 gives

$$\|D^2 v\|_{L^p(B_R)} \leq C\left(\sup_{z \in B_R} |a(0) - a(z)| \|D^2 v\|_{L^p(B_R)} + \|Lv\|_{L^p(B_R)}\right) \tag{11.8}$$

where $C > 0$ depends only on n and p. Since a is uniformly continuous in $\overline{B_{3/4}}$, there exists $0 < R < 1/4$ such that, for any $x \in B_{1/2}$,

$$\sup_{z \in B_R(x)} |a(x) - a(z)| < \frac{1}{2C}.$$

In particular, choosing $x = 0$, from (11.8) we get

$$\|D^2 v\|_{L^p(B_R)} \leq C\|Lv\|_{L^p(B_R)}. \tag{11.9}$$

The estimate for $D^2 v$ in (11.9) holds whenever $v \in W_0^{2,p}(B_1)$. To be able to use it for our strong solution $u \in W_{\mathrm{loc}}^{2,p}(B_1) \cap L^p(B_1)$, we need to localize u by using a cutoff. The tradeoff is the appearance of an extra term involving ∇u (which can be taken care of by means of the interpolation inequality of Lemma 11.7) and a discrepancy in the radii of the balls that prevents from directly absorbing the term with $D^2 u$ on the right hand side into the left hand side. To fix these issues, a parameter σ to be optimized is introduced.

Fix $0 < \sigma < 1$ and let $\sigma' = \frac{1+\sigma}{2}$. Let $\eta \in C_c^2(B_R)$ be a cutoff function satisfying

$$0 \leq \eta \leq 1 \text{ in } B_R, \qquad \eta = 1 \text{ in } B_{\sigma R}, \qquad \eta = 0 \text{ outside of } B_{\sigma' R}$$

and such that

$$|\nabla \eta| \leq \frac{c}{(1-\sigma)R} \qquad \text{and} \qquad |D^2 \eta| \leq \frac{c^2}{(1-\sigma)^2 R^2}$$

for some $c > 0$. Consider $v = \eta u \in W_0^{2,p}(B_1)$. Then, by (11.9), the boundedness of the coefficients and the properties of η,

$$\begin{aligned}\|D^2 u\|_{L^p(B_{\sigma R})} &\leq \|D^2(\eta u)\|_{L^p(B_R)} \leq C\|L(\eta u)\|_{L^p(B_R)} \\ &\leq C\Big(\|f\|_{L^p(B_R)} + \frac{1}{(1-\sigma)R}\|\nabla u\|_{L^p(B_{\sigma' R})} + \frac{1}{(1-\sigma)^2 R^2}\|u\|_{L^p(B_R)}\Big).\end{aligned}$$

Since $1 - \sigma = 2(1 - \sigma')$, we get

$$(1-\sigma)^2 R^2 \|D^2 u\|_{L^p(B_{\sigma R})} \leq C(R^2\|f\|_{L^p(B_R)} + (1-\sigma')R\|\nabla u\|_{L^p(B_{\sigma' R})} + \|u\|_{L^p(B_R)}).$$

If we call

$$\Phi_k = \sup_{0<\sigma<1} (1-\sigma)^k R^k \|D^k u\|_{L^p(B_{\sigma R})} \qquad \text{for } k = 0, 1, 2,$$

then the previous estimate implies that

$$\Phi_2 \leq C(R^2\|f\|_{L^p(B_R)} + \Phi_1 + \Phi_0). \tag{11.10}$$

Next we claim that the following interpolation inequality holds: for any $\varepsilon > 0$,

$$\Phi_1 \leq \varepsilon \Phi_2 + \frac{C}{\varepsilon}\Phi_0 \tag{11.11}$$

where $C > 0$ depends only on n. Indeed, by scaling, it is enough to prove it for $R = 1$. Given any $\delta > 0$, by definition of supremum, there exists $0 < \sigma_\delta < 1$ such that

$$\Phi_1 \le (1 - \sigma_\delta)\|\nabla u\|_{L^p(B_{\sigma_\delta})} + \delta.$$

By the interpolation inequality for $W^{2,p}$ functions of Lemma 11.7 with $\varepsilon(1 - \sigma_\delta) > 0$ in place of ε, we obtain

$$\begin{aligned}\Phi_1 &\le \varepsilon(1 - \sigma_\delta)^2\|D^2 u\|_{L^p(B_{\sigma_\delta})} + \frac{C}{\varepsilon}\|u\|_{L^p(B_{\sigma_\delta})} + \delta \\ &\le \varepsilon\Phi_2 + \frac{C}{\varepsilon}\Phi_0 + \delta\end{aligned}$$

and (11.11) follows by letting $\delta \to 0$. Now, applying (11.11) into (11.10) and choosing $\varepsilon > 0$ sufficiently small,

$$\Phi_2 \le C\big(R^2\|f\|_{L^p(B_R)} + \Phi_0\big).$$

This gives

$$\sup_{0<\sigma<1} (1 - \sigma)^2 R^2\|D^2 u\|_{L^p(B_{\sigma R})} \le C\big(R^2\|f\|_{L^p(B_R)} + \|u\|_{L^p(B_R)}\big)$$

which, in particular, implies that

$$\|D^2 u\|_{L^p(B_{R/2})} \le C\|f\|_{L^p(B_R)} + \frac{C}{R^2}\|u\|_{L^p(B_R)}$$

as desired. □

NOTES ON CHAPTER 11

The method we presented for the proof of Schauder estimates was originally developed by Caffarelli for fully nonlinear elliptic equations in [4], see also [7]. Caffarelli applied it to linear equations in [3]. Our presentation is based on these references. See also Han–Lin [44]. The classical proof of the Schauder estimate by freezing the coefficients with equations including lower-order terms can be found in Gilbarg–Trudinger [43].

The Calderón–Zygmund perturbation technique is classical. Here we followed Gilbarg–Trudinger [43], where equations with lower-order terms, interpolation inequalities and the method of continuity are also treated.

Another way of proving *a priori* $W^{2,p}$ estimates for nondivergence form elliptic equations, that works even for equations with discontinuous coefficients satisfying a suitable *vanishing mean oscillation* condition, is due to Chiarenza–Frasca–Longo [21]. Their method involves the "freezing the coefficient" idea but instead applies the boundedness of singular integrals and commutators in an essential way.

The method of continuity works for many scenarios, including fully nonlinear elliptic equations, see, for example, Caffarelli–Cabré [7] and Gilbarg–Trudinger [43].

IV

The fractional Laplacian

CHAPTER 12

Basic properties of the fractional Laplacian

In this chapter we introduce the method of semigroups to define the fractional Laplacian $(-\Delta)^s$ on $\mathbb{R}^n$, for $0 < s < 1$, and to derive pointwise formulas for $(-\Delta)^s u(x)$ with explicit constants, for a large class of functions u. We prove the validity of the pointwise limits

$$\lim_{s\to 0}(-\Delta)^s u(x) = u(x) \qquad \text{and} \qquad \lim_{s\to 1}(-\Delta)^s u(x) = -\Delta u(x)$$

under mild assumptions on u. The pointwise formulas also allow us to show maximum and comparison principles for classical solutions. The method of semigroups is applied again to find pointwise formulas for the inverse of the fractional Laplacian $(-\Delta)^{-s}$ and to prove the distributional solvability of the Poisson equation

$$(-\Delta)^s u = f \qquad \text{in } \mathbb{R}^n.$$

The weak definition of $(-\Delta)^s$ is given and suitable fractional Sobolev spaces are introduced to prove existence and uniqueness of weak solutions to the Poisson–Dirichlet problem

$$\begin{cases} (-\Delta)^s u = f & \text{in } \Omega \\ u = 0 & \text{in } \mathbb{R}^n \setminus \Omega. \end{cases}$$

Maximum and comparison principles for weak solutions are shown. Using the Fourier transform and properties of special functions, an explicit example of a solution to a Poisson–Dirichlet problem is computed. We introduce the notions of pointwise and viscosity solutions to $(-\Delta)^s u = f$ in Ω and show that they are actually equivalent whenever u can be touched by a test function. Finally, we prove that bounded viscosity solutions to $(-\Delta)^s u = 0$ in B_1 are locally Hölder continuous in B_1.

12.1 METHOD OF SEMIGROUPS AND POINTWISE FORMULAS

The reader should review the definition and properties of the Fourier transform and of the classical heat diffusion semigroup $\{e^{t\Delta}\}_{t\geq 0}$ given in Sections 3.1 and 3.3, and

DOI: 10.1201/9781032679853-12

the Gamma function at the end of the Introduction. We will repeatedly use them throughout this chapter without making any explicit further reference.

Let u be a function in the Schwartz class $\mathcal{S}$. The Fourier transform of u, $\widehat{u} = \mathcal{F}u$ is also in $\mathcal{S}$ and $\mathcal{F}^{-1}(\widehat{u}) = u$. Recall that

$$\widehat{(-\Delta)u}(\xi) = |\xi|^2\widehat{u}(\xi) \qquad \text{for every } \xi \in \mathbb{R}^n.$$

Given any $s > 0$, we define the **fractional Laplacian** $(-\Delta)^s u$ of u as the Fourier multiplier operator

$$\widehat{(-\Delta)^s u}(\xi) = |\xi|^{2s}\widehat{u}(\xi) \tag{12.1}$$

where we have raised the Fourier multiplier of the Laplacian to the power s. The name "fractional Laplacian" is used even when s is not necessarily a fraction. Equivalently,

$$(-\Delta)^s u(x) = \mathcal{F}^{-1}\big(|\xi|^{2s}\widehat{u}(\xi)\big)(x) \qquad \text{for } x \in \mathbb{R}^n.$$

It is obvious from the definition in (12.1) that

$$(-\Delta)^0 u = u \qquad \text{and} \qquad (-\Delta)^1 u = -\Delta u$$

and that, for any $s_1, s_2 > 0$,

$$(-\Delta)^{s_1} \circ (-\Delta)^{s_2} u = (-\Delta)^{s_1+s_2} u.$$

Moreover, the fractional Laplacian rescales as an operator of *fractional differential order* $2s$, in the sense that if $u_R(x) = u(Rx)$, where $R > 0$, then

$$(-\Delta)^s u_R(x) = R^{2s}[(-\Delta)^s u](Rx). \tag{12.2}$$

Observe that, for example,

$$(-\Delta)^{3/2} u = (-\Delta)^{1/2}(-\Delta u) \qquad \text{or} \qquad (-\Delta)^{\pi} u = (-\Delta)^{\pi-3}(-\Delta)^3 u$$

so, from now on, we restrict our attention to

$$0 < s < 1.$$

Although $|\xi|^{2s}\widehat{u}(\xi)$ is a well defined function of $\xi \in \mathbb{R}^n$, we still have

$$(-\Delta)^s u \notin \mathcal{S}$$

because differentiating $|\xi|^{2s}$ creates a singularity at $\xi = 0$. Therefore $(-\Delta)^s$ does not map $\mathcal{S}$ into $\mathcal{S}$. On the other hand, for any multi-index $\gamma \in \mathbb{N}_0^n$, by (12.1),

$$\mathcal{F}((-\Delta)^s D^\gamma u)(\xi) = |\xi|^{2s}(i\xi)^\gamma\widehat{u}(\xi) = (i\xi)^\gamma|\xi|^{2s}\widehat{u}(\xi) = \mathcal{F}(D^\gamma(-\Delta)^s u)(\xi)$$

which shows that if $u \in \mathcal{S}$ then $(-\Delta)^s u \in C^\infty$ and

$$D^\gamma(-\Delta)^s = (-\Delta)^s D^\gamma. \tag{12.3}$$

The next step is to derive explicit pointwise formulas for $(-\Delta)^s u(x)$ for each point $x \in \mathbb{R}^n$ that do not involve the Fourier transform of u. One can try to explicitly perform the inverse Fourier transform of (12.1). Since $|\xi|^{2s}\widehat{u}(\xi) \in L^1(\mathbb{R}^n)$, the formula $(-\Delta)^s u(x) = \mathcal{F}^{-1}(|\xi|^{2s}\widehat{u}(\xi))(x)$ makes sense for every $x \in \mathbb{R}^n$. But $|\xi|^{2s}\widehat{u}(\xi) \notin \mathcal{S}$, so the computation of this inverse Fourier transform must be done in an appropriate sense of distributions. However, we easily solve these issues by introducing the widely applicable **method of semigroups** for fractional power operators.

For the case of the fractional Laplacian, the method of semigroups begins with a numerical formula involving the Gamma function. Recall that, for any $0 < s < 1$, the Gamma function at $-s \in (-1, 0)$ is given by

$$\Gamma(-s) = \frac{\Gamma(1-s)}{-s} = \int_0^\infty (e^{-r} - 1)\,\frac{dr}{r^{1+s}} < 0.$$

Let $\lambda > 0$. By changing variables $r = \lambda t$, we get

$$\lambda^s = \frac{1}{\Gamma(-s)} \int_0^\infty (e^{-t\lambda} - 1)\,\frac{dt}{t^{1+s}}. \tag{12.4}$$

This identity is also true when $\lambda = 0$. Choose $\lambda = |\xi|^2$, multiply by $\widehat{u}(\xi)$ and recall the Fourier transform definition of the fractional Laplacian in (12.1) to obtain

$$\widehat{(-\Delta)^s u}(\xi) = |\xi|^{2s}\widehat{u}(\xi) = \frac{1}{\Gamma(-s)} \int_0^\infty (e^{-t|\xi|^2}\widehat{u}(\xi) - \widehat{u}(\xi))\,\frac{dt}{t^{1+s}}.$$

Therefore, after applying the inverse Fourier transform,

$$(-\Delta)^s u(x) = \frac{1}{(2\pi)^{n/2}\Gamma(-s)} \int_{\mathbb{R}^n} \int_0^\infty (e^{-t|\xi|^2}\widehat{u}(\xi)e^{ix\cdot\xi} - \widehat{u}(\xi)e^{ix\cdot\xi})\,\frac{dt}{t^{1+s}}\,d\xi. \tag{12.5}$$

The double integral in (12.5) is absolutely convergent because, by (12.4),

$$\int_{\mathbb{R}^n} \int_0^\infty (1 - e^{-t|\xi|^2})|\widehat{u}(\xi)|\,\frac{dt}{t^{1+s}}\,d\xi = |\Gamma(-s)| \int_{\mathbb{R}^n} |\xi|^{2s}|\widehat{u}(\xi)|\,d\xi < \infty.$$

Hence, by applying Fubini's theorem in (12.5), we obtain the **semigroup formula for the fractional Laplacian**

$$(-\Delta)^s u(x) = \frac{1}{\Gamma(-s)} \int_0^\infty (e^{t\Delta}u(x) - u(x))\,\frac{dt}{t^{1+s}} \qquad \text{for } x \in \mathbb{R}^n \tag{12.6}$$

and $u \in \mathcal{S}$. Recall that $\{e^{t\Delta}\}_{t\geq 0}$ is the classical heat diffusion semigroup given by

$$e^{t\Delta}u(x) = G_t * u(x) = \int_{\mathbb{R}^n} G_t(x - z)u(z)\,dz$$

where

$$G_t(x) = \frac{1}{(4\pi t)^{n/2}} e^{-|x|^2/(4t)} > 0 \tag{12.7}$$

is the Gauss–Weierstrass heat kernel. Moreover,

$$e^{t\Delta}1(x) = \int_{\mathbb{R}^n} G_t(x-z)\,dz \equiv 1 \qquad \text{for any } x \in \mathbb{R}^n,\ t>0. \tag{12.8}$$

Since $\|e^{t\Delta}u\|_{L^\infty(\mathbb{R}^n)} \le \|u\|_{L^\infty(\mathbb{R}^n)}$ and

$$|e^{t\Delta}u(x) - u(x)| \le \int_0^t |\partial_r e^{r\Delta}u(x)|\,dr = \int_0^t |e^{r\Delta}(\Delta u)(x)|\,dr \le t\|\Delta u\|_{L^\infty(\mathbb{R}^n)}$$

for all $t>0$, it is clear that (12.6) is an absolutely convergent integral.

By using the semigroup formula (12.6) and the expression of the heat kernel (12.7) we compute the pointwise formulas for the fractional Laplacian. Observe that the semigroup technique avoids the distributional calculation of the inverse Fourier transform in (12.1). In turn, it takes advantage of the formula for the heat kernel (12.7) and gives the explicit value of the constant $c_{n,s}>0$ in the following statement.

Theorem 12.1 (Pointwise formulas for $(-\Delta)^s u$). *Let $u \in \mathcal{S}$, $x \in \mathbb{R}^n$ and $0<s<1$.*

(1) *If $0<s<1/2$ then*

$$(-\Delta)^s u(x) = c_{n,s}\int_{\mathbb{R}^n} \frac{u(x)-u(z)}{|x-z|^{n+2s}}\,dz$$

where the integral is absolutely convergent.

(2) *If $1/2 \le s < 1$ then, for any $\delta>0$,*

$$\begin{aligned}(-\Delta)^s u(x) &= c_{n,s}\,\mathrm{P.V.}\int_{\mathbb{R}^n} \frac{u(x)-u(z)}{|x-z|^{n+2s}}\,dz \\ &= c_{n,s}\int_{\mathbb{R}^n} \frac{u(x)-u(z)-\nabla u(x)\cdot(x-z)\chi_{|x-z|<\delta}}{|x-z|^{n+2s}}\,dz\end{aligned}$$

where P. V. *means that the integral is understood in the principal value sense, that is,* $\mathrm{P.V.}\int_{\mathbb{R}^n} = \lim_{\varepsilon\to 0}\int_{|x-z|>\varepsilon}$, *and the second integral is absolutely convergent.*

(3) *For any $0<s<1$ we can also write*

$$-(-\Delta)^s u(x) = \frac{c_{n,s}}{2}\int_{\mathbb{R}^n} \frac{u(x+z)+u(x-z)-2u(x)}{|z|^{n+2s}}\,dz$$

where the integral is absolutely convergent.

The constant $c_{n,s}>0$ in the formulas above is explicitly given by

$$c_{n,s} = \frac{4^s\Gamma(n/2+s)}{|\Gamma(-s)|\pi^{n/2}} \qquad \textit{for } 0<s<1. \tag{12.9}$$

It is important to observe that, unlike differential operators, the fractional Laplacian is a **nonlocal operator**. The term "nonlocal" can be understood in two ways. Theorem 12.1 says that to compute the fractional Laplacian of u at a point x we need the values of $u(z)$ for all $z \in \mathbb{R}^n$. This is in contrast with the classical derivative operator, in which one only needs the values of $u(z)$ for z infinitesimally close to x. Next, if u has compact support then any derivative of u has support contained in the support of u. This is not the case in general for the fractional Laplacian: if u has compact support then $(-\Delta)^s u$ has noncompact support.

Proof of Theorem 12.1. We write down the convolution formula for $e^{t\Delta}u(x)$ in (12.6) and use (12.8) to get

$$(-\Delta)^s u(x) = \frac{1}{|\Gamma(-s)|}\int_0^\infty \int_{\mathbb{R}^n} G_t(x-z)(u(x)-u(z))\,dz\,\frac{dt}{t^{1+s}}. \tag{12.10}$$

The next step is to interchange the integrals in (12.10). We consider two cases.

Suppose first that $0 < s < 1/2$. We have

$$\int_0^\infty \int_{\mathbb{R}^n} G_t(x-z)|u(x)-u(z)|\,dz\,\frac{dt}{t^{1+s}} = \int_{\mathbb{R}^n} |u(x)-u(z)| \int_0^\infty G_t(x-z)\,\frac{dt}{t^{1+s}}\,dz.$$

By the formula for $G_t(x-z)$ in (12.7) and the change of variables $r = |x-z|^2/(4t)$,

$$\begin{aligned}\frac{1}{|\Gamma(-s)|}\int_0^\infty G_t(x-z)\,\frac{dt}{t^{1+s}} &= \frac{4^s}{|\Gamma(-s)|\pi^{n/2}}\int_0^\infty \frac{e^{-|x-z|^2/(4t)}}{(4t)^{n/2+s}}\,\frac{dt}{t} \\ &= \frac{4^s}{|\Gamma(-s)|\pi^{n/2}}\int_0^\infty e^{-r}\left(\frac{r}{|x-z|^2}\right)^{n/2+s}\frac{dr}{r} \\ &= \frac{4^s\Gamma(n/2+s)}{|\Gamma(-s)|\pi^{n/2}}\cdot\frac{1}{|x-z|^{n+2s}}.\end{aligned} \tag{12.11}$$

Therefore,

$$\frac{1}{|\Gamma(-s)|}\int_0^\infty \int_{\mathbb{R}^n} |G_t(x-z)||u(x)-u(z)|\,dz\,\frac{dt}{t^{1+s}} = c_{n,s}\int_{\mathbb{R}^n}\frac{|u(x)-u(z)|}{|x-z|^{n+2s}}\,dz$$

with $c_{n,s} > 0$ as in (12.9). The last integral is absolutely convergent because $u \in \mathcal{S}$. Indeed, on one hand, by the mean value theorem and passing to polar coordinates,

$$\begin{aligned}\int_{|x-z|<1}\frac{|u(x)-u(z)|}{|x-z|^{n+2s}}\,dz &\le \|\nabla u\|_{L^\infty(\mathbb{R}^n)}\int_{|x-z|<1}\frac{|x-z|}{|x-z|^{n+2s}}\,dz \\ &= C\int_0^1 r^{1-n-2s}r^{n-1}\,dr < \infty\end{aligned}$$

because $s < 1/2$. On the other hand,

$$\begin{aligned}\int_{|x-z|\ge 1}\frac{|u(x)-u(z)|}{|x-z|^{n+2s}}\,dz &\le 2\|u\|_{L^\infty(\mathbb{R}^n)}\int_{|x-z|\ge 1}\frac{1}{|x-z|^{n+2s}}\,dz \\ &= C\int_1^\infty r^{-n-2s}r^{n-1}\,dr < \infty\end{aligned} \tag{12.12}$$

since $s > 0$. Therefore, when $0 < s < 1/2$, we can apply Fubini's theorem in (12.10) and, by taking into account (12.11), the pointwise formula in (1) follows.

Next, suppose that $1/2 \leq s < 1$. Fix any $\delta > 0$. We write the double integral in (12.10) as

$$\begin{aligned} I + II := &\int_0^\infty \int_{|x-z|<\delta} G_t(x-z)(u(x)-u(z))\,dz\,\frac{dt}{t^{1+s}} \\ &+ \int_0^\infty \int_{|x-z|\geq\delta} G_t(x-z)(u(x)-u(z))\,dz\,\frac{dt}{t^{1+s}}. \end{aligned} \tag{12.13}$$

As in (12.12), the integral II in (12.13) is absolutely convergent. For the integral I in (12.13) we notice that, for any $i = 1, \ldots, n$ and $0 < \varepsilon < \delta$,

$$\int_{|z|<\delta} z_i G_t(z)\,dz = \int_{\varepsilon<|z|<\delta} z_i G_t(z)\,dz = 0 \tag{12.14}$$

for all $t > 0$. This is because $G_t(z)$ is a radially symmetric function of z and z_i is an odd function of the i-th variable. By changing variables and using (12.14),

$$\begin{aligned} I &= \int_0^\infty \int_{|z|<\delta} G_t(z)(u(x)-u(x-z))\,dz\,\frac{dt}{t^{1+s}} \\ &= \int_0^\infty \int_{|z|<\delta} G_t(z)(u(x)-u(x-z)-\nabla u(x)\cdot z)\,dz\,\frac{dt}{t^{1+s}}. \end{aligned}$$

Since $u \in \mathcal{S}$, the estimate $|u(x)-u(x-z)-\nabla u(x)\cdot z| \leq \|D^2u\|_{L^\infty(\mathbb{R}^n)}|z|^2$ holds for all $|z| < \delta$. Hence, by (12.11), the double integral in I is absolutely convergent:

$$\begin{aligned} \int_0^\infty \int_{|z|<\delta} G_t(z)|u(x)-u(x-z)-\nabla u(x)\cdot z|\,dz\,\frac{dt}{t^{1+s}} &\leq C \int_{|z|<\delta} \frac{|z|^2}{|z|^{n+2s}}\,dz \\ &= C \int_0^\delta r^{2-n-2s} r^{n-1}\,dr < \infty \end{aligned}$$

because $s < 1$. With this, from (12.10) and (12.13),

$$\begin{aligned} &(-\Delta)^s u(x) \\ &\quad = \frac{1}{|\Gamma(-s)|} \int_0^\infty \int_{|x-z|<\delta} G_t(x-z)(u(x)-u(z)-\nabla u(x)\cdot(x-z))\,dz\,\frac{dt}{t^{1+s}} \\ &\qquad + \frac{1}{|\Gamma(-s)|} \int_0^\infty \int_{|x-z|\geq\delta} G_t(x-z)(u(x)-u(z))\,dz\,\frac{dt}{t^{1+s}} \end{aligned}$$

where both integrals are absolutely convergent. Hence, by Fubini's theorem and (12.11), the second formula in (2) holds and gives an absolutely convergent integral. The principal value expression in (2) is a consequence of a cancelation like in

(12.14). Indeed, $z_i/|z|^{n+2s}$ is an odd function of z_i, so that

$$\begin{aligned}
&\int_{\mathbb{R}^n} \frac{u(x)-u(z)-\nabla u(x)\cdot(x-z)\chi_{|x-z|<\delta}}{|x-z|^{n+2s}}\,dz \\
&\quad= \lim_{\varepsilon\to 0}\int_{\varepsilon<|x-z|<\delta} \frac{u(x)-u(z)-\nabla u(x)\cdot(x-z)}{|x-z|^{n+2s}}\,dz + \int_{|x-z|\geq\delta} \frac{u(x)-u(z)}{|x-z|^{n+2s}}\,dz \\
&\quad= \lim_{\varepsilon\to 0}\int_{\varepsilon<|x-z|<\delta} \frac{u(x)-u(z)}{|x-z|^{n+2s}}\,dz + \int_{|x-z|\geq\delta} \frac{u(x)-u(z)}{|x-z|^{n+2s}}\,dz \\
&\quad= \lim_{\varepsilon\to 0}\int_{|x-z|>\varepsilon} \frac{u(x)-u(z)}{|x-z|^{n+2s}}\,dz = \mathrm{P.V.}\int_{\mathbb{R}^n} \frac{u(x)-u(z)}{|x-z|^{n+2s}}\,dz.
\end{aligned}$$

We give two proofs of the pointwise formula in (3). For the first proof, observe that, by changing variables in (12.10) and using that $G(z)$ is even in z, we can write

$$\begin{aligned}
-(-\Delta)^s u(x) &= \frac{1}{|\Gamma(-s)|}\int_0^\infty \int_{\mathbb{R}^n} G_t(z)(u(x-z)-u(x))\,dz\,\frac{dt}{t^{1+s}} \\
&= \frac{1}{|\Gamma(-s)|}\int_0^\infty \int_{\mathbb{R}^n} G_t(z)(u(x+z)-u(x))\,dz\,\frac{dt}{t^{1+s}}
\end{aligned}$$

Adding these two identities,

$$-(-\Delta)^s u(x) = \frac{1}{2|\Gamma(-s)|}\int_0^\infty \int_{\mathbb{R}^n} G_t(z)(u(x+z)+u(x-z)-2u(x))\,dz\,\frac{dt}{t^{1+s}}.$$

This double integral is absolutely convergent for any $0<s<1$. Indeed, this follows by using (12.11), the boundedness of u at infinity and the fact that, since $u\in\mathcal{S}$, we have $|u(x+z)+u(x-z)-2u(x)|\leq \|D^2u\|_{L^\infty(\mathbb{R}^n)}|z|^2$ for all $x,z\in\mathbb{R}^n$. Hence, by Fubini's theorem and (12.11), the formula in (3) holds.

The second proof of the pointwise formula in (3) uses the formulas we just proved in (1) and (2). We show the case $1/2\leq s<1$. By changing variables in the pointwise formula in (2),

$$\begin{aligned}
-(-\Delta)^s u(x) &= c_{n,s}\int_{\mathbb{R}^n} \frac{u(x-z)-u(x)+\nabla u(x)\cdot z\chi_{|z|<\delta}}{|z|^{n+2s}}\,dz \\
&= c_{n,s}\int_{\mathbb{R}^n} \frac{u(x+z)-u(x)-\nabla u(x)\cdot z\chi_{|z|<\delta}}{|z|^{n+2s}}\,dz
\end{aligned}$$

and (3) is obtained by adding these two identities. The same idea works for the case $0<s<1/2$. The theorem is proved. □

The pointwise formulas for the fractional Laplacian in Theorem 12.1

$$(-\Delta)^s u(x) = c_{n,s}\,\mathrm{P.V.}\int_{\mathbb{R}^n} \frac{u(x)-u(z)}{|x-z|^{n+2s}}\,dz$$

are valid for $u\in\mathcal{S}$ and all $x\in\mathbb{R}^n$. But the integral in the right hand side above is also well defined for less regular functions.

Observe that the kernel $1/|x|^{n+2s}$ is integrable at infinity. Hence we can relax the requirement on u at infinity by asking, say, that $u \in L^\infty(\mathbb{R}^n)$, or the sharp condition

$$\|u\|_{L_s} := \int_{\mathbb{R}^n} \frac{|u(x)|}{1+|x|^{n+2s}}\,dx < \infty. \tag{12.15}$$

Since the kernel is singular at the origin, we need some regularity of u at x for the integral formula for $(-\Delta)^s u(x)$ to be well defined.

When $0 < s < 1/2$, we only need u to be $C^{2s+\varepsilon}$ at x, for $2s+\varepsilon < 1$, for the singular part of the integral (that is, when z is near x) to be finite:

$$\begin{aligned}\int_{|x-z|<1} \frac{|u(x)-u(z)|}{|x-z|^{n+2s}}\,dz &\leq [u]_{C^{2s+\varepsilon}(x)} \int_{|x-z|<1} \frac{|x-z|^{2s+\varepsilon}}{|x-z|^{n+2s}}\,dz \\ &= C\int_0^1 r^{\varepsilon-n} r^{n-1}\,dr < \infty.\end{aligned}$$

When $1/2 \leq s < 1$, the requirement $u \in C^{1,2s+\varepsilon-1}$ at x, for $2s+\varepsilon-1 < 1$ will suffice to have a well defined singular part. Indeed, by the mean value theorem,

$$\begin{aligned}&\int_{|x-z|<\delta} \frac{|u(x)-u(z)-\nabla u(x)\cdot(x-z)|}{|x-z|^{n+2s}}\,dz \\ &\qquad \leq [\nabla u]_{C^{2s+\varepsilon-1}(x)} \int_{|x-z|<\delta} \frac{|x-z|^{2s+\varepsilon}}{|x-z|^{n+2s}}\,dz = C\int_0^\delta r^{\varepsilon-n} r^{n-1}\,dr < \infty.\end{aligned}$$

We then proceed to extend the definition of the fractional Laplacian to less regular functions and show that the pointwise formula still holds. For this, we begin by generalizing the definition of $(-\Delta)^s$ to the sense of distributions. If $u, f \in \mathcal{S}$,

$$\int_{\mathbb{R}^n} (-\Delta)^s u(x) f(x)\,dx = \int_{\mathbb{R}^n} |\xi|^{2s}\widehat{u}(\xi)\widehat{f}(\xi)\,d\xi = \int_{\mathbb{R}^n} u(x)(-\Delta)^s f(x)\,dx.$$

In other words, the fractional Laplacian is a symmetric operator on $L^2(\mathbb{R}^n)$. One may then reason in the following way. For $u \in \mathcal{S}'$ (a tempered distribution) and $f \in \mathcal{S}$ we could define the distribution $(-\Delta)^s u$ as

$$((-\Delta)^s u)(f) = u((-\Delta)^s f).$$

The problem here is that $(-\Delta)^s f \notin \mathcal{S}$, so the identity above makes no sense for $u \in \mathcal{S}'$. We need to understand first what is the image of $\mathcal{S}$ via the fractional Laplacian.

Lemma 12.2. *Let $f \in \mathcal{S}$. Then $(-\Delta)^s f$ belongs to the class $\mathcal{S}_s$ defined by*

$$\mathcal{S}_s = \left\{\varphi \in C^\infty(\mathbb{R}^n) : (1+|x|^{n+2s})D^\gamma\varphi(x) \in L^\infty(\mathbb{R}^n) \text{ for every } \gamma \in \mathbb{N}_0^n\right\}.$$

In fact, there is a constant $C > 0$ depending on n, s and Schwartz seminorms of f such that, for all $x \in \mathbb{R}^n$,

$$\int_{\mathbb{R}^n} \frac{|f(x+z)+f(x-z)-2f(x)|}{|z|^{n+2s}}\,dz \leq \frac{C}{1+|x|^{n+2s}}.$$

Proof. We already know that $(-\Delta)^s f \in C^\infty(\mathbb{R}^n)$. Since derivatives commute with the fractional Laplacian, see (12.3), it is enough to show the decay of $(-\Delta)^s f(x)$ at infinity. Assume that $|x| > 2$. Then

$$\begin{aligned}\Big|\int_{|x-z|>|x|/4} \frac{f(x)-f(z)}{|x-z|^{n+2s}}\,dz\Big| &\\ &\leq |f(x)|\int_{|x-z|>|x|/4}\frac{1}{|x-z|^{n+2s}}\,dz + \frac{4^{n+2s}}{|x|^{n+2s}}\int_{\mathbb{R}^n}|f(z)|\,dz \qquad (12.16)\\ &= C|f(x)||x|^n\frac{1}{|x|^{n+2s}} + C\|f\|_{L^1(\mathbb{R}^n)}\frac{1}{|x|^{n+2s}} \leq \frac{C}{|x|^{n+2s}}\end{aligned}$$

where $C > 0$ depends only on n, s and Schwartz seminorms of f. Thus, we are left to estimate the singular part of the integral defining $(-\Delta)^s f$. Consider two cases.

Suppose that $0 < s < 1/2$. Let z be such that $|x - z| < |x|/4$. For any point $w = tz + (1-t)x$, $0 \leq t \leq 1$, in between x and z, we have $x = w + t(x - z)$, so that $|x| \leq |w| + |x - z| \leq |w| + |x|/4$, which implies that $3|x|/4 \leq |w|$. Then, by the mean value theorem and the fact that $f \in \mathcal{S}$, for any $M > 0$ there exists a constant $C > 0$ depending only on M and Schwartz seminorms of f such that, if w is in between x and z,

$$|f(x) - f(z)| = |\nabla f(w)\cdot(x-z)| \leq \frac{|w|^M|\nabla f(w)|}{|w|^M}|x-z| \leq \frac{C}{|x|^M}|x-z|.$$

Now we can bound

$$\begin{aligned}\Big|\int_{|x-z|<|x|/4}\frac{f(x)-f(z)}{|x-z|^{n+2s}}\,dz\Big| &\leq \frac{C}{|x|^M}\int_{|x-z|<|x|}\frac{|x-z|}{|x-z|^{n+2s}}\,dz\\ &= \frac{C}{|x|^M}\int_0^{|x|} r^{1-n-2s}r^{n-1}\,dr = \frac{C}{|x|^M}|x|^{1-2s}.\end{aligned}$$

The result is proved by choosing $M = n + 1$.

Suppose next that $1/2 \leq s < 1$. In the same way as before, if $|x - z| < |x|/4$, then

$$|f(x) - f(z) - \nabla f(x)\cdot(x-z)| \leq \frac{|w|^M|D^2 f(w)|}{|w|^M}|x-z|^2 \leq \frac{C}{|x|^M}|x-z|^2.$$

with $C > 0$ depending only on M and Schwartz seminorms of f. Then

$$\begin{aligned}\Big|\int_{|x-z|<|x|/4}\frac{f(x)-f(z)-\nabla f(x)\cdot(x-z)}{|x-z|^{n+2s}}\,dz\Big| &\leq \frac{C}{|x|^M}\int_{|x-z|<|x|}\frac{|x-z|^2}{|x-z|^{n+2s}}\,dz\\ &= \frac{C}{|x|^M}\int_0^{|x|} r^{2-n-2s}r^{n-1}\,dr\\ &= \frac{C}{|x|^M}|x|^{2-2s}\end{aligned}$$

and it is enough to choose $M = n + 2$.

The last estimate in the statement follows by applying the same ideas. Indeed, for any $|x| > 2$, we have

$$\begin{aligned}\int_{|z|>|x|/4} & \frac{|f(x+z)+f(x-z)-2f(x)|}{|z|^{n+2s}}\,dz \\ & \leq \frac{C}{|x|^{n+2s}}\int_{|z|>|x|/4}|f(x+z)+f(x-z)|\,dx + C|f(x)|\int_{|z|>|x|/4}\frac{1}{|z|^{n+2s}}\,dz\end{aligned}$$

which, as in (12.16), is bounded by $C/|x|^{n+2s}$. On the other hand, with a similar reasoning using the mean value theorem as before, if $|z| < |x|/4$ then, for any $M > 0$ there is a constant $C > 0$ depending on M and Schwartz seminorms of f such that

$$|f(x+z)+f(x-z)-2f(x)| \leq \frac{C}{|x|^M}|z|^2.$$

With this,

$$\begin{aligned}\int_{|z|\leq|x|/4}\frac{|f(x+z)+f(x-z)-2f(x)|}{|z|^{n+2s}}\,dz &\leq \frac{C}{|x|^M}\int_{|z|\leq|x|/4}\frac{|z|^2}{|z|^{n+2s}}\,dz \\ &= \frac{C}{|x|^M}|x|^{2-2s}\end{aligned}$$

and it is enough to choose $M = n + 2$. □

A topology on the class of functions $\mathcal{S}_s$ of Lemma 12.2 can be defined through the family of seminorms

$$p_\gamma(\varphi) = \sup_{x\in\mathbb{R}^n}|(1+|x|^{n+2s})D^\gamma\varphi(x)| \qquad \text{for } \gamma \in \mathbb{N}_0^n.$$

Denote by $\mathcal{S}_s'$ the dual space of $\mathcal{S}_s$ under such topology. We have

$$\mathcal{S} \subset \mathcal{S}_s \qquad \text{so that} \qquad \mathcal{S}_s' \subset \mathcal{S}'.$$

The suitable space for the distributional definition of the fractional Laplacian is $\mathcal{S}_s'$.

Given $u \in \mathcal{S}_s'$, we define $(-\Delta)^s u$ **in the sense of distributions** as the tempered distribution $(-\Delta)^s u \in \mathcal{S}'$ given by

$$((-\Delta)^s u)(f) = u((-\Delta)^s f) \qquad \text{for every } f \in \mathcal{S}.$$

In terms of pairings, we write this as

$$\langle(-\Delta)^s u, f\rangle_{\mathcal{S}',\mathcal{S}} = \langle u, (-\Delta)^s f\rangle_{\mathcal{S}_s',\mathcal{S}_s}.$$

If $u \in \mathcal{S}$ then this distributional definition coincides with the one given in terms of the Fourier transform. From the proof of Lemma 12.2, it is also clear that

$$(-\Delta)^s : \mathcal{S}_s' \to \mathcal{S}' \text{ continuously.}$$

Next, we consider locally integrable functions in $\mathcal{S}_s'$. For this, define the space

$$L_s = \left\{ u \in L^1_{\text{loc}}(\mathbb{R}^n) : \|u\|_{L_s} = \int_{\mathbb{R}^n} \frac{|u(x)|}{1+|x|^{n+2s}}\, dx < \infty \right\} \tag{12.17}$$

see (12.15). Then $L_s \subset \mathcal{S}_s'$ in the sense that if $u \in L_s$ then

$$u(\varphi) = \int_{\mathbb{R}^n} u(x)\varphi(x)\, dx \qquad \text{for } \varphi \in \mathcal{S}_s$$

defines a bounded linear functional on $\mathcal{S}_s$. Indeed, it is clear that $u(\varphi)$ is a linear functional and $|u(\varphi)| \leq \|u\|_{L_s} p_0(\varphi)$. Therefore, $(-\Delta)^s u$ is well defined in the distributional sense for all functions u in L_s.

Let $\Omega \subset \mathbb{R}^n$, $n \geq 1$, be a domain and consider $f \in L^1_{\text{loc}}(\Omega)$. We say that $u \in L_s$ is a **distributional solution** to

$$(-\Delta)^s u = f \qquad \text{in } \Omega$$

if for any $\varphi \in C_c^\infty(\Omega)$ we have

$$(-\Delta)^s u(\varphi) = \int_{\mathbb{R}^n} u(-\Delta)^s \varphi\, dx = \int_\Omega f\varphi\, dx.$$

In this case, we say that the tempered distribution $(-\Delta)^s u$ **coincides with** f **in** Ω.

If $u \in L_s$ has enough regularity to guarantee that the pointwise formulas for $(-\Delta)^s u(x)$ are well defined at the point x, then the pointiwse formulas for the fractional Laplacian define continuous functions that actually coincide with the distributional definition. This is made precise in the following statement.

Theorem 12.3 (Pointwise formulas for less regular functions). *Let Ω be a domain of $\mathbb{R}^n$, $n \geq 1$, and let $u \in L_s$, $0 < s < 1$. Suppose that*

$$u \in C^{2s+\varepsilon}(\Omega) \qquad \textit{if } s < 1/2$$

or that

$$u \in C^{1,2s+\varepsilon-1}(\Omega) \qquad \textit{if } s \geq 1/2$$

for some $\varepsilon > 0$. Then $(-\Delta)^s u$, which is initially defined in the sense of distributions, coincides with a continuous function in Ω and, for every $x \in \Omega$, it is given by the pointwise formulas of Theorem 12.1, that is,

(1) *if $0 < s < 1/2$ then*

$$(-\Delta)^s u(x) = c_{n,s} \int_{\mathbb{R}^n} \frac{u(x)-u(z)}{|x-z|^{n+2s}}\, dz$$

where the integral is absolutely convergent;

(2) *if* $1/2 \leq s < 1$ *then, for any* $\delta > 0$,

$$(-\Delta)^s u(x) = c_{n,s}\,\mathrm{P.V.}\int_{\mathbb{R}^n} \frac{u(x)-u(z)}{|x-z|^{n+2s}}\,dz$$
$$= c_{n,s}\int_{\mathbb{R}^n} \frac{u(x)-u(z)-\nabla u(x)\cdot(x-z)\chi_{|x-z|<\delta}}{|x-z|^{n+2s}}\,dz$$

where the second integral is absolutely convergent;

(3) *for any* $0 < s < 1$,

$$-(-\Delta)^s u(x) = \frac{c_{n,s}}{2}\int_{\mathbb{R}^n} \frac{u(x+z)+u(x-z)-2u(x)}{|z|^{n+2s}}\,dz$$

where the integral is absolutely convergent.

The constant $c_{n,s} > 0$ *in the formulas above is explicitly given by* (12.9).

Furthermore, for any $0 < s < 1$ *and* $x \in \Omega$, *the semigroup formula*

$$(-\Delta)^s u(x) = \frac{1}{\Gamma(-s)}\int_0^\infty \left(e^{t\Delta}u(x) - u(x)\right)\frac{dt}{t^{1+s}}$$

holds, where the integral is absolutely convergent.

Proof. The proof is by an approximation argument. To construct a suitable approximating sequence, let $\psi \in C_c^\infty(B_1)$ such that $\int_{\mathbb{R}^n}\psi\,dx = 1$ and take a cutoff $\eta \in C_c^\infty(B_2)$ such that $0 \leq \eta \leq 1$ and $\eta \equiv 1$ in B_1. Denote $\psi_k(x) = k^n\psi(kx)$ and $\eta_k(x) = \eta(x/k)$, for $k \geq 1$. Consider family of functions

$$u_k(x) = \eta_k(x)(\psi_k * u)(x) \in C_c^\infty(\mathbb{R}^n).$$

By construction, $\psi_k * u$ defines an approximation of the identity, so $u_k \to u$ uniformly on compact subsets of Ω as $k \to \infty$. Fix any bounded subdomain $\Omega' \subset\subset \Omega$. If $s < 1/2$, then

$$[u_k]_{C^{2s+\varepsilon}(\Omega')} \leq [u]_{C^{2s+\varepsilon}(\Omega)}$$

for all k sufficiently large. Indeed, for any $x_1, x_2 \in \Omega'$ and all k sufficiently large, $\eta_k(x_1) = \eta_k(x_2) = 1$, $\mathrm{supp}(\psi_k) \subset \Omega$ and

$$|u_k(x_1) - u_k(x_2)| \leq \int_{\mathrm{supp}(\psi_k)} |u(x_1-z) - u(x_2-z)|\psi_k(z)\,dz$$
$$\leq [u]_{C^{2s+\varepsilon}(\Omega)}|x_1-x_2|^{2s+\varepsilon}\int_{\mathrm{supp}(\psi_k)}\psi_k(z)\,dz$$
$$= [u]_{C^{2s+\varepsilon}(\Omega)}|x_1-x_2|^{2s+\varepsilon}.$$

Similarly, when $s \geq 1/2$, for all k sufficiently large, we have

$$[\nabla u_k]_{C^{2s+\varepsilon-1}(\Omega')} \leq [\nabla u]_{C^{2s+\varepsilon-1}(\Omega)}.$$

Furthermore, $u_k \to u$ as $k \to \infty$ in the norm of L_s given in (12.17). To see this, write

$$\begin{aligned}\|u_k - u\|_{L_s} &= \int_{\mathbb{R}^n} \frac{|u_k(x) - u(x)|}{1+|x|^{n+2s}}\,dx \\ &\le \int_{\mathbb{R}^n} \frac{|\psi_k * u(x) - u(x)|}{1+|x|^{n+2s}}\,dx + \int_{\mathbb{R}^n} |\eta_k(x) - 1|\frac{|u(x)|}{1+|x|^{n+2s}}\,dx \\ &=: I_k + II_k\end{aligned}$$

where in I_k we used that $0 \le \eta_k \le 1$ in $\mathbb{R}^n$. Since $u \in L_s$, the dominated convergence theorem gives that $II_k \to 0$ as $k \to \infty$. For the first term, using that $\int_{\mathbb{R}^n} \psi_k\,dz = 1$ and changing variables,

$$\begin{aligned}I_k &\le \int_{\mathrm{supp}(\psi_k)} |\psi_k(z)| \int_{\mathbb{R}^n} \frac{|u(x-z) - u(x)|}{1+|x|^{n+2s}}\,dx\,dz \\ &= \int_{B_1} |\psi(z)| \int_{\mathbb{R}^n} \frac{|u(x-z/k) - u(x)|}{1+|x|^{n+2s}}\,dx\,dz.\end{aligned} \tag{12.18}$$

Notice that the integrand above converges pointwise a.e. to 0 as $k \to \infty$. To apply the generalized dominated convergence theorem, define

$$v_k(z,x) = |\psi(z)|\left[\frac{|u(x-z/k)|}{1+|x-z/k|^{n+2s}} + \frac{|u(x)|}{1+|x|^{n+2s}}\right] \qquad \text{for } (z,x) \in B_1 \times \mathbb{R}^n.$$

As $k \to \infty$, we have the a.e. pointwise convergence

$$v_k(z,x) \to |\psi(z)|\frac{2|u(x)|}{1+|x|^{n+2s}} =: v(z,x) \in L^1(B_1 \times \mathbb{R}^n).$$

Also, by the translation invariance of the Lebesgue measure,

$$\begin{aligned}\|v_k\|_{L^1(B_1\times\mathbb{R}^n)} &= \int_{B_1} |\psi(z)| \int_{\mathbb{R}^n} \frac{|u(x-z/k)|}{1+|x-z/k|^{n+2s}}\,dx\,dz \\ &\quad + \int_{B_1} |\psi(z)| \int_{\mathbb{R}^n} \frac{|u(x)|}{1+|x|^{n+2s}}\,dx\,dz \\ &= \|v\|_{L^1(B_1\times\mathbb{R}^n)}.\end{aligned}$$

Moreover, since $1+|x-z/k|^{n+2s} \le d_{n,s}(1+|x|^{n+2s})$ for all $(z,x) \in B_1 \times \mathbb{R}^n$ and $k \ge 1$, for some constant $d_{n,s} > 1$ depending only on n and s, we see that, for the integrand in (12.18),

$$|\psi(z)|\frac{|u(x-z/k) - u(x)|}{1+|x|^{n+2s}} \le d_{n,s} v_k(z,x) \qquad \text{for all } k \ge 1.$$

Therefore, the generalized dominated convergence theorem can be applied in (12.18) to get that $I_k \to 0$ as $k \to \infty$. In conclusion, $u_k \to u$ as $k \to \infty$ in the norm of L_s.

Assume that $s < 1/2$. We first verify that the pointwise formula for $(-\Delta)^s u_k(x)$, for $x \in \Omega$, converges to the pointwise formula with u in place of u_k, that is,

$$\lim_{k\to\infty} (-\Delta)^s u_k(x) = \lim_{k\to\infty} c_{n,s} \int_{\mathbb{R}^n} \frac{u_k(x) - u_k(z)}{|x-z|^{n+2s}}\,dz = c_{n,s} \int_{\mathbb{R}^n} \frac{u(x) - u(z)}{|x-z|^{n+2s}}\,dz.$$

Let $\varepsilon > 0$. Fix $x \in \Omega'$ and let $\delta > 0$ small such that $B_\delta(x) \subset \Omega'$. We write

$$(-\Delta)^s u_k(x) = c_{n,s} \int_{B_\delta(x)} \frac{u_k(x) - u_k(z)}{|x-z|^{n+2s}}\, dz + c_{n,s} \int_{\mathbb{R}^n \setminus B_\delta(x)} \frac{u_k(x) - u_k(z)}{|x-z|^{n+2s}}\, dz.$$

Let us choose $\delta > 0$ sufficiently small such that

$$c_{n,s} \int_{B_\delta} \frac{[u_k]_{C^{\alpha+2\varepsilon}(\Omega')}}{|z|^{n-\varepsilon}}\, dz \le c_{n,s} \int_{B_\delta} \frac{[u]_{C^{\alpha+2\varepsilon}(\Omega)}}{|z|^{n-\varepsilon}}\, dz < \frac{\varepsilon}{3}. \tag{12.19}$$

Then

$$\left| c_{n,s} \int_{B_\delta(x)} \frac{u_k(x) - u_k(z)}{|x-z|^{n+2s}}\, dz \right| \le \frac{\varepsilon}{3}$$

and

$$\left| c_{n,s} \int_{B_\delta(x)} \frac{u(x) - u(z)}{|x-z|^{n+2s}}\, dz \right| \le \frac{\varepsilon}{3}.$$

Let $R > 0$ such that $\Omega' \subset B_R$. If $|x - z| \ge \delta$ and $|z| < 2R$, then

$$\frac{1 + |z|^{n+2s}}{|x-z|^{n+2s}} \le C_{n,s,\delta,R}.$$

If $|x - z| \ge \delta$ and $|z| \ge 2R$ then $|z| \le |x - z| + |x| \le |x - z| + R$, so that $|z|/2 \le |z| - R \le |x - z|$ and again

$$\frac{1 + |z|^{n+2s}}{|x-z|^{n+2s}} \le C_{n,s,\delta,R}.$$

With this, for all k sufficiently large,

$$\begin{aligned}
&\left| (-\Delta)^s u_k(x) - c_{n,s} \int_{\mathbb{R}^n} \frac{u(x) - u(z)}{|x-z|^{n+2s}}\, dz \right| \\
&\quad \le \frac{2\varepsilon}{3} + \left| c_{n,s} \int_{\mathbb{R}^n \setminus B_\delta(x)} \frac{u_k(x) - u_k(z)}{|x-z|^{n+2s}}\, dz - c_{n,s} \int_{\mathbb{R}^n \setminus B_\delta(x)} \frac{u(x) - u(z)}{|x-z|^{n+2s}}\, dz \right| \\
&\quad \le \frac{2\varepsilon}{3} + c_{n,s} \int_{\mathbb{R}^n \setminus B_\delta(x)} \frac{|u_k(x) - u(x)|}{|x-z|^{n+2s}}\, dz + c_{n,s} \int_{\mathbb{R}^n \setminus B_\delta(x)} \frac{|u_k(z) - u(z)|}{|x-z|^{n+2s}}\, dz \\
&\quad \le \frac{2\varepsilon}{3} + C_{n,s,\delta} \|u_k - u\|_{L^\infty(\Omega')} + C_{n,s,\delta,R} \|u_k - u\|_{L_s} \\
&\quad < \varepsilon
\end{aligned}$$

uniformly in x such that $B_\delta(x) \subset \Omega'$, where we used that $u_k \to u$ uniformly in Ω' and $u_k \to u$ in the norm of L_s, as $k \to \infty$. In particular, since $(-\Delta)^s u_k \in C(\Omega)$ and Ω' was arbitrary, the pointwise formula for u defines a continuous function in Ω. To see that it coincides with $(-\Delta)^s u$ in Ω in the sense of distributions, note that the convergence of u_k to u in $L_s \subset \mathcal{S}_s$ and the continuity of $(-\Delta)^s$ from $\mathcal{S}'_s$ into $\mathcal{S}'$ give

$$(-\Delta)^s u_k \to (-\Delta)^s u \qquad \text{in } \mathcal{S}'_s.$$

Thus, for any $\varphi \in C_c^\infty(\Omega')$,

$$
\begin{aligned}
((-\Delta)^s u)(\varphi) &= u((-\Delta)^s \varphi) \\
&= \lim_{k\to\infty} u_k((-\Delta)^s\varphi) \\
&= \lim_{k\to\infty} \int_{\mathbb{R}^n} u_k(x)(-\Delta)^s\varphi(x)\,dx \\
&= \lim_{k\to\infty} \int_{\mathbb{R}^n} (-\Delta)^s u_k(x)\varphi(x)\,dx \\
&= \int_{\mathbb{R}^n} \Big[\lim_{k\to\infty} (-\Delta)^s u_k(x)\Big]\varphi(x)\,dx \\
&= \int_{\mathbb{R}^n} \Big[c_{n,s}\int_{\mathbb{R}^n} \frac{u(x)-u(z)}{|x-z|^{n+2s}}\,dz\Big]\varphi(x)\,dx
\end{aligned}
$$

where in the second to last line we applied the dominated convergence therorem. Therefore, $(-\Delta)^s u$ coincides with the function

$$
c_{n,s}\int_{\mathbb{R}^n} \frac{u(x)-u(z)}{|x-z|^{n+2s}}\,dz \qquad \text{in } \Omega'.
$$

Again, since Ω' was arbitrary, the conclusion holds for all $x \in \Omega$.

The proof for the case $s \geq 1/2$ follows the same lines. The only differences are that (12.19) must be replaced by

$$
c_{n,s}\int_{B_\delta} \frac{[\nabla u_k]_{C^{\alpha+2\varepsilon-1}(\Omega')}}{|z|^{n-\varepsilon}}\,dz \leq c_{n,s}\int_{B_\delta} \frac{[\nabla u]_{C^{\alpha+2\varepsilon-1}(\Omega)}}{|z|^{n-\varepsilon}}\,dz < \frac{\varepsilon}{3}
$$

and the second formula in Theorem 12.1(2) must now be used with this $\delta > 0$.

Exactly as we did at the end of the proof of Theorem 12.1, the formula

$$
-(-\Delta)^s u(x) = \frac{c_{n,s}}{2}\int_{\mathbb{R}^n} \frac{u(x+z)+u(x-z)-2u(x)}{|z|^{n+2s}}\,dz
$$

holds for any $x \in \Omega$ and $0 < s < 1$. Observe that if $s \geq 1/2$ the integral is in fact absolutely convergent because, for all $|z|$ small,

$$
|u(x+z)+u(x-z)-2u(x)| \leq C[\nabla u]_{C^{2s+\varepsilon-1}(\Omega)}|z|^{2s+\varepsilon}.
$$

For the semigroup formula, notice that, by the exponential decay of $G_t(x)$,

$$
e^{t\Delta}u(x) = \int_{\mathbb{R}^n} G_t(x-z)u(z)\,dz \qquad \text{for } u \in L_s
$$

is well defined as an absolutely convergent integral. Therefore, all the steps done in the proof of Theorem 12.1 can be reversed to show that $(-\Delta)^s u(x)$ coincides with the semigroup formula for any $x \in \Omega$. Notice that the semigroup formula is absolutely convergent because in (12.10) one can use the regularity of u around x, (12.11) and the fact that $u \in L_s$. □

Let Ω be a domain of $\mathbb{R}^n$, $n \geq 1$, and consider $f \in C(\Omega)$. We say that $u \in L_s$ is a **classical solution** to

$$(-\Delta)^s u = f \qquad \text{in } \Omega$$

if $u \in C^{2s+\varepsilon}(\Omega)$ if $s < 1/2$, or $u \in C^{1,2s+\varepsilon-1}(\Omega)$ if $s \geq 1/2$, for some $\varepsilon > 0$, and

$$(-\Delta)^s u(x) = f(x) \qquad \text{for every } x \in \Omega$$

where $(-\Delta)^s u(x)$ is given by the pointwise formulas of Theorem 12.3. With this notion, Theorem 12.3 says that sufficiently regular distributional solutions are, in fact, classical solutions.

12.2 POINTWISE LIMITS

It is clear from the Fourier transform definition (12.1) that if $u \in \mathcal{S}$ then

$$\lim_{s \to 1} (-\Delta)^s u = -\Delta u \qquad \text{and} \qquad \lim_{s \to 0} (-\Delta)^s u = u.$$

However, by looking at the pointwise formulas in Theorem 12.3, it is not obvious why these limits should hold for less regular functions. As a matter of fact, the constant $c_{n,s} > 0$ that appears in front of the integral formulas for the fractional Laplacian, see (12.9), can be written as

$$c_{n,s} = \frac{4^s \Gamma(n/2+s)}{|\Gamma(-s)| \pi^{n/2}} = \frac{s 4^s \Gamma(n/2+s)}{\Gamma(1-s) \pi^{n/2}} = \frac{s(1-s) 4^s \Gamma(n/2+s)}{\Gamma(2-s) \pi^{n/2}}. \tag{12.20}$$

It follows that

$$c_{n,s} \sim s(1-s) \qquad \text{as } s \to 0 \text{ or } s \to 1$$

that is, $c_{n,s}$ degenerates linearly when $s \to 0$ or $s \to 1$. This asymptotic behavior of $c_{n,s}$ must be combined with appropriate cancelations in order to prove the following two results.

Theorem 12.4 (Poinwise limit as $s \to 1$). *Suppose that $u \in L^\infty(\mathbb{R}^n) \cap C^2(B_2(x))$ for some $x \in \mathbb{R}^n$. Then*

$$\lim_{s \to 1} (-\Delta)^s u(x) = -\Delta u(x).$$

Proof. By Theorem 12.3, $(-\Delta)^s u(x)$ is well defined and, for any $\delta > 0$,

$$\begin{aligned}(-\Delta)^s u(x) &= c_{n,s} \int_{|x-z|>\delta} \frac{u(x)-u(z)}{|x-z|^{n+2s}}\,dz \\ &\quad + c_{n,s} \int_{|x-z|<\delta} \frac{u(x)-u(z)-\nabla u(x)\cdot(x-z)}{|x-z|^{n+2s}}\,dz \\ &=: I + II.\end{aligned}$$

Clearly, by (12.20) and passing to polar coordinates,

$$|I| \leq C_n (1-s) \|u\|_{L^\infty(\mathbb{R}^n)} \int_\delta^\infty \frac{r^{n-1}}{r^{n+2s}}\,dr = \frac{C_n}{(2s)\delta^{2s}} (1-s) \|u\|_{L^\infty(\mathbb{R}^n)} \to 0$$

as $s \to 1$. Let $\varepsilon > 0$ and choose $0 < \delta = \delta(\varepsilon) < 1$ such that

$$|D^2u(z) - D^2u(z')| < \varepsilon \qquad \text{for all } z, z' \in B_1(x) \text{ with } |z - z'| < \delta. \tag{12.21}$$

By changing variables and using polar coordinates, we can write

$$\begin{aligned}
II &= c_{n,s} \int_0^\delta \frac{r^{n-1}}{r^{n+2s}} \int_{|w|=1} (u(x) - u(x - rw) - \nabla u(x) \cdot (rw))\, dS_w\, dr \\
&= c_{n,s} \int_0^\delta r^{-1-2s} \int_{|w|=1} \left[\frac{1}{2}(rw)^T D^2u(x)(rw) - R_1u(x, rw)\right] dS_w\, dr \\
&\quad - c_{n,s} \int_0^\delta r^{-1-2s} \int_{|w|=1} \frac{1}{2}(rw)^T D^2u(x)(rw)\, dS_w\, dr \\
&=: II_1 - II_2
\end{aligned}$$

where $R_1u(x, rw) = u(x - rw) - u(x) + \nabla u(x) \cdot (rw)$. Observe that, for some intermediate point ξ in between x and $x - rw$, $|w| = 1$, by the formula for the Taylor remainder $R_1u(x, rw)$ and (12.21),

$$\begin{aligned}
\left|\frac{1}{2}(rw)^T D^2u(x)(rw) - R_1u(x, rw)\right| &= \frac{r^2}{2}|w^T(D^2u(x) - D^2u(\xi))w| \\
&\leq \frac{r^2}{2}|w|^2\varepsilon = \varepsilon\frac{r^2}{2}.
\end{aligned}$$

Hence, by (12.20),

$$|II_1| \leq C_n\varepsilon(1 - s) \int_0^\delta r^{1-2s}\, dr = \frac{C_n}{2}\varepsilon\delta^{2(1-s)} \leq C\varepsilon.$$

Next, since w_iw_j is an odd function whenever $i \neq j$,

$$\begin{aligned}
\int_{|w|=1} w^T D^2u(x)w\, dS_w &= \sum_{i,j=1}^n \frac{\partial^2u(x)}{\partial x_i \partial x_j} \int_{|w|=1} w_iw_j\, dw \\
&= \sum_{i=1}^n \frac{\partial^2u(x)}{\partial x_i^2} \int_{|w|=1} (w_i)^2\, dw \\
&= \sum_{i=1}^n \frac{\partial^2u(x)}{\partial x_i^2} \int_{|w|=1} (w_1)^2\, dw \\
&= \frac{(n/2 + 1)\pi^{n/2}}{\Gamma(n/2 + 2)} \Delta u(x)
\end{aligned}$$

where in the last line we applied (3.19). Therefore, with this, (12.20) and the functional identity for the Gamma function,

$$\begin{aligned}
II_2 &= c_{n,s} \int_0^\delta r^{-1-2s}\frac{r^2}{2}\frac{(n/2 + 1)\pi^{n/2}}{\Gamma(n/2 + 2)} \Delta u(x)\, dr \\
&= \Delta u(x)\frac{s4^{s-1}\Gamma(n/2 + s)}{\Gamma(2 - s)\Gamma(n/2 + 1)}\delta^{2-2s} \\
&\to \Delta u(x)
\end{aligned}$$

as $s \to 1$. Since $\varepsilon > 0$ was arbitrary, the conclusion follows. □

For the next result, let L_0 be defined as the space L_s in (12.17) with $s = 0$.

Theorem 12.5 (Pointwise limit as $s \to 0$). *Suppose that $u \in L_0 \cap C^\alpha(B_2(x))$ for some $x \in \mathbb{R}^n$ and some $0 < \alpha < 1$. Then*

$$\lim_{s \to 0} (-\Delta)^s u(x) = u(x).$$

Proof. Observe that $L_0 \subset L_s$ for any $0 < s < 1$. Since we are interested in the limit as $s \to 0$, we can assume that $2s < \alpha$. By Theorem 12.3, $(-\Delta)^s u(x)$ is well defined and, for any $R > 0$, we can write

$$\begin{aligned}(-\Delta)^s u(x) &= c_{n,s} \int_{|x-z|<R} \frac{u(x) - u(z)}{|x-z|^{n+2s}}\, dz \\ &\quad + u(x) c_{n,s} \int_{|x-z|>R} \frac{1}{|x-z|^{n+2s}}\, dz - c_{n,s} \int_{|x-z|>R} \frac{u(z)}{|x-z|^{n+2s}}\, dz \\ &=: I + II - III.\end{aligned}$$

By (12.20) and the regularity of u,

$$|I| \leq s C_n [u]_{C^\alpha(x)} \int_{|x-z|<R} \frac{|x-z|^\alpha}{|x-z|^{n+2s}}\, dz = s C_n [u]_{C^\alpha(x)} \frac{R^{\alpha - 2s}}{\alpha - 2s} \to 0$$

as $s \to 0$. Next, by using polar coordinates, the formula for the surface area of ∂B_1, see (2.7), (12.20) and the functional identity for the Gamma function,

$$II = u(x) c_{n,s} \frac{|\partial B_1|}{(2s) R^{2s}} = u(x) \frac{n 4^{s-1/2} \Gamma(n/2 + s)}{\Gamma(n/2+1)\Gamma(1-s) R^{2s}} \to u(x)$$

as $s \to 0$. For the integral in III, we choose R conveniently so we can replace $|x-z|$ in the denominator by $1 + |z|$. Let $R = 1 + |x|$. Then, for $|x - z| > R$ we can write $1 + |z| \leq 1 + |x-z| + |x| \leq 2|x-z|$. Thus, by (12.20) and using that $(1+|z|)^n \leq (1+|z|)^{n+2s}$,

$$|III| \leq 2^{n+2s} c_{n,s} \int_{\mathbb{R}^n} \frac{|u(z)|}{(1+|z|)^{n+2s}}\, dz \leq s C_n \|u\|_{L_0} \to 0$$

as $s \to 0$. □

Remark 12.6. If u is a constant then $(-\Delta)^s u = 0$ in $\mathbb{R}^n$, so in this case it is not possible that $(-\Delta)^s u(x) \to u(x)$ as $s \to 0$ unless $u \equiv 0$. But this is not a contradiction with Theorem 12.5 for two reasons. First, the result requires that $u \in L_0$, so constants are ruled out. Second, the only constant function u in the Schwartz class $\mathcal{S}$ is $u = 0$, so the Fourier transform proof for this case still works.

12.3 MAXIMUM AND COMPARISON PRINCIPLES

Unlike the case of the Laplacian or more general nondivergence form elliptic operators, the maximum principle for the fractional Laplacian is quite easy to prove. In some sense, this is true because $(-\Delta)^s u(x)$ already carries global information on u.

Theorem 12.7 (Maximum and comparsion principles). *Let* $0 < s < 1$.

(1) *Let* $u \in L_s$ *such that*

$$u \geq 0 \text{ in } \mathbb{R}^n \qquad \text{and} \qquad u(x_0) = 0 \text{ for some } x_0 \in \mathbb{R}^n.$$

Suppose that

$$u \in C^{2s+\varepsilon} \text{ if } s < 1/2 \qquad \text{or} \qquad u \in C^{1,2s+\varepsilon-1} \text{ if } s \geq 1/2$$

in a neighborhood of x_0, *for some* $\varepsilon > 0$. *Then*

$$(-\Delta)^s u(x_0) \leq 0.$$

Moreover, $(-\Delta)^s u(x_0) = 0$ *if and only if* $u \equiv 0$.

In other words, if x_0 *is a point where* u *attains a global minimum then, at* x_0, *"the fractional Laplacian has the correct sign," namely,* $-(-\Delta)^s u(x_0) \geq 0$, *as* u *is "bending upwards" at* x_0.

(2) *Let* $u, v \in L_s$ *such that*

$$u \geq v \text{ in } \mathbb{R}^n \qquad \text{and} \qquad u(x_0) = v(x_0) \text{ for some } x_0 \in \mathbb{R}^n.$$

Suppose that

$$u, v \in C^{2s+\varepsilon} \text{ if } s < 1/2 \qquad \text{or} \qquad u, v \in C^{1,2s+\varepsilon-1} \text{ if } s \geq 1/2$$

in a neighborhood of x_0, *for some* $\varepsilon > 0$. *Then*

$$(-\Delta)^s u(x_0) \leq (-\Delta)^s v(x_0).$$

Moreover, $(-\Delta)^s u(x_0) = (-\Delta)^s v(x_0)$ *if and only if* $u \equiv v$.

Proof. For (1), under the given hypotheses of u, by Theorem 12.3,

$$(-\Delta)^s u(x_0) = c_{n,s} \int_{\mathbb{R}^n} \frac{-u(z)}{|x_0 - z|^{n+2s}} \, dz \leq 0.$$

For the second part of the statement in (1), if u is not identically zero then $u(z) > 0$ in a set of positive measure, and then the last inequality is strict.

For (2) we apply (1) to the difference $u - v \geq 0$. □

In view of the pointwise formulas for $(-\Delta)^s u(x)$ (see Theorems 12.1 and 12.3), it becomes evident that the adequate **Poisson–Dirichlet problem** for the fractional Laplacian in a bounded domain Ω with "homogeneous boundary condition" is

$$\begin{cases} (-\Delta)^s u = f & \text{in } \Omega \\ u = 0 & \text{in } \mathbb{R}^n \setminus \Omega. \end{cases}$$

Indeed, we need to prescribe the values of u outside of Ω, as it is not enough to know u in $\overline{\Omega}$ in order to compute $(-\Delta)^s u(x)$ for $x \in \Omega$. Furthermore, changing the values of u only on $\partial\Omega$ does not affect the value of $(-\Delta)^s u$ because $\partial\Omega$ is a set of measure zero in $\mathbb{R}^n$. A consequence of these observations is that, unlike the case of the Laplacian (see Corollary 2.7), the weak maximum principle for the fractional Laplacian must involve the values of u outside of Ω.

For the sake of the following result, let us introduce the terminology of classical subsolutions and supersolutions. Recall Theorem 12.3. Let $\Omega \subset \mathbb{R}^n$, $n \geq 1$, be a domain and consider $f \in C(\Omega)$. Let $u \in L_s$ such that $u \in C^{2s+\varepsilon}(\Omega)$ if $s < 1/2$, or $u \in C^{1,2s+\varepsilon-1}(\Omega)$ if $s \geq 1/2$, for some $\varepsilon > 0$. We say that u is a **classical subsolution** to $(-\Delta)^s u = f$ in Ω if

$$(-\Delta)^s u(x) \leq f(x) \qquad \text{for every } x \in \Omega.$$

Similarly, we say that u is a **classical supersolution** to $(-\Delta)^s u = f$ in Ω if

$$(-\Delta)^s u(x) \geq f(x) \qquad \text{for every } x \in \Omega.$$

Clearly, u is a classical solution to $(-\Delta)^s u = f$ in Ω if it is both a classical subsolution and a classical supersolution to $(-\Delta)^s u = f$ in Ω.

The reader may notice that the inequalities are reversed with respect to the definition of subsolutions and supersolutions to $\Delta u = f$, where we require $\Delta u \geq f$ and $\Delta u \leq f$, respectively. However, the inequalities in the definitions above are consistent with the Laplacian because $-(-\Delta)^s u \to \Delta u$ as $s \to 1$, and consistent as well with the maximum principle, see Theorem 12.7.

The following should be compared with the Laplacian case in Corollary 2.7.

Corollary 12.8 (Direct consequences of the maximum principle)**.** *Let Ω be a bounded domain and $0 < s < 1$. Let $u, v \in L_s \cap C(\overline{\Omega})$ such that*

$$u, v \in C^{2s+\varepsilon}(\Omega) \text{ if } s < 1/2 \qquad \text{or} \qquad u, v \in C^{1,2s+\varepsilon-1}(\Omega) \text{ if } s \geq 1/2$$

for some $\varepsilon > 0$.

(1) (Sign of supersolutions). *If*

$$\begin{cases} (-\Delta)^s u \geq 0 & \text{in } \Omega \\ u = 0 & \text{in } \mathbb{R}^n \setminus \Omega, \end{cases}$$

then $u \geq 0$ in $\mathbb{R}^n$.

(2) (Weak maximum principle for subsolutions). *If*

$$\begin{cases} (-\Delta)^s u \leq 0 & \text{in } \Omega \\ u \leq 0 & \text{in } \mathbb{R}^n \setminus \Omega, \end{cases}$$

then

$$\sup_{\mathbb{R}^n} u = \sup_{\mathbb{R}^n \setminus \Omega} u.$$

(3) (Weak maximum principle for supersolutions). *If*

$$\begin{cases} (-\Delta)^s u \geq 0 & \text{in } \Omega \\ u \geq 0 & \text{in } \mathbb{R}^n \setminus \Omega, \end{cases}$$

then

$$\inf_{\mathbb{R}^n} u = \inf_{\mathbb{R}^n \setminus \Omega} u.$$

(4) (Ordering). *If*

$$\begin{cases} (-\Delta)^s u \geq (-\Delta)^s v & \text{in } \Omega \\ u \geq v & \text{in } \mathbb{R}^n \setminus \Omega, \end{cases}$$

then

$$u \geq v \qquad \text{in } \mathbb{R}^n.$$

(5) (Uniqueness for the Poisson–Dirichlet problem). *If*

$$\begin{cases} (-\Delta)^s u = (-\Delta)^s v & \text{in } \Omega \\ u = v & \text{in } \mathbb{R}^n \setminus \Omega, \end{cases}$$

then

$$u = v \qquad \text{in } \mathbb{R}^n.$$

Proof. For (1), notice that, by our assumptions, $u \in C(\mathbb{R}^n)$. By contradiction, suppose that $x_0 \in \Omega$ is a point where u attains a negative global minimum, that is, $u(x_0) = \min_{\mathbb{R}^n} u < 0$. We apply Theorem 12.7(1) to the vertical translation $\tilde{u}$ of u given by $\tilde{u} = u - u(x_0) \in L_s$. We have $(-\Delta)^s \tilde{u}(x_0) = (-\Delta)^s u(x_0) \leq 0$. If $(-\Delta)^s u(x_0) = 0$ then $u \equiv u(x_0) < 0$, which is a contradiction to $u = 0$ in $\mathbb{R}^n \setminus \Omega$. Then we must have $(-\Delta)^s u(x_0) < 0$, which is now a contraction to $(-\Delta)^s u(x_0) \geq 0$. Hence $u \geq 0$ in $\mathbb{R}^n$.

The proof of (2) is by contradiction. We can assume that u is not identically constant. Let $x_0 \in \Omega$ such that $u(x_0) = \sup_{\mathbb{R}^n} u$. We apply Theorem 12.7(1) to $\tilde{u} = u(x_0) - u \in L_s$. We obtain $-(-\Delta)^s \tilde{u}(x_0) = (-\Delta)^s u(x_0) \geq 0$. If $(-\Delta)^s u(x_0) = 0$ then $u \equiv u(x_0)$ and this is a contradiction to our assumption that u is not identically constant. If $(-\Delta)^s u(x_0) > 0$ then we are in contradiction with the equation $(-\Delta)^s u(x_0) \leq 0$. Hence, x_0 cannot be in Ω.

For (3), apply (2) to $-u$ and use that $\sup(-u) = -\inf u$.

The proof of (4) follows by applying (3) to the difference $u - v$.

Finally, (5) is a direct consequence of (4). □

12.4 THE INVERSE FRACTIONAL LAPLACIAN

In this section we consider the global **Poisson equation** for the fractional Laplacian

$$(-\Delta)^s u = f \qquad \text{in } \mathbb{R}^n.$$

Since the problem is posed on $\mathbb{R}^n$, we can apply the Fourier transform to find that $|\xi|^{2s}\widehat{u}(\xi) = \widehat{f}(\xi)$ and, therefore, $\widehat{u}(\xi) = |\xi|^{-2s}\widehat{f}(\xi)$.

Given $s > 0$, the **inverse of the fractional Laplacian**, or **negative power of the Laplacian**, $(-\Delta)^{-s}$ is defined for $f \in \mathcal{S}$ as

$$\widehat{(-\Delta)^{-s} f}(\xi) = |\xi|^{-2s}\widehat{f}(\xi) \qquad \xi \neq 0. \tag{12.22}$$

Note that in (12.22) we are raising the Fourier multiplier of the Laplacian to the power $-s$. Equivalently,

$$(-\Delta)^{-s} f(x) = \mathcal{F}^{-1}\big(|\xi|^{-2s}\widehat{f}(\xi)\big)(x) \qquad \text{for } x \in \mathbb{R}^n.$$

Similarly to the case of the fractional Laplacian, see (12.2), the inverse fractional Laplacian rescales as an operator of *fractional differential order* $-2s$ or, equivalently, as an operator of *fractional integral order* $2s$, in the sense that if $f_R(x) = f(Rx)$, where $R > 0$, then

$$(-\Delta)^{-s} f_R(x) = R^{-2s}[(-\Delta)^{-s} f](Rx).$$

In principle, definition (12.22) works for any $s > 0$. However, we need the restriction $0 < s < n/2$ because when $s \geq n/2$ the Fourier multiplier $|\xi|^{-2s} \notin L^1_{\mathrm{loc}}(\mathbb{R}^n)$ and, therefore, does not even define a tempered distribution. In the harmonic analysis literature, $(-\Delta)^{-s}$ is known as the **fractional integral operator of order** $2s$.

The next step is to find a pointwise expression of $(-\Delta)^{-s} f(x)$ at a point $x \in \mathbb{R}^n$ that does not involve the Fourier transform of f. As we did for the case of the fractional Laplacian, we apply the general **method of semigroups** for fractional operators. For the inverse fractional Laplacian, and parallel to Section 12.1, the method begins with a numerical formula involving the Gamma function. Recall the definition of the Gamma function: for any $s > 0$,

$$\Gamma(s) = \int_0^\infty e^{-r}\,\frac{dr}{r^{1-s}}.$$

Let $\lambda > 0$ and perform the change of variables $r = t\lambda$ to get

$$\lambda^{-s} = \frac{1}{\Gamma(s)}\int_0^\infty e^{-t\lambda}\,\frac{dt}{t^{1-s}}. \tag{12.23}$$

Now choose $\lambda = |\xi|^2$ in the identity above, multiply by $\widehat{f}(\xi)$ and recall (12.22) to get

$$\widehat{(-\Delta)^{-s} f}(\xi) = |\xi|^{-2s}\widehat{f}(\xi) = \frac{1}{\Gamma(s)}\int_0^\infty e^{-t|\xi|^2}\widehat{f}(\xi)\,\frac{dt}{t^{1-s}}.$$

Therefore, after applying the inverse Fourier transform,

$$(-\Delta)^{-s} f(x) = \frac{1}{(2\pi)^{n/2}\Gamma(s)}\int_{\mathbb{R}^n}\int_0^\infty e^{-t|\xi|^2}\widehat{f}(\xi)e^{ix\cdot\xi}\,\frac{dt}{t^{1-s}}\,d\xi. \tag{12.24}$$

If $0 < s < n/2$, then the double integral above is absolutely convergent. By (12.23),

$$\frac{1}{\Gamma(s)}\int_{\mathbb{R}^n}\int_0^\infty e^{-t|\xi|^2}|\widehat{f}(\xi)|\,\frac{dt}{t^{1-s}}\,d\xi = \int_{\mathbb{R}^n}|\xi|^{-2s}|\widehat{f}(\xi)|\,dx < \infty$$

because $f \in \mathcal{S}$ and $|\xi|^{-2s} \in L^1_{\text{loc}}(\mathbb{R}^n)$. We can also take $s = n/2$ under the additional assumption that $\widehat{f}(0) = 0$ because, in this case, by the mean value theorem,

$$\begin{aligned}\int_{|\xi|<1} |\xi|^{-n}|\widehat{f}(\xi)|\,dx &= \int_{|\xi|<1} |\xi|^{-n}|\widehat{f}(\xi) - \widehat{f}(0)|\,dx \\ &\leq \|\nabla \widehat{f}\|_{L^\infty(\mathbb{R}^n)} \int_{|\xi|<1} |\xi|^{-n+1}\,d\xi < \infty.\end{aligned}$$

Hence, after applying Fubini's theorem in (12.24), we arrive to the **semigroup formula for the inverse fractional Laplacian**:

$$(-\Delta)^{-s} f(x) = \frac{1}{\Gamma(s)} \int_0^\infty e^{t\Delta} f(x)\,\frac{dt}{t^{1-s}} \qquad \text{for } x \in \mathbb{R}^n$$

and $f \in \mathcal{S}$. Here, as in Section 12.1, $\{e^{t\Delta}\}_{t\geq 0}$ is the classical heat diffusion semigroup.

We are ready to give the pointwise formula for $(-\Delta)^{-s} f(x)$. The reader should compare the proof of the following result with that of Theorem 12.1. Indeed, by means of the semigroup methodology, we avoid the direct computation of the inverse Fourier transform of (12.22). Instead, we take advantage of the formula for the heat kernel (12.7) and obtain the explicit value of the constant $c_{n,-s} > 0$ (which is consistent with $c_{n,s}$ in (12.9)) in the following statement. In addition, the methodology was already encountered in the proof of the formula for the inverse Laplacian (fundamental solution), see Section 3.4.

Theorem 12.9 (Pointwise formula for $(-\Delta)^{-s}$). *Let $f \in \mathcal{S}$, $x \in \mathbb{R}^n$ and $0 < s < n/2$. Then*

$$(-\Delta)^{-s} f(x) = c_{n,-s} \int_{\mathbb{R}^n} \frac{f(z)}{|x-z|^{n-2s}}\,dz \tag{12.25}$$

where the integral is absolutely convergent. The constant $c_{n,-s} > 0$ is given by replacing s by $-s$ in the constant $c_{n,s}$ of Theorem 12.1, see (12.9), *namely,*

$$c_{n,-s} = \frac{\Gamma(n/2 - s)}{4^s \Gamma(s) \pi^{n/2}} \qquad \text{for } 0 < s < n/2. \tag{12.26}$$

If, in addition, $\int_{\mathbb{R}^n} f\,dz = 0$ then

$$(-\Delta)^{-n/2} f(x) = \frac{-2}{(4\pi)^{n/2}\Gamma(n/2)} \int_{\mathbb{R}^n} f(z) \log|x-z|\,dz$$

where the integral is absolutely convergent.

Proof. By writing $e^{t\Delta} f(x)$ in terms of the heat kernel (12.7),

$$(-\Delta)^{-s} f(x) = \frac{1}{\Gamma(s)} \int_0^\infty \int_{\mathbb{R}^n} G_t(x-z) f(z)\,\frac{dt}{t^{1-s}}. \tag{12.27}$$

Suppose that $0 < s < n/2$. Then the double integral in (12.27) is absolutely convergent. Indeed, we have

$$\int_0^\infty \int_{\mathbb{R}^n} G_t(x-z)|f(z)|\,\frac{dt}{t^{1-s}} = \int_{\mathbb{R}^n} |f(z)| \int_0^\infty G_t(x-z)\,\frac{dt}{t^{1-s}}\,dz.$$

To compute the integral in t, by the change of variables $r = |x-z|^2/(4t)$,

$$\begin{aligned}\frac{1}{\Gamma(s)}\int_0^\infty G_t(x-z)\,\frac{dt}{t^{1-s}} &= \frac{1}{\Gamma(s)(4\pi)^{n/2}}\int_0^\infty \frac{e^{-|x-z|^2/(4t)}}{t^{n/2-s}}\,\frac{dt}{t}\\ &= \frac{1}{\Gamma(s)(4\pi)^{n/2}}\int_0^\infty e^{-r}\left(\frac{4r}{|x-z|^2}\right)^{n/2-s}\frac{dr}{r}\\ &= \frac{\Gamma(n/2-s)}{4^s\Gamma(s)\pi^{n/2}}\cdot\frac{1}{|x-z|^{n-2s}}.\end{aligned} \tag{12.28}$$

Hence, since $f \in \mathcal{S}$ and $1/|x-z|^{n-2s} \in L^1_{\mathrm{loc}}(\mathbb{R}^n)$, we find that (12.27) is an absolutely convergent integral. Therefore, by Fubini's theorem and (12.28), we obtain (12.25) with $c_{n,-s} > 0$ as in (12.26).

Consider the case $2s = n$. We perform the change of variables $r = 1/(4t)$ in (12.27) and use that $\int_{\mathbb{R}^n} f(x-z)\,dz = 0$ to write

$$\begin{aligned}(-\Delta)^{-n/2}f(x) &= \frac{1}{(4\pi)^{n/2}\Gamma(n/2)}\int_0^\infty\int_{\mathbb{R}^n} e^{-|z|^2/(4t)}f(x-z)\,dz\,\frac{dt}{t}\\ &= \frac{1}{(4\pi)^{n/2}\Gamma(n/2)}\int_0^\infty\int_{\mathbb{R}^n} e^{-r|z|^2}f(x-z)\,dz\,\frac{dr}{r}\\ &= \frac{1}{(4\pi)^{n/2}\Gamma(n/2)}\int_0^\infty\int_{\mathbb{R}^n} (e^{-r|z|^2}-\chi_{(0,1)}(r))f(x-z)\,dz\,\frac{dr}{r}.\end{aligned}$$

We see that the last double integral is absolutely convergent because $f \in \mathcal{S}$ and, by the change of variables $t = r|z|^2$ and integration by parts,

$$\begin{aligned}\int_0^\infty \left|e^{-r|z|^2}-\chi_{(0,1)}(r)\right|\frac{dr}{r} &= \int_0^{|z|^2}(1-e^{-t})\,\frac{dt}{t} + \int_{|z|^2}^\infty e^{-t}\,\frac{dt}{t}\\ &= (1-e^{-|z|^2})\log(|z|^2) - \int_0^{|z|^2} e^{-t}\log t\,dt\\ &\quad - e^{-|z|^2}\log(|z|^2) + \int_{|z|^2}^\infty e^{-t}\log t\,dt\\ &\le C(1+\log|z|)\end{aligned} \tag{12.29}$$

which is locally integrable. Therefore, by Fubini's theorem,

$$(-\Delta)^{-n/2}f(x) = \frac{1}{(4\pi)^{n/2}\Gamma(n/2)}\int_{\mathbb{R}^n} f(x-z)\left[\int_0^\infty (e^{-r|z|^2}-\chi_{(0,1)}(r))\,\frac{dr}{r}\right]dz.$$

The integral in brackets can be computed as done in (12.29) (see also (3.16)) to get

$$\int_0^1 (e^{-r|z|^2}-1)\,\frac{dr}{r} + \int_1^\infty e^{-r|z|^2}\,\frac{dr}{r} = -\log(|z|^2) - \gamma,$$

where $\gamma = -\int_0^\infty e^{-r}\log r\,dr > 0$ is the Euler–Mascheroni constant. Hence

$$\begin{aligned}(-\Delta)^{-n/2}f(x) &= \frac{1}{(4\pi)^{n/2}\Gamma(n/2)}\int_{\mathbb{R}^n} f(x-z)\big(-\log(|z|^2)-\gamma\big)\,dz\\ &= \frac{-2}{(4\pi)^{n/2}\Gamma(n/2)}\int_{\mathbb{R}^n} f(x-z)\log|z|\,dz.\end{aligned}$$

□

Next we turn to showing that, under appropriate conditions on f, $u = (-\Delta)^{-s}f$ solves the global Poisson equation for the fractional Laplacian $(-\Delta)^s u = f$ in $\mathbb{R}^n$ in a suitable sense. The reader should recall the strategy we implemented in Section 3.5 to solve the Poisson equation for the Laplacian. In that case, after the integral formula for the inverse Laplacian $u = (-\Delta)^{-1}f$ was found, see Theorem 3.3, the problem reduced to that of finding suitable conditions on f in order to differentiate inside the integral sign. That will not be so for our present situation: directly computing the fractional Laplacian of a function is a rather complicated task. In the following result we use the method of semigroups in combination with the Fourier transform to show that the Poisson problem for the fractional Laplacian is solvable in the distributional sense (for comparison with the Laplacian case, see Section 4.1).

Observe that the kernel $\Phi(x) = c_{n,-s}/|x|^{n-2s}$ in the pointwise formula (12.25) is integrable at the origin but not at infinity. Therefore, the integral in (12.25) is well defined whenever f has some decay at infinity. To simplify matters, we next assume that f has compact support. We will see in Chapter 13 that, under suitable conditions on f, the distributional solution u given by Theorem 12.10 is regular and the Poisson equation can actually be understood in the classical sense.

Theorem 12.10 (Distributional solvability). *Let $f \in L^\infty(\mathbb{R}^n)$ with compact support and $0 < s < \min\{1, n/2\}$. Define*

$$u(x) = (-\Delta)^{-s}f(x) = \frac{1}{\Gamma(s)} \int_0^\infty e^{t\Delta} f(x) \frac{dt}{t^{1-s}} \qquad \text{for } x \in \mathbb{R}^n.$$

Then u coincides with the integral formula in (12.25), that is,

$$u(x) = c_{n,-s} \int_{\mathbb{R}^n} \frac{f(z)}{|x-z|^{n-2s}}\, dz \qquad \text{for all } x \in \mathbb{R}^n$$

where the integral is absolutely convergent and the constant $c_{n,-s} > 0$ is given by (12.26). Moreover, $u \in L^\infty(\mathbb{R}^n)$ with

$$\|u\|_{L^\infty(\mathbb{R}^n)} \le C \|f\|_{L^\infty(\mathbb{R}^n)}$$

where $C > 0$ is a constant depending only on n and s, and $|u(x)| \to 0$ as $|x| \to \infty$. Furthermore,

$$(-\Delta)^s u = f \qquad \text{in } \mathbb{R}^n \text{ in the sense of distributions.}$$

Proof. The arguments in the proof of Theorem 12.9 continue to hold for f bounded with compact support, so $u(x)$ coincides with the integral formula given in (12.25). For simplicity, assume that $\operatorname{supp}(f) \subset B_1$. Then, for any $x \in \mathbb{R}^n$,

$$\begin{aligned} |u(x)| &\le C\|f\|_{L^\infty(\mathbb{R}^n)} \int_{B_1} \frac{1}{|x-z|^{n-2s}}\, dz \\ &\le C\|f\|_{L^\infty(\mathbb{R}^n)} \int_{B_1(x)} \frac{1}{|x-z|^{n-2s}}\, dz = C\|f\|_{L^\infty(\mathbb{R}^n)}. \end{aligned}$$

Next, if $|z| < 1$ and $|x| > 2$ then $|x| - 1 \leq |x-z| + |z| - 1 \leq |x-z|$, so that

$$|u(x)| \leq C\|f\|_{L^\infty(\mathbb{R}^n)} \int_{B_1} \frac{1}{|x-z|^{n-2s}}\,dz \leq C\|f\|_{L^\infty(\mathbb{R}^n)} \frac{|B_1|}{(|x|-1)^{n-2s}} \to 0$$

as $|x| \to \infty$. We are left to show that $(-\Delta)^s u = f$ in $\mathbb{R}^n$ in the sense of distributions. Since u is bounded, $u \in L_s \subset \mathcal{S}_s'$ and $(-\Delta)^s u \in \mathcal{S}'$ is well defined as

$$(-\Delta)^s u(\varphi) = u((-\Delta)^s \varphi) = \int_{\mathbb{R}^n} u(x)(-\Delta)^s \varphi(x)\,dx \qquad \text{for } \varphi \in \mathcal{S}.$$

Recall that $(-\Delta)^s \varphi \in \mathcal{S}_s$, see Lemma 12.2. We need to prove that

$$\int_{\mathbb{R}^n} u(x)(-\Delta)^s \varphi(x)\,dx = \int_{\mathbb{R}^n} f(x)\varphi(x)\,dx$$

which shows that $(-\Delta)^s u = f$ in the sense of distributions. For this, notice that both integrals are absolutely convergent. Let us call $\psi = (-\Delta)^s \varphi$. We perform a Fourier transform computation and use the method of semigroups. Recall that, for the heat kernel G_t, $\widehat{G_t}(\xi) = e^{-t|\xi|^2}/(2\pi)^{n/2}$. By the Gamma function formula (12.23) with $\lambda = |\xi|^2$, Fubini's theorem, Parseval's identity and (12.28),

$$\begin{aligned} \frac{1}{(2\pi)^{n/2}} \int_{\mathbb{R}^n} |\xi|^{-2s}\overline{\widehat{\psi}(\xi)}\,d\xi &= \frac{1}{(2\pi)^{n/2}} \int_{\mathbb{R}^n} \left[\frac{1}{\Gamma(s)} \int_0^\infty e^{-t|\xi|^2} \frac{dt}{t^{1-s}}\right] \overline{\widehat{\psi}(\xi)}\,d\xi \\ &= \frac{1}{\Gamma(s)} \int_0^\infty \int_{\mathbb{R}^n} \frac{e^{-t|\xi|^2}}{(2\pi)^{n/2}} \overline{\widehat{\psi}(\xi)}\,d\xi \frac{dt}{t^{1-s}} \\ &= \frac{1}{\Gamma(s)} \int_0^\infty \int_{\mathbb{R}^n} G_t(x)\psi(x)\,dx \frac{dt}{t^{1-s}} \\ &= c_{n,-s} \int_{\mathbb{R}^n} \frac{\psi(x)}{|x|^{n-2s}}\,dx. \end{aligned} \tag{12.30}$$

Next, for any fixed $z \in \mathbb{R}^n$, by the translation property of the Fourier transform,

$$\frac{1}{(2\pi)^{n/2}} \int_{\mathbb{R}^n} |\xi|^{-2s}\overline{\widehat{\psi}(-\xi)} e^{iz\cdot\xi}\,d\xi = c_{n,-s} \int_{\mathbb{R}^n} \frac{\psi(z-x)}{|x|^{n-2s}}\,dx.$$

Multiplying both sides by $f(z)$ and integrating, we get

$$\begin{aligned} \int_{\mathbb{R}^n} |\xi|^{-2s} \widehat{f}(-\xi)\overline{\widehat{\psi}(-\xi)}\,d\xi &= \frac{1}{(2\pi)^{n/2}} \int_{\mathbb{R}^n}\int_{\mathbb{R}^n} |\xi|^{-2s}\overline{\widehat{\psi}(-\xi)} f(z) e^{iz\cdot\xi}\,dz\,d\xi \\ &= c_{n,-s} \int_{\mathbb{R}^n}\int_{\mathbb{R}^n} \frac{f(z)\psi(z-x)}{|x|^{n-2s}}\,dz\,dx. \end{aligned}$$

Using this, Parseval's identity and recalling that $\psi = (-\Delta)^s\varphi$,

$$\begin{aligned}\int_{\mathbb{R}^n} u(x)(-\Delta)^s\varphi(x)\,dx &= c_{n,-s}\int_{\mathbb{R}^n}\psi(x)\int_{\mathbb{R}^n}\frac{f(z)}{|x-z|^{n-2s}}\,dz\,dx\\ &= c_{n,-s}\int_{\mathbb{R}^n} f(z)\int_{\mathbb{R}^n}\frac{\psi(x)}{|z-x|^{n-2s}}\,dx\,dz\\ &= c_{n,-s}\int_{\mathbb{R}^n} f(z)\int_{\mathbb{R}^n}\frac{\psi(z-x)}{|x|^{n-2s}}\,dx\,dz\\ &= \int_{\mathbb{R}^n}|\xi|^{-2s}\widehat{f}(-\xi)\overline{\widehat{\psi}(-\xi)}\,d\xi\\ &= \int_{\mathbb{R}^n}|\xi|^{-2s}\widehat{f}(\xi)\overline{|\xi|^{2s}\widehat{\varphi}(\xi)}\,d\xi = \int_{\mathbb{R}^n} f(x)\varphi(x)\,dx.\end{aligned}$$

□

With the Fourier transform and semigroup methods of the proof of Theorem 12.10 we also obtain the following fundamental property of the kernel of the inverse fractional Laplacian.

Proposition 12.11 (Fractional Laplacian of the kernel of the inverse fractional Laplacian). *For $0 < s < \min\{1, n/2\}$,*

$$\Phi(x) = \frac{c_{n,-s}}{|x|^{n-2s}} \in L_s$$

and

$$(-\Delta)^s\Phi(x) = 0 \qquad \textit{for all } x \in \mathbb{R}^n \setminus \{0\}.$$

Proof. It is clear that Φ is locally integrable and belongs to L_s. Hence, $(-\Delta)^s\Phi$ is well defined in the sense of distributions. To see that $(-\Delta)^s\Phi(x) = 0$ for $x \neq 0$, let $\varphi \in C_c^\infty(\mathbb{R}^n)$ such that $0 \notin \operatorname{supp}(\varphi)$. Then, if $\psi = (-\Delta)^s\varphi$, by (12.30),

$$\begin{aligned}(-\Delta)^s\Phi(\varphi) &= \int_{\mathbb{R}^n}\Phi(x)(-\Delta)^s\varphi(x)\,dx\\ &= c_{n,-s}\int_{\mathbb{R}^n}\frac{\psi(x)}{|x|^{n-2s}}\,dx\\ &= \frac{1}{(2\pi)^{n/2}}\int_{\mathbb{R}^n}|\xi|^{-2s}\overline{\widehat{\psi}(\xi)}\,d\xi\\ &= \frac{1}{(2\pi)^{n/2}}\int_{\mathbb{R}^n}\overline{\widehat{\varphi}(\xi)}\,d\xi = \varphi(0) = 0.\end{aligned}$$

In other words, the distribution $(-\Delta)^s\Phi$ coincides with 0 outside of the origin. □

12.5 WEAK SOLUTIONS AND FRACTIONAL SOBOLEV SPACES

In this section we introduce a suitable fractional Sobolev space and an appropriate concept of weak solution to the Poisson–Dirichlet problem

$$\begin{cases}(-\Delta)^s u = f & \text{in } \Omega\\ u = 0 & \text{in } \mathbb{R}^n \setminus \Omega\end{cases} \tag{12.31}$$

where Ω, is a bounded domain of $\mathbb{R}^n$, $n \geq 1$, and $f \in L^2(\Omega)$. By applying the fractional Poincaré inequality (that we prove here) and the Riesz representation theorem, existence and uniqueness of weak solutions will be obtained. Furthermore, we show maximum and comparison principles for weak solutions.

To motivate the definition of weak solution to (12.31), observe the following. Let $u, v \in \mathcal{S}$. Then, by using Parseval's identity for the Fourier transform, we can always "pass half fractional Laplacian" of a "full fractional Laplacian" from u to v:

$$\begin{aligned}\int_{\mathbb{R}^n} (-\Delta)^s u(x) v(x)\, dx &= \int_{\mathbb{R}^n} |\xi|^{2s} \widehat{u}(\xi) \overline{\widehat{v}(\xi)}\, d\xi \\ &= \int_{\mathbb{R}^n} (|\xi|^s \widehat{u}(\xi)) (\overline{|\xi|^s \widehat{v}(\xi)})\, d\xi \\ &= \int_{\mathbb{R}^n} (-\Delta)^{s/2} u(x) (-\Delta)^{s/2} v(x)\, dx \\ &= \int_{\mathbb{R}^n} u(x) (-\Delta)^s v(x)\, dx.\end{aligned} \tag{12.32}$$

This formula is the parallel to the integration by parts identity

$$\int_\Omega (-\Delta u) v\, dx = \int_\Omega \nabla u \nabla v\, dx = \int_\Omega u(-\Delta v)\, dx \qquad \text{for } u, v \in H_0^1(\Omega).$$

Now we use the pointwise formula for the fractional Laplacian to derive a more explicit, simplified integral expression for (12.32) that will motivate the definitions of weak solutions and fractional Sobolev space.

Lemma 12.12. *Let $u, v \in \mathcal{S}$ and $0 < s < 1$. Then*

$$\int_{\mathbb{R}^n} (-\Delta)^{s/2} u (-\Delta)^{s/2} v\, dx = \frac{c_{n,s}}{2} \int_{\mathbb{R}^n} \int_{\mathbb{R}^n} \frac{(u(x) - u(z))(v(x) - v(z))}{|x - z|^{n+2s}}\, dz\, dx.$$

Proof. By Theorem 12.1(3), we can write

$$\int_{\mathbb{R}^n} u(-\Delta)^s v\, dx = -\frac{c_{n,s}}{2} \int_{\mathbb{R}^n} u(x) \int_{\mathbb{R}^n} \frac{v(x+z) + v(x-z) - 2v(x)}{|z|^{n+2s}}\, dz\, dx$$

and

$$\int_{\mathbb{R}^n} v(-\Delta)^s u\, dx = -\frac{c_{n,s}}{2} \int_{\mathbb{R}^n} v(x) \int_{\mathbb{R}^n} \frac{u(x+z) + u(x-z) - 2u(x)}{|z|^{n+2s}}\, dz\, dx$$

where the integrals are absolutely convergent. We add these two identities, use (12.32) and apply Fubini's theorem to get

$$2 \int_{\mathbb{R}^n} (-\Delta)^{s/2} u (-\Delta)^{s/2} v\, dx = -\frac{c_{n,s}}{2} \int_{\mathbb{R}^n} \frac{F(z)}{|z|^{n+2s}}\, dz \tag{12.33}$$

where, by the translation invariance of the Lebesgue measure,

$$\begin{aligned} F(z) &= \int_{\mathbb{R}^n} \Big[u(x)[v(x+z) + v(x-z)] + v(x)[u(x+z) + u(x-z)] - 4u(x)v(x) \Big]\, dx \\ &= \int_{\mathbb{R}^n} \Big[u(x)[v(x+z) + v(x-z)] + v(x)[u(x+z) + u(x-z)] \\ &\qquad - 2u(x)v(x) - u(x-z)v(x-z) - u(x+z)v(x+z) \Big]\, dx. \end{aligned}$$

On the other hand, by changing variables,

$$\begin{aligned}
\frac{c_{n,s}}{2}\int_{\mathbb{R}^n}\int_{\mathbb{R}^n}&\frac{(u(x)-u(z))(v(x)-v(z))}{|x-z|^{n+2s}}\,dz\,dx\\
&=\frac{c_{n,s}}{2}\int_{\mathbb{R}^n}\int_{\mathbb{R}^n}\frac{(u(x)-u(x-z))(v(x)-v(x-z))}{|z|^{n+2s}}\,dx\,dz\\
&=\frac{c_{n,s}}{2}\int_{\mathbb{R}^n}\int_{\mathbb{R}^n}\frac{(u(x)-u(x+z))(v(x)-v(x+z))}{|z|^{n+2s}}\,dx\,dz.
\end{aligned}$$

After adding the last two formulas we arrive at (12.33). The lemma is proved. □

We now give the definition of the **fractional Laplacian in** $L^2(\mathbb{R}^n)$ by means of the Fourier transform and extend Lemma 12.12 to this new setting. Let $0 < s < 1$. For $u \in L^2(\mathbb{R}^n)$ such that $|\xi|^s\widehat{u}(\xi) \in L^2(\mathbb{R}^n)$, we define $(-\Delta)^{s/2}u \in L^2(\mathbb{R}^n)$ as

$$(-\Delta)^{s/2}u(x) = \mathcal{F}^{-1}(|\xi|^s\widehat{u}(\xi))(x) \qquad \text{in } L^2(\mathbb{R}^n).$$

Observe that the fractional Laplacian defined in this L^2 way coincides with the distributional definition given in Section 12.1. Indeed, if $u \in L^2(\mathbb{R}^n)$ then, by Hölder's inequality, $u \in L_{s/2}$ and $(-\Delta)^{s/2}u$ is well defined in the sense of distributions. Then, for any $\varphi \in \mathcal{S}$, by Parseval's identity,

$$\begin{aligned}
(-\Delta)^{s/2}u(\varphi) &= \int_{\mathbb{R}^n} u(x)(-\Delta)^{s/2}\varphi(x)\,dx\\
&= \int_{\mathbb{R}^n} \widehat{u}(\xi)\overline{|\xi|^s\widehat{\varphi}(\xi)}\,d\xi\\
&= \int_{\mathbb{R}^n} |\xi|^s\widehat{u}(\xi)\overline{\widehat{\varphi}(\xi)}\,d\xi = \int_{\mathbb{R}^n}(-\Delta)^{s/2}u(x)\varphi(x)\,dx
\end{aligned}$$

and the distribution $(-\Delta)^{s/2}u$ coincides with $(-\Delta)^{s/2}u(x)$ a.e. in $\mathbb{R}^n$.

Lemma 12.13. *Let $u, v \in L^2(\mathbb{R}^n)$ such that $(-\Delta)^{s/2}u, (-\Delta)^{s/2}v \in L^2(\mathbb{R}^n)$. Then*

$$\int_{\mathbb{R}^n}(-\Delta)^{s/2}u(-\Delta)^{s/2}v\,dx = \frac{c_{n,s}}{2}\int_{\mathbb{R}^n}\int_{\mathbb{R}^n}\frac{(u(x)-u(z))(v(x)-v(z))}{|x-z|^{n+2s}}\,dz\,dx.$$

Proof. Notice that, unless u and v are sufficiently regular, we cannot use the pointwise formulas for the fractional Laplacian nor "pass half fractional Laplacian" from u to v. We overcome these issues by using the method of semigroups. Suppose first that $u = v$. By the Gamma function formula (12.4) with $\lambda = |\xi|^2$, Parseval's identity, the

convolution formula for $e^{t\Delta}u(x)$ involving the heat kernel (12.7) and (12.8),

$$\begin{aligned}
\int_{\mathbb{R}^n} |(-\Delta)^{s/2}u|^2\,dx &= \int_{\mathbb{R}^n} |\xi|^{2s}\widehat{u}(\xi)\overline{\widehat{u}(\xi)}\,d\xi \\
&= \frac{1}{\Gamma(-s)}\int_0^\infty \int_{\mathbb{R}^n} \left[e^{-t|\xi|^2}\widehat{u}(\xi)\overline{\widehat{u}(\xi)} - \widehat{u}(\xi)\overline{\widehat{u}(\xi)}\right] d\xi\,\frac{dt}{t^{1+s}} \\
&= \frac{1}{\Gamma(-s)}\int_0^\infty \int_{\mathbb{R}^n} \left[e^{t\Delta}u(x)u(x) - u(x)u(x)\right] dx\,\frac{dt}{t^{1+s}} \\
&= \frac{1}{\Gamma(-s)}\int_0^\infty \int_{\mathbb{R}^n}\int_{\mathbb{R}^n} G_t(x-z)\big[u(z)u(x) - u(x)u(x)\big]\,dz\,dx\,\frac{dt}{t^{1+s}} \\
&= \frac{1}{\Gamma(-s)}\int_0^\infty \int_{\mathbb{R}^n}\int_{\mathbb{R}^n} G_t(x-z)\big[u(x)u(z) - u(z)u(z)\big]\,dz\,dx\,\frac{dt}{t^{1+s}}
\end{aligned}$$

where in the last line we interchanged the roles of x and z and used that the heat kernel $G_t(z)$ is even in z. By adding the last two identities, applying Tonelli's theorem and using (12.11),

$$\begin{aligned}
&\int_{\mathbb{R}^n} |(-\Delta)^{s/2}u|^2\,dx \\
&\quad = \frac{1}{2|\Gamma(-s)|}\int_0^\infty \int_{\mathbb{R}^n}\int_{\mathbb{R}^n} G_t(x-z)(u(x)-u(z))^2\,dz\,dx\,\frac{dt}{t^{1+s}} \\
&\quad = \int_{\mathbb{R}^n}\int_{\mathbb{R}^n}\left[\frac{1}{2|\Gamma(-s)|}\int_0^\infty G_t(x-z)\,\frac{dt}{t^{1+s}}\right](u(x)-u(z))^2\,dz\,dx \\
&\quad = \frac{c_{n,s}}{2}\int_{\mathbb{R}^n}\int_{\mathbb{R}^n} \frac{(u(x)-u(z))^2}{|x-z|^{n+2s}}\,dz\,dx.
\end{aligned}$$

The general case when u and v are not necessarily the same follows from this one by polarization, that is, by using the Fourier transform definitions to write

$$\begin{aligned}
&\int_{\mathbb{R}^n} (-\Delta)^{s/2}u(-\Delta)^{s/2}v\,dx \\
&\qquad = \frac{1}{2}\int_{\mathbb{R}^n}\left[|(-\Delta)^{s/2}u|^2 + |(-\Delta)^{s/2}v|^2 - |(-\Delta)^{s/2}(u-v)|^2\right] dx
\end{aligned}$$

and applying the previous case to the formulas in the right hand side. □

For $0 < s < 1$, the **fractional Sobolev space** $H^s(\mathbb{R}^n)$ is defined as the set

$$H^s(\mathbb{R}^n) = \{u \in L^2(\mathbb{R}^n) : (-\Delta)^{s/2}u \in L^2(\mathbb{R}^n)\}.$$

We endow $H^s(\mathbb{R}^n)$ with the inner product (see Lemma 12.13)

$$\begin{aligned}
\langle u, v\rangle_{H^s(\mathbb{R}^n)} &= \int_{\mathbb{R}^n} uv\,dx + \int_{\mathbb{R}^n} (-\Delta)^{s/2}u(-\Delta)^{s/2}v\,dx \\
&= \int_{\mathbb{R}^n} uv\,dx + \frac{c_{n,s}}{2}\int_{\mathbb{R}^n}\int_{\mathbb{R}^n} \frac{(u(x)-u(z))(v(x)-v(z))}{|x-z|^{n+2s}}\,dz\,dx.
\end{aligned}$$

Then the associated induced norm is

$$\begin{aligned}\|u\|_{H^s(\mathbb{R}^n)}^2 &= \int_{\mathbb{R}^n} u^2\,dx + \int_{\mathbb{R}^n} |(-\Delta)^{s/2}u|^2\,dx \\ &= \int_{\mathbb{R}^n} u^2\,dx + \frac{c_{n,s}}{2}\int_{\mathbb{R}^n}\int_{\mathbb{R}^n}\frac{|u(x)-u(z)|^2}{|x-z|^{n+2s}}\,dz\,dx.\end{aligned}$$

It is known that $H^s(\mathbb{R}^n)$ is a Hilbert space and that $C_c^\infty(\mathbb{R}^n)$ is dense in $H^s(\mathbb{R}^n)$.

The fractional Sobolev space $H^s(\mathbb{R}^n)$ is interpreted as the set of those functions in $L^2(\mathbb{R}^n)$ that have "s derivatives" in $L^2(\mathbb{R}^n)$. Indeed, $-\Delta$ is an operator of order 2, so $(-\Delta)^{s/2}$ is an operator of fractional order $2\cdot(s/2)=s$. This interpretation of the fractional Laplacian as a *fractional derivative operator* will also be apparent at the scale of Hölder spaces in Chapter 13.

To encode the exterior Dirichlet condition in (12.31), we introduce the space

$$H_0^s(\Omega) = \{u \in H^s(\mathbb{R}^n) : u = 0 \text{ in } \mathbb{R}^n \setminus \Omega\}. \tag{12.34}$$

Under the same inner product as that of $H^s(\mathbb{R}^n)$, $H_0^s(\Omega)$ becomes a Hilbert space. It is known that $C_c^\infty(\Omega)$ is dense in $H_0^s(\Omega)$ whenever Ω is a bounded domain with **continuous boundary**, that is, such that $\partial\Omega$ can be locally represented as the graph of a continuous function.

Let $f \in L^2(\Omega)$. We say that $u \in H_0^s(\Omega)$ is a **weak solution** to the Poisson–Dirichlet problem (12.31) if

$$\int_{\mathbb{R}^n} (-\Delta)^{s/2}u(-\Delta)^{s/2}v\,dx = \int_\Omega fv\,dx \qquad \text{for all } v \in H_0^s(\Omega)$$

or, equivalently,

$$\frac{c_{n,s}}{2}\int_{\mathbb{R}^n}\int_{\mathbb{R}^n}\frac{(u(x)-u(z))(v(x)-v(z))}{|x-z|^{n+2s}}\,dz\,dx = \int_\Omega fv\,dx \qquad \text{for all } v \in H_0^s(\Omega).$$

With the same reasoning we used before the statement of Lemma 12.13, this time by taking $\varphi \in C_c^\infty(\Omega) \subset H_0^s(\Omega)$, it is clear that if $u \in H_0^s(\Omega)$ is a weak solution to (12.31) then u is a distributional solution to $(-\Delta)^s u = f$ in Ω. For the converse, if the bounded domain Ω has continuous boundary and $u \in H_0^s(\Omega)$ is a distributional solution to $(-\Delta)^s u = f$ in Ω then, by using the density of $C_c^\infty(\Omega)$ in $H_0^s(\Omega)$, we conclude that u is a weak solution to (12.31).

To prove existence and uniqueness of weak solutions we need a fractional Poincaré inequality (see Theorem 12.14 with $p=2$) that, in particular, implies that

$$\langle u, v\rangle_{H_0^s(\Omega)} := \int_{\mathbb{R}^n} (-\Delta)^{s/2}u(-\Delta)^{s/2}v\,dx$$

is an equivalent inner product to the one in $H_0^s(\Omega)$. It is clear that

$$F(v) = \int_\Omega fv\,dx$$

is a bounded linear functional on $H_0^s(\Omega)$. Therefore, the weak formulation of (12.31) reads as follows: find $u \in H_0^s(\Omega)$ such that

$$F(v) = \langle u, v \rangle_{H_0^s(\Omega)} \qquad \text{for all } v \in H_0^s(\Omega).$$

By the Riesz representation theorem, u exists and is unique and, moreover,

$$\int_{\mathbb{R}^n} |(-\Delta)^{s/2} u|^2 \, dx = \frac{c_{n,s}}{2} \int_{\mathbb{R}^n} \int_{\mathbb{R}^n} \frac{|u(x) - u(z)|^2}{|x - z|^{n+2s}} \, dz \, dx = \|f\|_{L^2(\Omega)}^2.$$

We are left to show the fractional Poincaré inequality. In fact, we will give an elementary proof of the fractional Sobolev embedding, from which the fractional Poincaré inequality follows as a particular case.

Theorem 12.14 (Fractional Sobolev embedding and fractional Poincaré inequality). *Fix $0 < s < \min\{1, n/2\}$. There is a constant $C > 0$ depending only on n and s such that, for any $u \in H^s(\mathbb{R}^n)$,*

$$\|u\|_{L^{\frac{2n}{n-2s}}(\mathbb{R}^n)}^2 \le C \int_{\mathbb{R}^n} \int_{\mathbb{R}^n} \frac{|u(x) - u(z)|^2}{|x - z|^{n+2s}} \, dz \, dx.$$

In particular, for any $1 \le p \le \frac{2n}{n-2s}$ and $u \in H_0^s(\Omega)$,

$$\|u\|_{L^p(\Omega)}^2 \le C |\Omega|^{\frac{2}{p} - \frac{n-2s}{n}} \int_{\mathbb{R}^n} \int_{\mathbb{R}^n} \frac{|u(x) - u(z)|^2}{|x - z|^{n+2s}} \, dz \, dx.$$

Proof. By the density of $C_c^\infty(\mathbb{R}^n)$ in $H^s(\mathbb{R}^n)$, we can assume that $u \in C_c^\infty(\mathbb{R}^n)$. For any $q \ge 1$, $x \in \mathbb{R}^n$ and $r > 0$ we have, by Hölder's inequality and using the fact that if $z \in B_r(x)$ then $1 < 2r/|x - z|$,

$$\begin{aligned}
|u(x)| &\le \frac{1}{|B_r(x)|} \int_{B_r(x)} |u(x) - u(z)| \, dz + \frac{1}{|B_r(x)|} \int_{B_r(x)} |u(z)| \, dz \\
&\le \left(\frac{1}{|B_r(x)|} \int_{B_r(x)} |u(x) - u(z)|^2 \, dz \right)^{1/2} + \left(\frac{1}{|B_r(x)|} \int_{B_r(x)} |u(z)|^q \, dz \right)^{1/q} \\
&\le \left(\frac{(2r)^{n+2s}}{|B_r(x)|} \int_{B_r(x)} \frac{|u(x) - u(z)|^2}{|x - z|^{n+2s}} \, dz \right)^{1/2} + \left(\frac{1}{|B_r(x)|} \int_{\mathbb{R}^n} |u(z)|^q \, dz \right)^{1/q} \\
&\le C_{n,s,q} \left[r^s \left(\int_{\mathbb{R}^n} \frac{|u(x) - u(z)|^2}{|x - z|^{n+2s}} \, dz \right)^{1/2} + r^{-n/q} \|u\|_{L^q(\mathbb{R}^n)} \right].
\end{aligned}$$

The function of $r > 0$ defined as $\phi(r) = Ar^s + Br^{-n/q}$, where $A, B > 0$, attains its minimum at $r_0 = (nB/(Asq))^{q/(n+sq)}$. Evaluating at this r_0 and raising to the power q, we obtain

$$|u(x)|^q \le C_{n,s,q} \|u\|_{L^q(\mathbb{R}^n)}^{sq^2/(n+sq)} \left(\int_{\mathbb{R}^n} \frac{|u(x) - u(z)|^2}{|x - z|^{n+2s}} \, dz \right)^{nq/[2(n+sq)]}.$$

Now we choose q so as to make the exponent of the last integral equal to 1:

$$\frac{nq}{2(n + sq)} = 1 \qquad \text{if and only if} \qquad q = \frac{2n}{n - 2s}.$$

With this choice of q we integrate the inequality above over $\mathbb{R}^n$ to get

$$\int_{\mathbb{R}^n} |u(x)|^{\frac{2n}{n-2s}}\,dx \le C_{n,s}\|u\|^{4s/(n-2s)}_{L^{\frac{2n}{n-2s}}(\mathbb{R}^n)} \int_{\mathbb{R}^n}\int_{\mathbb{R}^n} \frac{|u(x)-u(z)|^2}{|x-z|^{n+2s}}\,dz\,dx.$$

From here, the conclusion follows.

If $u \in H^s_0(\Omega)$ and $1 \le p \le \frac{2n}{n-2s}$ then, by Hölder's inequality,

$$\|u\|^2_{L^p(\Omega)} \le |\Omega|^{\frac{2}{p}-\frac{n-2s}{n}} \|u\|^2_{L^{\frac{2n}{n-2s}}(\Omega)}$$

and the conclusion follows by applying the inequality we just proved. □

We can extend the weak maximum principle of Corollary 12.8 to weak solutions $u \in H^s_0(\Omega)$ to (12.31). For this, we need the simple observation that, for the negative part $u^- = -\min\{u, 0\} \ge 0$ of u,

$$(u^-(x) - u^-(z))^2 \le (u(x) - u(z))^2 \qquad \text{for all } x, z \in \mathbb{R}^n.$$

Thus, if $u \in H^s(\mathbb{R}^n)$ then $u^- \in H^s(\mathbb{R}^n)$ and, similarly, $u^+ = \max\{u, 0\} \in H^s(\mathbb{R}^n)$.

Theorem 12.15 (Weak maximum principle for weak solutions). *Let $f \in L^2(\Omega)$ and let $u \in H^s_0(\Omega)$ be the unique weak solution to*

$$\begin{cases} (-\Delta)^s u = f & \text{in } \Omega \\ u = 0 & \text{in } \mathbb{R}^n \setminus \Omega. \end{cases}$$

If $f \ge 0$ a.e. in Ω then $u \ge 0$ a.e. in $\mathbb{R}^n$.

Proof. Suppose, by way of contradiction, that $|\{x \in \Omega : u^- \neq 0\}| > 0$. We can choose $v = u^- \in H^s_0(\Omega)$ as test function in the weak formulation. On one hand,

$$\int_\Omega f u^-\,dx \ge 0.$$

On the other hand, observe that

$$(u^+(x) - u^+(z))(u^-(x) - u^-(z)) \le 0 \qquad \text{for all } x, z \in \mathbb{R}^n$$

and that, by our assumption,

$$-(u^-(x) - u^-(z))^2 < 0 \qquad \text{on a set of positive measure in } \mathbb{R}^n \times \mathbb{R}^n.$$

Therefore, by writing $u = u^+ - u^-$, we find that

$$\int_{\mathbb{R}^n}\int_{\mathbb{R}^n} \frac{(u(x)-u(z))(u^-(x)-u^-(z))}{|x-z|^{n+2s}}\,dx\,dz < \int_{\mathbb{R}^n}\int_{\mathbb{R}^n} \frac{(u^+(x)-u^+(z))(u^-(x)-u^-(z))}{|x-z|^{n+2s}}\,dx\,dz \le 0$$

which is a contradiction to the weak formulation of the problem. □

The following comparison principle is obtained immediately by applying the weak maximum principle of Theorem 12.15 to the difference $u - v$.

Corollary 12.16 (Comparison principle for weak solutions). *Let $f, g \in L^2(\Omega)$ and let $u, v \in H_0^s(\Omega)$ be the unique weak solutions to*

$$\begin{cases} (-\Delta)^s u = f & \text{in } \Omega \\ u = 0 & \text{in } \mathbb{R}^n \setminus \Omega \end{cases} \quad \text{and} \quad \begin{cases} (-\Delta)^s v = g & \text{in } \Omega \\ v = 0 & \text{in } \mathbb{R}^n \setminus \Omega, \end{cases}$$

respectively. If $f \geq g$ a.e. in Ω then $u \geq v$ a.e. in $\mathbb{R}^n$.

12.6 AN EXPLICIT EXAMPLE

At this point, we pause for a moment with the theory and consider an example of an explicit solution to the Poisson–Dirichlet problem (12.31) for the fractional Laplacian. As we will see, computing the fractional Laplacian of a given function is not an easy task. In the example, it involves several properties of special functions. The example will also highlight several features in terms of regularity of solutions to (12.31).

Example 12.17 (An explicit solution to the Poisson–Dirichlet problem). The unique weak solution $u \in H_0^s(B_1)$ to the Poisson–Dirichlet problem

$$\begin{cases} (-\Delta)^s u = 1 & \text{in } B_1 \\ u = 0 & \text{in } \mathbb{R}^n \setminus B_1 \end{cases}$$

is

$$u(x) = d_{n,s}[(1 - |x|^2)^+]^s$$

where

$$d_{n,s} = \frac{\Gamma(n/2)}{4^s \Gamma(s+1)\Gamma(n/2+s)} > 0. \tag{12.35}$$

Since u is continuous with compact support in $\mathbb{R}^n$, we have $u \in L_s$ so that its fractional Laplacian $(-\Delta)^s u$ is well defined in the sense of distributions and, as we already know, $(-\Delta)^s u = f$ in the distributional sense in B_1. Moreover, u is smooth in the interior of B_1, so that, by Theorem 12.3, $(-\Delta)^s u(x)$ is a well defined continuous function of $x \in B_1$ that is given by the pointwise formulas for the fractional Laplacian. In particular, by Corollary 12.8, u is the unique classical solution to the problem.

To prove that u is a weak solution, we use the Fourier transform, the symmetries of the Laplacian and properties of Bessel and hypergeometric functions.

Clearly, $u \in L^2(\mathbb{R}^n)$. We must show that $(-\Delta)^{s/2} u \in L^2(\mathbb{R}^n)$, which gives that $u \in H_0^s(B_1)$, and also that, by the density of $C_c^\infty(B_1)$ in $H_0^s(B_1)$,

$$\int_{\mathbb{R}^n} (-\Delta)^{s/2} u (-\Delta)^{s/2} \varphi \, dx = \int_{B_1} \varphi \, dx \qquad \text{for any } \varphi \in C_c^\infty(B_1). \tag{12.36}$$

We begin with the computation of $\widehat{u}(\xi)$. Since $u \in L^1(\mathbb{R}^n)$ is a radial function, that is, $u(x) = u_0(|x|)$ for $u_0(r) = d_{n,s}[(1-r^2)^+]^s$, $r > 0$, its Fourier transform is a radial function given by

$$\begin{aligned}\widehat{u}(\xi) &= |\xi|^{1-n/2} \int_0^\infty u_0(r) J_{\frac{n-2}{2}}(r|\xi|) r^{n/2}\, dr \\ &= d_{n,s}|\xi|^{1-n/2} \int_0^1 (1-r^2)^s J_{\frac{n-2}{2}}(r|\xi|) r^{n/2}\, dr\end{aligned}$$

see, for instance, [37, 66]. Here $J_\nu(r)$, $\nu \in \mathbb{R}$, is the **Bessel function of the first kind of order** ν, which is defined via the power series

$$J_\nu(r) = \sum_{k=0}^\infty \frac{(-1)^k}{k!\Gamma(k+\nu+1)} \left(\frac{r}{2}\right)^{\nu+2k}$$

see [37, 51]. For $\nu, \mu > -1$ and $y > 0$, the following integral identity holds:

$$\int_0^1 r^{\nu+1/2}(1-r^2)^\mu J_\nu(ry)(ry)^{1/2}\, dr = 2^\mu \Gamma(\mu+1) y^{-\mu-1/2} J_{\nu+\mu+1}(y)$$

see [66] or [31, Section 8.5, eq. (33)]. If in this formula we let $y = |\xi|$, $\nu = \frac{n-2}{2}$ and $\mu = s$, we obtain

$$\widehat{u}(\xi) = d_{n,s} 2^s \Gamma(s+1) |\xi|^{-n/2-s} J_{\frac{n}{2}+s}(|\xi|).$$

It is known that $J_{\frac{n}{2}+s}(|\xi|) \sim |\xi|^{\frac{n}{2}+s}$ as $|\xi| \to 0$ and that $J_{\frac{n}{2}+s}(|\xi|) \sim |\xi|^{-1/2}$ as $|\xi| \to \infty$, see, for example, [37, 51]. With these estimates and the formula for $\widehat{u}$ above, we conclude that $|\xi|^s \widehat{u}(\xi) \in L^2(\mathbb{R}^n)$. Therefore, $u \in H_0^s(B_1)$.

To show that (12.36) holds, fix $\varphi \in C_c^\infty(B_1)$ and denote by I the integral in the left hand side of (12.36). By the formula for $\widehat{u}$ above and passing to polar coordinates,

$$\begin{aligned}I &= \int_{\mathbb{R}^n} |\xi|^s \widehat{u}(\xi) \overline{|\xi|^s \widehat{\varphi}(\xi)}\, d\xi \\ &= d_{n,s} 2^s \Gamma(s+1) \int_{\mathbb{R}^n} |\xi|^{-n/2+s} J_{\frac{n}{2}+s}(|\xi|) \overline{\widehat{\varphi}(\xi)}\, d\xi \\ &= d_{n,s} 2^s \Gamma(s+1) \int_0^\infty r^{n/2+s-1} J_{\frac{n}{2}+s}(r) \left[\int_{|\theta|=1} \overline{\widehat{\varphi}(r\theta)}\, dS_\theta \right] dr.\end{aligned}$$

Using the definition of the Fourier transform and Fubini's theorem, the integral in brackets above can be written as

$$\begin{aligned}\int_{|\theta|=1} \overline{\widehat{\varphi}(r\theta)}\, dS_\theta &= \int_{B_1} \varphi(x) \left[\int_{|\theta|=1} \frac{e^{i(rx)\cdot\theta}}{(2\pi)^{n/2}}\, dS_\theta \right] dx \\ &= \int_{B_1} \varphi(x)(r|x|)^{1-n/2} J_{\frac{n-2}{2}}(r|x|)\, dx\end{aligned}$$

where we used the fact that the Fourier transform of the surface area measure of the unit sphere is given in terms of the Bessel function $J_{\frac{n-2}{2}}$, see [37, 65]. Thus,

$$I = d_{n,s} 2^s \Gamma(s+1) \int_{B_1} \varphi(x)|x|^{1-n/2} \int_0^\infty r^s J_{\frac{n}{2}+s}(r) J_{\frac{n-2}{2}}(r|x|)\, dr\, dx. \tag{12.37}$$

Another integral formula is used, this time involving the **Gauss hypergeometric function** $F \equiv {}_2F_1$, which is defined as

$$F(\alpha, \beta; \gamma; r) = \sum_{k=0}^{\infty} \frac{(\alpha)_k (\beta)_k}{(\gamma)_k k!} r^k$$

for $|r| < 1$, $\alpha, \beta, \gamma \in \mathbb{R}$ with $\gamma \neq 0, -1, -2, \ldots$, where

$$(\lambda)_0 = 1 \qquad \text{and} \qquad (\lambda)_k = \lambda(\lambda+1)\cdots(\lambda+k-1) \text{ for } k \geq 1$$

see [51]. It follows directly from the definition that

$$F(\alpha, 0; \gamma; r) = 1. \tag{12.38}$$

We have

$$\begin{aligned} &\int_0^{\infty} r^{-\lambda-1/2} J_\mu(r) J_\nu(ry)(ry)^{1/2}\, dr \\ &= y^{\nu+1/2} \frac{\Gamma[\frac{1}{2}(\mu+\nu-\lambda+1)]}{2^\lambda \Gamma(\nu+1)\Gamma[\frac{1}{2}(\lambda+\mu-\nu+1)]} F\Big(\tfrac{\mu+\nu-\lambda+1}{2}, \tfrac{\nu-\lambda-\mu+1}{2}; \nu+1; y^2\Big) \end{aligned}$$

where $\mu + \nu + 1 > \lambda > -1$ and $0 < y < 1$, see [31, Section 8.11, eq. (9)]. By choosing $\mu = \frac{n}{2} + s$, $\nu = \frac{n-2}{2}$, $\lambda = -s > -1$ and $y = |x| < 1$, and using (12.38), we get

$$\int_0^{\infty} r^s J_{\frac{n}{2}+s}(r) J_{\frac{n-2}{2}}(r|x|)\, dr = |x|^{n/2-1} \frac{2^s \Gamma(n/2+s)}{\Gamma(n/2)}.$$

Hence, by plugging this identity into I in (12.37) and taking into account the value of the constant $d_{n,s}$ in (12.35), we arrive at (12.36).

It is important to underscore a couple of features of Example 12.17. Observe that

$$\lim_{s\to 1} u(x) = \lim_{s \to 1} d_{n,s}[(1-|x|^2)^+]^s = \frac{1}{2n}(1-|x|^2)^+ = w(x).$$

In other words, as $s \to 1$, the solution u converges to the unique classical solution w to the Poisson–Dirichlet problem for the Laplacian, that is,

$$\begin{cases} -\Delta w = 1 & \text{in } B_1 \\ w = 0 & \text{on } \partial B_1. \end{cases}$$

The verification that w satisfies this equation is much simpler than the one we did in Example 12.17 for u and the fractional Laplacian. Next, and more importantly, notice that the solution w is smooth all the way up to the boundary of B_1. However, $u(x) = d_{n,s}[(1-|x|^2)^+]^s$, although smooth in the interior of B_1 and in $\mathbb{R}^n \setminus B_1$, it is only $C^s(\overline{B_1})$ and no better. The domain B_1, the right hand side $f = 1$ in B_1 and the exterior data $g = 0$ in $\mathbb{R}^n \setminus B_1$ are all smooth, but the solution is not globally smooth, only $C^s(\mathbb{R}^n)$. This is in high contrast with the Laplacian case.

12.7 VISCOSITY AND POINTWISE SOLUTIONS, HÖLDER REGULARITY

In this section we give the definitions of viscosity and pointwise solutions to the fractional Laplacian equation

$$-(-\Delta)^s u = f \qquad \text{in } \Omega \tag{12.39}$$

where $f \in C(\Omega)$ and $\Omega \subset \mathbb{R}^n$, $n \geq 1$, is a bounded domain, and analyze their relation. Furthermore, we show a Hölder regularity result: if u is a bounded viscosity solution to $-(-\Delta)^s u = 0$ in B_1, then u is locally Hölder continuous in B_1.

Both definitions of solutions are based on the pointwise integral formula for $-(-\Delta)^s u$ given in Theorems 12.1(3) and 12.3(3). Observe that such a formula is valid for all $0 < s < 1$ and does not involve principal values nor derivatives of u.

As we mentioned in the Introduction, unlike the case of local elliptic PDEs (see Chapter 8), the concept of viscosity solution for the fractional Laplacian must take into account the values of u at infinity. More importantly, we prove that whenever a viscosity subsolution (supersolution) u to (12.39) is touched from above (below) by a test function at a point x_0 then the integral formula for the fractional Laplacian of u at x_0 is in fact convergent and that the equation can be understood in the pointwise sense. This is in high contrast with local PDEs, where the equation cannot be applied directly to u even at points where it can be touched by test functions.

Let $u \in L_s$. We say that u is a **pointwise subsolution** to (12.39) if, for every point $x_0 \in \Omega$,

$$\frac{c_{n,s}}{2} \int_{\mathbb{R}^n} \frac{u(x_0+z) + u(x_0-z) - 2u(x_0)}{|z|^{n+2s}}\, dz \geq f(x_0)$$

where the integral is convergent. Similarly, we say that u is a **pointwise supersolution** to (12.39) if $-u$ is a pointwise subsolution to (12.39), that is, if for every point $x_0 \in \Omega$ we have

$$\frac{c_{n,s}}{2} \int_{\mathbb{R}^n} \frac{u(x_0+z) + u(x_0-z) - 2u(x_0)}{|z|^{n+2s}}\, dz \leq f(x_0)$$

where the integral is convergent. Finally, we say that u is a **pointwise solution** to (12.39) if it is both a pointwise subsolution and a pointwise supersolution to (12.39), namely, if at every point $x_0 \in \Omega$ we have

$$\frac{c_{n,s}}{2} \int_{\mathbb{R}^n} \frac{u(x_0+z) + u(x_0-z) - 2u(x_0)}{|z|^{n+2s}}\, dz = f(x_0)$$

and the integral is convergent.

Clearly, a classical solution $u \in L_s$ to (12.39) is also a pointwise solution. Conversely, by Theorem 12.3, if $u \in L_s$ is a pointwise solution to (12.39) such that $u \in C^{2s+\varepsilon}(\Omega)$ if $s < 1/2$, or $u \in C^{1,2s+\varepsilon-1}(\Omega)$ if $s \geq 1/2$, for some $\varepsilon > 0$, then u is a classical solution and, in this case, $-(-\Delta)^s u(x_0) = f(x_0)$ for every $x_0 \in \Omega$.

Let $u \in L_s \cap C(\Omega)$. We say that u is a **viscosity subsolution** to (12.39), written

$$-(-\Delta)^s u \geq f \qquad \text{in the viscosity sense in } \Omega$$

if whenever φ is a C^2 function in a neighborhood N of a point $x_0 \in \Omega$ that touches u from above at x_0, that is, $\varphi(x_0) = u(x_0)$ and $\varphi \geq u$ in N, and v is defined as

$$v = \begin{cases} \varphi & \text{in } N \\ u & \text{in } \mathbb{R}^n \setminus N, \end{cases} \tag{12.40}$$

then we have

$$-(-\Delta)^s v(x_0) = \frac{c_{n,s}}{2} \int_{\mathbb{R}^n} \frac{v(x_0+z) + v(x_0-z) - 2v(x_0)}{|z|^{n+2s}}\, dz \geq f(x_0).$$

Similarly, we say u is a **viscosity supersolution** to (12.39), written

$$-(-\Delta)^s u \leq f \qquad \text{in the viscosity sense in } \Omega$$

if $-u$ is a viscosity subsolution to (12.39) or, equivalently, if whenever φ is a C^2 function in a neighborhood N of a point $x_0 \in \Omega$ that touches u from below at x_0, that is, $\varphi(x_0) = u(x_0)$ and $\varphi \leq u$ in N, and v is defined as in (12.40), we have

$$-(-\Delta)^s v(x_0) = \frac{c_{n,s}}{2} \int_{\mathbb{R}^n} \frac{v(x_0+z) + v(x_0-z) - 2v(x_0)}{|z|^{n+2s}}\, dz \leq f(x_0).$$

Finally, we say that u is a **viscosity solution** to (12.39), written

$$-(-\Delta)^s u = f \qquad \text{in the viscosity sense in } \Omega$$

if it is both a viscosity subsolution and a viscosity supersolution to (12.39).

Notice here that v defined as in (12.40) is in the space L_s and is also C^2 in a neighborhood of x_0. Therefore, by Theorem 12.3, $-(-\Delta)^s v(x_0)$ is well defined and coincides with the absolutely convergent integral formulas given above.

Let $u \in L_s \cap C(\Omega)$ be a pointwise subsolution to (12.39). Suppose that φ is a C^2 function in a neighborhood N of a point $x_0 \in \Omega$ that touches u from above at x_0 and define v as in (12.40). Then, since $v \geq u$ in $\mathbb{R}^n$ and $v(x_0) = u(x_0)$,

$$v(x_0+z) + v(x_0-z) - 2v(x_0) \geq u(x_0+z) + u(x_0-z) - 2u(x_0)$$

for all $z \in \mathbb{R}^n$. Therefore,

$$-(-\Delta)^s v(x_0) \geq \frac{c_{n,s}}{2} \int_{\mathbb{R}^n} \frac{u(x_0+z) + u(x_0-z) - 2u(x_0)}{|z|^{n+2s}}\, dz \geq f(x_0).$$

Hence, u is a viscosity subsolution to (12.39). Surprisingly enough, the converse is also true in the following sense.

Theorem 12.18. *Let $u \in L_s \cap C(\Omega)$ satisfy $-(-\Delta)^s u \geq f$ in the viscosity sense in Ω. If φ is a C^2 function in a neighborhood N of a point $x_0 \in \Omega$ that touches u from above at x_0 in N, then*

$$\frac{c_{n,s}}{2} \int_{\mathbb{R}^n} \frac{u(x_0+z) + u(x_0-z) - 2u(x_0)}{|z|^{n+2s}}\, dz \geq f(x_0)$$

where the integral is finite and, in this case, the left hand side is denoted by $-(-\Delta)^s u(x_0)$.

Proof. For simplicity with the notation, given a function w, let us write

$$\delta(w, x, z) = w(x+z) + w(x-z) - 2w(x). \tag{12.41}$$

Let $r_0 > 0$ such that $B_r(x_0) \subset N$ for all $r < r_0$. For any $r \le r_0$, define

$$v_r = \begin{cases} \varphi & \text{in } B_r(x_0) \\ u & \text{in } \mathbb{R}^n \setminus B_r(x_0). \end{cases}$$

By definition of viscosity subsolution, $-(-\Delta)^s v_r(x_0) \ge f(x_0)$, that is,

$$\frac{c_{n,s}}{2}\int_{\mathbb{R}^n} \frac{\delta(v_r, x_0, z)^+}{|z|^{n+2s}}\,dz - \frac{c_{n,s}}{2}\int_{\mathbb{R}^n} \frac{\delta(v_r, x_0, z)^-}{|z|^{n+2s}}\,dz \ge f(x_0)$$

where both integrals are finite. Since $v_r \ge u$ in $\mathbb{R}^n$ and $v_r(x_0) = u(x_0)$,

$$\delta(v_r, x_0, z) \ge \delta(u, x_0, z) \qquad \text{for all } z \in \mathbb{R}^n.$$

Then, as $v_r \in L_s \cap C^2(B_r(x_0))$,

$$0 \le \int_{\mathbb{R}^n} \frac{\delta(u, x_0, z)^+}{|z|^{n+2s}}\,dz \le \int_{\mathbb{R}^n} \frac{|\delta(v_r, x_0, z)|}{|z|^{n+2s}}\,dz < \infty.$$

We now consider $\delta(u, x_0, z)^-/|z|^{n+2s}$. Note that

$$\frac{c_{n,s}}{2}\int_{\mathbb{R}^n} \frac{\delta(v_r, x_0, z)^+}{|z|^{n+2s}}\,dz - f(x_0) \ge \frac{c_{n,s}}{2}\int_{\mathbb{R}^n} \frac{\delta(v_r, x_0, z)^-}{|z|^{n+2s}}\,dz.$$

Since φ touches u from above at x_0, we have $\delta(v_r, x_0, z) \le \delta(v_{r_0}, x_0, z)$ for every $r < r_0$ and all $z \in \mathbb{R}^n$, so that,

$$\frac{c_{n,s}}{2}\int_{\mathbb{R}^n} \frac{\delta(v_{r_0}, x_0, z)^+}{|z|^{n+2s}}\,dz - f(x_0) \ge \frac{c_{n,s}}{2}\int_{\mathbb{R}^n} \frac{\delta(v_r, x_0, z)^-}{|z|^{n+2s}}\,dz.$$

Now, for every $z \in \mathbb{R}^n$, $\delta(v_r, x_0, z)^-$ increases to $\delta(u, x_0, z)^-$ as $r \to 0$. Hence, by the monotone convergence theorem,

$$\frac{c_{n,s}}{2}\int_{\mathbb{R}^n} \frac{\delta(v_{r_0}, x_0, z)^+}{|z|^{n+2s}}\,dz - f(x_0) \ge \frac{c_{n,s}}{2}\int_{\mathbb{R}^n} \frac{\delta(u, x_0, z)^-}{|z|^{n+2s}}\,dz \ge 0.$$

Thus, $\delta(u, x_0, z)/|z|^{n+2s}$ is integrable. Next, for any $0 < r < r_0$,

$$\begin{aligned} \frac{c_{n,s}}{2}\int_{\mathbb{R}^n} \frac{\delta(u, x_0, z)}{|z|^{n+2s}}\,dz &= \frac{c_{n,s}}{2}\int_{\mathbb{R}^n} \frac{\delta(v_r, x_0, z)}{|z|^{n+2s}}\,dz - \frac{c_{n,s}}{2}\int_{\mathbb{R}^n} \frac{\delta(v_r - u, x_0, z)}{|z|^{n+2s}}\,dz \\ &\ge f(x_0) - \frac{c_{n,s}}{2}\int_{\mathbb{R}^n} \frac{\delta(v_r - u, x_0, z)}{|z|^{n+2s}}\,dz. \end{aligned}$$

Since $\delta(v_r - u, x_0, z)/|z|^{n+2s}$ decreases to 0 as $r \to 0$ and, for all $r < r_0$, is bounded by $\delta(v_{r_0} - u, x_0, z)/|z|^{n+2s} \in L^1(\mathbb{R}^n)$, the dominated convergence theorem gives that the last integral vanishes as $r \to 0$, and the theorem is proved. □

Next we show that bounded viscosity solutions to $-(-\Delta)^s u = 0$ in B_1 are locally Hölder continuous in B_1.

Let us geometrically understand how the nonlocality comes into play in the method of proof. The idea is to show the Hölder regularity of u at, say, the origin, by *zooming out* the solution from the origin to B_1. Observe that the rescaling of u from a small ball around the origin to B_1 will also rescale the whole solution outside of the ball at the same time. Therefore, if, say, $u(0) = 0$, and we were to have the estimate $|u(x)| \leq C|x|^\alpha$ near the origin, for some $C > 0$ and $0 < \alpha < 1$, then the behavior around the origin would propagate all the way to infinity. In other words, the rescalings of u will have a growth like $|x|^\alpha$ outside of B_1. However, as α becomes small, these $|x|^\alpha$ *wings* outside of B_1 become flatter and flatter and, consequently, the integral part of the fractional Laplacian of u outside of B_1 can be maintained under control and also be made arbitrarily small.

Theorem 12.19 (Hölder regularity of viscosity solutions). *Let $0 < s < 1$. There exist $0 < \alpha < 1$ and a constant $C > 0$ depending only on n and s such that for any $u \in C(\mathbb{R}^n) \cap L^\infty(\mathbb{R}^n)$ viscosity solution to*

$$-(-\Delta)^s u = 0 \qquad \textit{in } B_1$$

we have $u \in C^\alpha(B_{1/2})$ with

$$\|u\|_{C^{0,\alpha}(B_{1/2})} \leq C\|u\|_{L^\infty(\mathbb{R}^n)}.$$

Proof. We prove that u is C^α at the origin. The proof at any other point in $B_{1/2}$ follows the same arguments. By considering $u/(\operatorname{osc}_{\mathbb{R}^n} u)$, we can assume, without loss of generality, that

$$\operatorname*{osc}_{\mathbb{R}^n} u = 1.$$

We must show that there exists $0 < \theta < 1/2$ depending only on n and s such that

$$\operatorname*{osc}_{B_{1/2^k}} u \leq (1-\theta)^k \tag{12.42}$$

for all $k \geq 0$. From here the Hölder regularity of u at 0 follows. Indeed, let $\alpha \in (0, 1)$ such that $(1 - \theta) = (1/2)^\alpha$, that is, $\alpha = \frac{\log(1-\theta)}{\log(1/2)}$. Fix $|x| < 1$ and let $k \geq 0$ such that $1/2^{k+1} \leq |x| < 1/2^k$. Then, by (12.42),

$$|u(x) - u(0)| \leq \operatorname*{osc}_{B_{1/2^k}} u \leq (1-\theta)^k = \frac{1}{(1-\theta)}\left(\frac{1}{2^{k+1}}\right)^\alpha \leq C|x|^\alpha$$

where $C = (1 - \theta)^{-1} > 0$.

To prove (12.42), we show that there exist an increasing sequence b_k and a decreasing sequence a_k such that

$$b_k \leq u \leq a_k \text{ in } B_{1/2^k} \qquad \text{and} \qquad a_k - b_k = (1-\theta)^k \tag{12.43}$$

for all $k \geq 0$. The construction is by induction. Clearly, (12.43) is true for all $k \leq 0$ by choosing $b_k = \inf_{\mathbb{R}^n} u$, $a_k = b_k + (1-\theta)^k$ and our assumption that $\operatorname{osc}_{\mathbb{R}^n} u = 1$. For the induction step, fix $k \geq 0$ and suppose that b_j and a_j have been constructed for every $j \leq k$. We need to prove that (12.43) holds for $k+1$ in place of k. For this, we rescale u from $B_{1/2^k}$ to B_1 and perform a vertical translation to center it in between $-1/2$ and $1/2$. Let

$$m = (a_k + b_k)/2 \tag{12.44}$$

and define

$$v(x) = \frac{1}{(1-\theta)^k}(u(x/2^k) - m) \qquad \text{for } x \in \mathbb{R}^n.$$

Then, by (12.43), $-1/2 \leq v \leq 1/2$ in B_1. It is easy to see that, by the scaling of the fractional Laplacian in (12.2),

$$-(-\Delta)^s v = 0 \qquad \text{in the viscosity sense in } B_1.$$

The induction hypothesis says that

$$a_\ell - b_\ell = (1-\theta)^\ell \qquad \text{for all } \ell \leq k.$$

Fix any $|x| \geq 1$. Let $j \geq 0$ such that $2^j \leq |x| < 2^{j+1}$. Using the induction hypothesis with $\ell = k - j - 1$ and the definition of m in (12.44),

$$\begin{aligned}
v(x) &= \frac{1}{(1-\theta)^k}(u(x/2^k) - m) \\
&\leq \frac{1}{(1-\theta)^k}\left(\sup_{B_{1/2^{k-j-1}}} u - m\right) \\
&\leq \frac{1}{(1-\theta)^k}(a_{k-j-1} - m) \\
&\leq \frac{1}{(1-\theta)^k}((a_{k-j-1} - b_{k-j-1}) + (b_k - m)) \\
&= \frac{1}{(1-\theta)^{j+1}} - \frac{a_k - b_k}{2(1-\theta)^k} \\
&= \frac{1}{(1-\theta)^{j+1}} - 1/2.
\end{aligned}$$

Since $2^j \leq |x| < 2^{j+1}$ and $1/(1-\theta) = 2^\alpha$, we then obtain

$$v(x) \leq (22^j)^\alpha - 1/2 \leq (2|x|)^\alpha - 1/2.$$

Suppose now that v is mostly negative in B_1, that is,

$$|\{v \leq 0\} \cap B_1| \geq \frac{1}{2}|B_1|.$$

We show that

$$v \leq (1/2 - \theta) \qquad \text{in } B_{1/2}. \tag{12.45}$$

In this case, we obtain (12.43) with $k+1$ in place of k. Indeed, rescaling back to u, (12.45) says that

$$u \le (1/2-\theta)(1-\theta)^k + m \qquad \text{in } B_{1/2^{k+1}}.$$

We then choose $a_{k+1} = (1/2-\theta)(1-\theta)^k + m$ and $b_{k+1} = b_k$. Hence, by the definition of m in (12.44),

$$\begin{aligned} a_{k+1} - b_{k+1} &= (1/2-\theta)(1-\theta)^k + \frac{a_k - b_k}{2} \\ &= (1-\theta)^k[(1/2-\theta)+1/2] = (1-\theta)^{k+1} \end{aligned}$$

and $a_{k+1} \le a_k$, as desired. In the case when v is mostly positive, that is,

$$|\{v \ge 0\} \cap B_1| \ge \frac{1}{2}|B_1|$$

we apply the same reasoning to $-v$ to again get (12.43) with $k+1$ in place of k.

We prove (12.45) by contradiction. Fix a radially symmetric smooth function $\eta \in C_c^\infty(B_{3/4})$ such that $0 \le \eta \le 1$ and $\eta \equiv 1$ in $B_{1/2}$. Notice that $-(-\Delta)^s\eta$ is bounded on $\mathbb{R}^n$. Define the bounded, continuous function

$$w = v + \theta\eta \qquad \text{in } \mathbb{R}^n.$$

Since $\eta = 1$ in $B_{1/2}$, $w = v + \theta\eta = v + \theta$ in $B_{1/2}$. Suppose that (12.45) does not hold, that is, there is a point in $B_{1/2}$ where $v + \theta > 1/2$. Then w has a local maximum in $B_{3/4}$ attained at a point $x_0 \in B_{3/4}$ such that $w(x_0) \ge 1/2$. In particular, since η is supported in $B_{3/4}$ and $v \le 1/2$,

$$\sup_{B_1} w = w(x_0) > 1/2.$$

We now compute the equation for w at x_0 and choose θ (and, therefore, α) to reach a contradiction.

Since x_0 is a point of local maximum for w, it is clear that w (and, as a consequence, v) can be touched from above at x_0 by a C^2 function in a neighborhood of x_0. Therefore, by Theorem 12.18, the equation can be computed in the pointwise sense at x_0 and we have

$$\begin{aligned} -(-\Delta)^s w(x_0) &= -(-\Delta)^s v(x_0) - \theta(-\Delta)^s\eta(x_0) \\ &\ge \theta \inf_{B_{3/4}}[-(-\Delta)^s\eta] = -\theta C_1 \end{aligned}$$

where $C_1 > 0$ is depends only on n and s. Now we bound $I := -(-\Delta)^s w(x_0)$ from above. Using the notation in (12.41) and changing variables, we write

$$\begin{aligned} I &= \frac{c_{n,s}}{2}\int_{\mathbb{R}^n} \frac{\delta(w,x_0,z)^+ - \delta(w,x_0,z)^-}{|z|^{n+2s}}\,dz \\ &\le c_{n,s}\int_{\mathbb{R}^n} \frac{(w(x_0+z)-w(x_0))^+ - (w(x_0+z)-w(x_0))^-}{|z|^{n+2s}}\,dz \\ &= c_{n,s}(I_1 + I_2) \end{aligned}$$

where

$$I_1 = \int_{|x_0+z|\geq 1} \frac{(w(x_0+z)-w(x_0))^+ - (w(x_0+z)-w(x_0))^-}{|z|^{n+2s}}\, dz$$

and

$$I_2 = \int_{|x_0+z|<1} \frac{(w(x_0+z)-w(x_0))^+ - (w(x_0+z)-w(x_0))^-}{|z|^{n+2s}}\, dz.$$

Let us see that I_1 can be made arbitrarily small as $\alpha \to 0$ and that I_2 is bounded above by a fixed, negative constant.

For I_1, if $|x_0+z| \geq 1$, then $v(x_0+z) \leq (2|x_0+z|)^\alpha - 1/2$ and, also, $w(x_0) > 1/2$. Therefore, since η is supported in $B_{3/4}$,

$$\begin{aligned} I_1 &\leq \int_{|x_0+z|\geq 1} \frac{(w(x_0+z)-w(x_0))^+}{|z|^{n+2s}}\, dz \\ &= \int_{|x_0+z|\geq 1} \frac{(v(x_0+z)-w(x_0))^+}{|z|^{n+2s}}\, dz \\ &\leq \int_{|x_0+z|\geq 1} \frac{(2|x_0+z|)^\alpha - 1}{|z|^{n+2s}}\, dz =: I_\alpha. \end{aligned}$$

The region of integration in I_α does not contain the origin because $x_0 \in B_{3/4}$ and I_α is finite if $\alpha \leq \alpha_0 < 2s$. Moreover, its integrand decreases to 0 when $\alpha \to 0$. Hence, by the dominated convergence theorem,

$$I_\alpha \to 0 \qquad \text{as } \alpha \to 0.$$

Consider now I_2. For any $|x_0+z| < 1$, since x_0 is a point of maximum of w in B_1, we have $(w(x_0+z)-w(x_0))^+ = 0$ and $(w(x_0+z)-w(x_0))^- = w(x_0)-w(x_0+z)$, and I_2 is negative. However, a refined negative upper bound can be obtained by considering the subset $A = \{v \leq 0\} \cap B_1$ of B_1 where v is negative. Indeed, if $x_0 + z \in A$ then, since $\eta(x_0+z) \leq 1$,

$$w(x_0) - w(x_0+z) \geq 1/2 - v(x_0+z) - \theta\eta(x_0+z) \geq 1/2 - \theta.$$

Therefore, by using these observations together with the facts that $1/|z|^{n+2s} \geq 1$ in B_1 and, by hypothesis, $|A| \geq |B_1|/2$,

$$\begin{aligned} I_2 &\leq -\int_{|x_0+z|\in B_1} \frac{(w(x_0+z)-w(x_0))^-}{|z|^{n+2s}}\, dz \\ &\leq -\int_{|x_0+z|\in A} \frac{w(x_0)-w(x_0+z)}{|z|^{n+2s}}\, dz \\ &\leq -\int_{|x_0+z|\in A} \frac{1/2-\theta}{|z|^{n+2s}}\, dz \\ &\leq -(1/2-\theta)|A| \leq -C_2 \end{aligned}$$

for all $\theta \leq \theta_0 < 1/2$, where $C_2 > 0$ depends only on n.

By collecting all the estimates, we obtain that, for all $\theta \leq \theta_0$ and $\alpha \leq \alpha_0$,

$$-\theta C_1 \leq c_{n,s}(I_\alpha - C_2).$$

By letting θ (and, as a consequence, α) tend to 0, we reach a contradiction. Thus, (12.45) holds and the result is proved. □

NOTES ON CHAPTER 12

One of the most classical potential theory references for the fractional Laplacian is the book by Landkof [50].

The method of semigroups for fractional power operators and, in particular, for the fractional Laplacian, was developed by Stinga–Torrea in [67, 69]. The results presented here using this method, as well as the pointwise limits, can be found in those references. We also follow Stinga *users' guide* for the fractional Laplacian and the method of semigroups [68]. The method of semigroups was extended to fractional powers of other, more general linear operators by Galé–Miana–Stinga in [40].

The distributional definition and the maximum and comparison principles for classical solutions to fractional Laplacian equations were given by Silvestre [59, 61].

The proof of the density of smooth functions with compact support in the space $H^s(\mathbb{R}^n)$ can be found in Adams [1]. The proof of the fractional Sobolev embedding (Theorem 12.14) we presented here is due to Brezis. See also Di Nezza–Palatucci–Valdinoci [28] for more about fractional Sobolev spaces and their relation with the fractional Laplacian. We warn the reader about the use of the notation $H_0^s(\Omega)$. In our case, this is the subspace of $H^s(\mathbb{R}^n)$ given in (12.34). However, other classical references such as Adams [1] use $H_0^s(\Omega)$ to denote the closure of $C_c^\infty(\Omega)$ with respect to the norm of a fractional Sobolev space $H^s(\Omega)$ defined only for functions in $L^2(\Omega)$. We do not use the latter space in this book. Some authors denote the space $H_0^s(\Omega)$ defined in (12.34) by $X_0^s(\Omega)$. The density of $C_c^\infty(\Omega)$ in $H_0^s(\Omega) = X_0^s(\Omega)$ when the boundary of the bounded domain Ω is continuous was proved by Fiscella–Servadei–Valdinoci in [36].

The explicit example of Section 12.6 was computed by Getoor [42]. Bessel and hypergeometric functions, other special functions and integral identities can be seen in Lebedev [51] or Erdélyi *et al.* [31]. Folland [37], Stein [65] and Stein–Weiss [66] consider the Fourier transforms of radial functions and of the measure of surface area on the unit sphere.

The notion of viscosity solutions and the proof of Theorem 12.18 are from Caffarelli–Silvestre [12]. The technique to prove the interior C^α estimate for viscosity solutions is due to Silvestre [60]. The regularity theory for fully nonlinear nonlocal elliptic equations of fractional order was initiated by Caffarelli and Silvestre in their series of groundbreaking papers [12–14].

CHAPTER 13

Hölder and Schauder estimates

In this chapter we prove the Hölder and Schauder estimates

$$(-\Delta)^{\pm s} : C^{\alpha} \to C^{\alpha \mp 2s}.$$

These are valid when $\alpha \mp 2s$ is not an integer. At integer values, we obtain estimates in Zygmund spaces. We also show that

$$(-\Delta)^{-s} : L^{\infty} \to C^{2s}$$

when $s \neq 1/2$. For $s = 1/2$, C^{2s} must be replaced by the Zygmund space Λ_*. The proofs are presented with two different techniques: through pointwise integral methods (leaving out the endpoint cases) and with the method of semigroups (which gives a unified approach to all the estimates).

13.1 HÖLDER ESTIMATES

Given any $\beta > 0$, we denote by $[\beta] \geq 0$ the **integer part** of β, that is, $[\beta]$ is the largest integer k such that $\beta \geq k$. For convenience, we sometimes use the notation C^{β} to denote the space $C^{[\beta],[\beta]-\beta}$, for $\beta > 0$.

In this section we show, through various cases, that, for any $\beta > 2s > 0$,

$$(-\Delta)^s : C^{\beta}(\mathbb{R}^n) \to C^{\beta-2s}(\mathbb{R}^n)$$

whenever $\beta - 2s$ is not an integer. In particular, this says that $(-\Delta)^s$ behaves as a *fractional differential operator of order* $2s$ at the scale of Hölder spaces.

Our first result is the global (that is, in the whole space $\mathbb{R}^n$) Hölder estimate

$$(-\Delta)^s : C^{0,\alpha}(\mathbb{R}^n) \to C^{0,\alpha-2s}(\mathbb{R}^n)$$

for the case when $0 < \alpha \leq 1$ and the fractional order of derivative is $2s < \alpha$ (notice that if $0 < \alpha < 1$ and $0 < s < 1$ then $-2 < \alpha - 2s < 1$).

The reader should recall the pointwise formulas for the fractional Laplacian from Chapter 12, see Theorem 12.3.

DOI: 10.1201/9781032679853-13

Theorem 13.1 (Hölder estimates). *Let $u \in C^{0,\alpha}(\mathbb{R}^n)$, for some $0 < \alpha \leq 1$. If $0 < \alpha - 2s < 1$, then $(-\Delta)^s u \in C^{0,\alpha-2s}(\mathbb{R}^n)$ with*

$$\|(-\Delta)^s u\|_{L^\infty(\mathbb{R}^n)} \leq C\|u\|_{C^{0,\alpha}(\mathbb{R}^n)}$$

and

$$[(-\Delta)^s u]_{C^{\alpha-2s}(\mathbb{R}^n)} \leq C[u]_{C^\alpha(\mathbb{R}^n)}$$

where the constant $C > 0$ depends only on n, s and α.

Proof. For the L^∞ estimate notice that, for any $x \in \mathbb{R}^n$,

$$\begin{aligned}
|(-\Delta)^s u(x)| &\leq c_{n,s} \int_{|x-z|<1} \frac{[u]_{C^\alpha(x)}|x-z|^\alpha}{|x-z|^{n+2s}}\,dz + c_{n,s}\int_{|x-z|\geq 1} \frac{2\|u\|_{L^\infty(\mathbb{R}^n)}}{|x-z|^{n+2s}}\,dz \\
&\leq C\big([u]_{C^\alpha(\mathbb{R}^n)} + \|u\|_{L^\infty(\mathbb{R}^n)}\big) = C\|u\|_{C^{0,\alpha}(\mathbb{R}^n)}.
\end{aligned}$$

For the Hölder estimate, let $x_1, x_2 \in \mathbb{R}^n$. Then

$$\begin{aligned}
&|(-\Delta)^s u(x_1) - (-\Delta)^s u(x_2)| \\
&\quad = c_{n,s}\bigg| \int_{|z|<|x_1-x_2|} \frac{(u(x_1) - u(x_1+z)) - (u(x_2) - u(x_2+z))}{|z|^{n+2s}}\,dz \\
&\qquad + \int_{|z|\geq|x_1-x_2|} \frac{(u(x_1) - u(x_2)) - (u(x_1+z) - u(x_2+z))}{|z|^{n+2s}}\,dz\bigg| \\
&\quad \leq C[u]_{C^\alpha(\mathbb{R}^n)}\bigg[\int_{|z|<|x_1-x_2|} \frac{|z|^\alpha}{|z|^{n+2s}}\,dz + |x_1-x_2|^\alpha \int_{|z|\geq|x_1-x_2|} \frac{1}{|z|^{n+2s}}\,dz\bigg] \\
&\quad = C[u]_{C^\alpha(\mathbb{R}^n)}|x_1-x_2|^{\alpha-2s}.
\end{aligned}$$

□

The next $C^{1,\alpha}$ result follows from the Hölder estimate in Theorem 13.1 and the fact that derivatives commute with the fractional Laplacian, see (12.3). The reader should recall the distributional definition of the fractional Laplacian given in Chapter 12.

Theorem 13.2 ($C^{1,\alpha}$ estimates). *Let $u \in C^{1,\alpha}(\mathbb{R}^n)$ for some $0 < \alpha \leq 1$.*

(1) *If $0 < \alpha - 2s < 1$ then $(-\Delta)^s u \in C^{1,\alpha-2s}(\mathbb{R}^n)$ and*

$$[(-\Delta)^s u]_{C^{1,\alpha-2s}(\mathbb{R}^n)} \leq C[u]_{C^{1,\alpha}(\mathbb{R}^n)}.$$

(2) *If $0 < \alpha - 2s + 1 < 1$ then $(-\Delta)^s u \in C^{0,\alpha-2s+1}(\mathbb{R}^n)$ and*

$$[(-\Delta)^s u]_{C^{0,\alpha-2s+1}(\mathbb{R}^n)} \leq C[u]_{C^{1,\alpha}(\mathbb{R}^n)}.$$

The constant $C > 0$ above depends only on n, s and α.

Proof. We know that $\nabla(-\Delta)^s u(x) = (-\Delta)^s \nabla u(x)$ whenever $u \in \mathcal{S}$. The same is true for $u \in \mathcal{S}_s'$ in the sense of distributions after we define $\nabla u \in \mathcal{S}_s'$ for $u \in \mathcal{S}_s'$ in the natural way. Indeed, if $u \in \mathcal{S}_s'$ then, for any $\varphi \in \mathcal{S}$,

$$\begin{aligned}\langle \nabla(-\Delta)^s u, \varphi\rangle_{\mathcal{S}',\mathcal{S}} &= -\langle(-\Delta)^s u, \nabla\varphi\rangle_{\mathcal{S}',\mathcal{S}} = -\langle u, (-\Delta)^s \nabla\varphi\rangle_{\mathcal{S}_s',\mathcal{S}_s} \\ &= -\langle u, \nabla(-\Delta)^s\varphi\rangle_{\mathcal{S}_s',\mathcal{S}_s} = \langle \nabla u, (-\Delta)^s\varphi\rangle_{\mathcal{S}_s',\mathcal{S}_s} \\ &= \langle(-\Delta)^s\nabla u, \varphi\rangle_{\mathcal{S}',\mathcal{S}}.\end{aligned}$$

In particular for (1), if $u \in C^{1,\alpha}(\mathbb{R}^n)$ and $0 < 2s < \alpha$, then $\nabla(-\Delta)^s u = (-\Delta)^s \nabla u$. Moreover, by Theorem 13.1, since $\nabla u \in C^\alpha(\mathbb{R}^n)$ and $2s < \alpha$,

$$\begin{aligned}[(-\Delta)^s u]_{C^{1,\alpha-2s}(\mathbb{R}^n)} &= [\nabla(-\Delta)^s u]_{C^{\alpha-2s}(\mathbb{R}^n)} = [(-\Delta)^s \nabla u]_{C^{\alpha-2s}(\mathbb{R}^n)} \\ &\leq C[\nabla u]_{C^\alpha(\mathbb{R}^n)} = C[u]_{C^{1,\alpha}(\mathbb{R}^n)}.\end{aligned}$$

For (2), note that $2s > \alpha$. Recall that the fact that $u \in C^{1,\alpha}(\mathbb{R}^n)$ implies that

$$|u(x) - u(x+z) + \nabla u(x)\cdot z| \leq [u]_{C^{1,\alpha}(\mathbb{R}^n)}|z|^{1+\alpha}$$

for any $x, z \in \mathbb{R}^n$. Fix $x_1, x_2 \in \mathbb{R}^n$ and choose $\delta = |x_1 - x_2|$ in the pointwise formula for the fractional Laplacian of Theorem 12.3(2). On one hand, for $i = 1, 2$,

$$\begin{aligned}\left|\int_{|z|<|x_1-x_2|} \frac{u(x_i) - u(x_i+z) + \nabla u(x_i)\cdot z}{|z|^{n+2s}}\,dz\right| &\leq \int_{|z|<|x_1-x_2|} \frac{[u]_{C^{1,\alpha}(\mathbb{R}^n)}|z|^{1+\alpha}}{|z|^{n+2s}}\,dz \\ &= C[u]_{C^{1,\alpha}(\mathbb{R}^n)}|x_1-x_2|^{\alpha-2s+1}.\end{aligned}$$

On the other hand, let $z \in \mathbb{R}^n$ such that $|z| \geq |x_1 - x_2|$. For some $\xi = tx_1 + (1-t)x_2$, $0 \leq t \leq 1$, in between x_1 and x_2, and for some $\eta = r(x_1+z) + (1-r)(x_2+z)$, $0 \leq r \leq 1$, in between $x_1 + z$ and $x_2 + z$, we have

$$|\xi - \eta| = |(t-r)(x_1-x_2) - z| \leq |x_1 - x_2| + |z| \leq 2|z|.$$

Now, for some ξ and η as above we have

$$u(x_1) - u(x_2) = \nabla u(\xi)\cdot(x_1 - x_2)$$

and

$$u(x_1+z) - u(x_2+z) = \nabla u(\eta)\cdot(x_1-x_2).$$

Hence, if $|z| \geq |x_1 - x_2|$,

$$\begin{aligned}|(u(x_1) - u(x_2)) - (u(x_1+z) - u(x_2+z)| &\leq [\nabla u]_{C^\alpha(\mathbb{R}^n)}|\xi-\eta|^\alpha|x_1-x_2| \\ &\leq 2^\alpha[\nabla u]_{C^\alpha(\mathbb{R}^n)}|x_1-x_2||z|^\alpha.\end{aligned}$$

With this we estimate

$$\begin{aligned}&\left|\int_{|z|\geq|x_1-x_2|} \frac{(u(x_1)-u(x_2)) - (u(x_1+z)-u(x_2+z))}{|z|^{n+2s}}\,dz\right| \\ &\qquad \leq 2^\alpha[\nabla u]_{C^\alpha(\mathbb{R}^n)}|x_1-x_2|\int_{|z|\geq|x_1-x_2|}\frac{|z|^\alpha}{|z|^{n+2s}}\,dz \\ &\qquad = C[\nabla u]_{C^\alpha(\mathbb{R}^n)}|x_1-x_2|^{\alpha-2s+1}\end{aligned}$$

and (2) follows. □

We iterate Theorems 13.1 and 13.2 to obtain the following general $C^{k,\alpha}$ result.

Theorem 13.3 ($C^{k,\alpha}$ estimates). *Let $u \in C^{k,\alpha}(\mathbb{R}^n)$ for some $k \geq 0$ and some $0 < \alpha \leq 1$. Suppose that $k + \alpha - 2s > 0$ and that $\alpha - 2s$ is not an integer. Then*

$$(-\Delta)^s u \in C^{[k+\alpha-2s],k+\alpha-2s-[k+\alpha-2s]}(\mathbb{R}^n).$$

Proof. For $k = 0, 1$, see Theorems 13.1 and 13.2, respectively. The case $k \geq 2$ follows by induction and using that $D^k(-\Delta)^s u = (-\Delta)^s D^k u$. Say that $u \in C^{2,\alpha}(\mathbb{R}^n)$. Then $\nabla u \in C^{1,\alpha}(\mathbb{R}^n)$, so that $(-\Delta)^s \nabla u \in C^{1+\alpha-2s}(\mathbb{R}^n)$ if $1 + \alpha - 2s$ is not an integer. Therefore $\nabla(-\Delta)^s u \in C^{1+\alpha-2s}(\mathbb{R}^n)$, which is the same as saying that $(-\Delta)^s u \in C^{2+\alpha-2s}(\mathbb{R}^n)$ and $2 + \alpha - 2s$ is not an integer. □

13.2 SCHAUDER ESTIMATES

In this section we use pointwise integral techniques in combination with the method of semigroups to prove the global Schauder estimate

$$(-\Delta)^{-s} : C^\alpha(\mathbb{R}^n) \to C^{\alpha+2s}(\mathbb{R}^n)$$

whenever $\alpha + 2s < 1$. The reader should recall that the case $s = 1$ was proved with pointwise integral techniques in Chapter 3, see Remark 3.11.

Theorem 13.4 (Schauder estimates for $(-\Delta)^s$). *Suppose that $0 < \alpha + 2s < 1$. Let $f \in C^{0,\alpha}(\mathbb{R}^n)$ with compact support in a bounded domain Ω. Define*

$$u(x) = (-\Delta)^{-s} f(x) = \frac{1}{\Gamma(s)} \int_0^\infty e^{t\Delta} f(x) \, \frac{dt}{t^{1-s}} = c_{n,-s} \int_\Omega \frac{f(z)}{|x - z|^{n-2s}} \, dz.$$

Then $u \in C^{0,\alpha+2s}(\mathbb{R}^n)$ with

$$\|u\|_{L^\infty(\mathbb{R}^n)} \leq C \|f\|_{L^\infty(\mathbb{R}^n)}$$

and

$$[u]_{C^{0,\alpha+2s}(\mathbb{R}^n)} \leq C [f]_{C^{0,\alpha}(\mathbb{R}^n)}.$$

where $C > 0$ is a constant depending only on n, s and α. As a consequence, by Theorems 12.10 and 12.3, u satisfies $(-\Delta)^s u = f$ in $\mathbb{R}^n$ in the classical sense.

Proof. By Theorem 12.10, we already know that $\|u\|_{L^\infty(\mathbb{R}^n)} \leq C \|f\|_{L^\infty(\mathbb{R}^n)}$, where $C > 0$ depends only on n and s.

For the Hölder estimate of u, fix any $x_1, x_2 \in \mathbb{R}^n$. We begin by claiming that

$$I := \int_{\mathbb{R}^n} \left(\frac{1}{|x_1 - z|^{n-2s}} - \frac{1}{|x_2 - z|^{n-2s}} \right) dz = 0. \tag{13.1}$$

To see this, we use the method of semigroups. Recall that, by (12.28), up to a multiplicative constant that depends only on n and s,

$$\frac{1}{|x - z|^{n-2s}} = \int_0^\infty G_t(x - z) \, \frac{dt}{t^{1-s}} \tag{13.2}$$

where $G_t(x) = \frac{1}{(4\pi t)^{n/2}} e^{-|x|^2/(4t)}$ is the heat kernel. Then, up to a multiplicative constant, we can rewrite (13.1) as

$$I = \int_{\mathbb{R}^n} \int_0^\infty \left(G_t(x_1 - z) - G_t(x_2 - z)\right) \frac{dt}{t^{1-s}}\, dz. \tag{13.3}$$

The double integral in (13.3) is absolutely convergent. Indeed, on one hand, since $G_t(x) > 0$ and

$$\int_{\mathbb{R}^n} G_t(x - z)\, dz \equiv 1 \qquad \text{for any } x \in \mathbb{R}^n,\ t > 0 \tag{13.4}$$

we get

$$\int_{\mathbb{R}^n} \int_0^1 |G_t(x_1 - z) - G_t(x_2 - z)| \frac{dt}{t^{1-s}}\, dz \le 2 \int_0^1 \frac{dt}{t^{1-s}} < \infty.$$

On the other hand, first note that, for some constants $C, c > 0$ depending only on n,

$$|\nabla G_t(x)| \le \frac{C}{t^{(n+1)/2}} e^{-|x|^2/(ct)} \qquad \text{for all } x \in \mathbb{R}^n,\ t > 0.$$

Second, suppose that $|z| > 2(|x_1| + |x_2|) =: r_0$. Then, for any intermediate point $\xi = t(x_1 - z) + (1 - t)(x_2 - z) = tx_1 + (1 - t)x_2 - z$, $0 \le t \le 1$, in between $x_1 - z$ and $x_2 - z$, we have $|z| \le |x_1| + |x_2| + |\xi|$, so that $|z|/2 \le |z| - (|x_1| + |x_2|) \le |\xi|$. Hence, by the mean value theorem, for any $|z| > 2(|x_1| + |x_2|) = r_0$,

$$|G_t(x_1 - z) - G_t(x_2 - z)| \le |\nabla G_t(\xi)||x_1 - x_2| \le \frac{C}{t^{(n+1)/2}} e^{-|z|^2/(4ct)} |x_1 - x_2|.$$

From these two facts and (13.4),

$$\begin{aligned}
&\int_{\mathbb{R}^n} \int_1^\infty |G_t(x_1 - z) - G_t(x_2 - z)| \frac{dt}{t^{1-s}}\, dz \\
&\quad \le C|x_1 - x_2| \int_1^\infty \left[\int_{|z| \le r_0} \frac{1}{t^{(n+1)/2}}\, dz + \int_{|z| > r_0} \frac{e^{-|z|^2/(4ct)}}{t^{(n+1)/2}}\, dz \right] \frac{dt}{t^{1-s}} \\
&\quad \le C_{n,r_0} |x_1 - x_2| \int_1^\infty \left[\frac{1}{t^{(n+1)/2}} + \frac{1}{t^{1/2}} \right] \frac{dt}{t^{1-s}} < \infty
\end{aligned}$$

because, by hypothesis, $s < 1/2$. We have proved that the double integral in (13.3) is absolutely convergent. Therefore, applying Fubini's theorem in (13.3) and using (13.4), we get (13.1).

Now, since (13.1) holds, we can write

$$\begin{aligned}
u(x_1) - u(x_2) &= c_{n,-s} \int_{\mathbb{R}^n} f(z) \left[\frac{1}{|x_1 - z|^{n-2s}} - \frac{1}{|x_2 - z|^{n-2s}} \right] dz \\
&= c_{n,-s} \int_{\mathbb{R}^n} (f(z) - f(x_1)) \left[\frac{1}{|x_1 - z|^{n-2s}} - \frac{1}{|x_2 - z|^{n-2s}} \right] dz.
\end{aligned}$$

We write the last integral as the sum of two integrals $I_1 + I_2$, where the region of integration for I_1 is $B = \{z : |x_1 - z| < 2|x_1 - x_2|\}$ and the region of integration for I_2 is $B^c = \{z : |x_1 - z| \geq 2|x_1 - x_2|\}$. If $z \in B$ then $|x_2 - z| < 4|x_1 - x_2|$ and

$$\begin{aligned}|I_1| &\leq [f]_{C^\alpha(\mathbb{R}^n)}\left[\int_B \frac{|z-x_1|^\alpha}{|x_1-z|^{n-2s}}\,dz + \int_B \frac{|z-x_1|^\alpha}{|x_2-z|^{n-2s}}\,dz\right] \\ &\leq C[f]_{C^\alpha(\mathbb{R}^n)}\left[\int_B \frac{|z-x_1|^\alpha}{|x_1-z|^{n-2s}}\,dz + \int_{|x_2-z|<4|x_1-x_2|} \frac{|x_1-x_2|^\alpha}{|x_2-z|^{n-2s}}\,dz\right] \\ &= C_{n,s,\alpha}[f]_{C^\alpha(\mathbb{R}^n)}|x_1-x_2|^{\alpha+2s}.\end{aligned}$$

For I_2 we need a preliminary estimate. Fix $z \in B^c$. If ξ is any intermediate point between x_1 and x_2, then $|x_1-z| \leq |x_1-\xi|+|\xi-z| \leq |x_1-x_2|+|\xi-z| \leq \frac{1}{2}|x_1-z|+|\xi-z|$ and, as a consequence, $|x_1 - z| \leq 2|\xi - z|$. Thus, for $z \in B^c$, by using (13.2), the mean value theorem and the change of variables $|\xi - z|^2/(2ct) = r$,

$$\begin{aligned}\left|\frac{1}{|x_1-z|^{n-2s}} - \frac{1}{|x_2-z|^{n-2s}}\right| &\leq C\int_0^\infty |G_t(x_1-z) - G_t(x_2-z)|\,\frac{dt}{t^{1-s}} \\ &\leq C|x_1-x_2|\int_0^\infty \frac{e^{-|\xi-z|^2/(ct)}}{t^{(n+1)/2}}\,\frac{dt}{t^{1-s}} \\ &= C|x_1-x_2|\frac{1}{|\xi-z|^{n-2s+1}} \\ &\leq C|x_1-x_2|\frac{1}{|x_1-z|^{n-2s+1}}.\end{aligned}$$

With this we can now estimate (notice that in this step we need $\alpha + 2s < 1$)

$$\begin{aligned}|I_2| &\leq C[f]_{C^\alpha(\mathbb{R}^n)}|x_1-x_2|\int_{B^c} \frac{|z-x_1|^\alpha}{|x_1-z|^{n-2s+1}}\,dz \\ &= C_{n,s,\alpha}[f]_{C^\alpha(\mathbb{R}^n)}|x_1-x_2|^{\alpha+2s}.\end{aligned}$$

We have shown that $[u]_{C^{\alpha+2s}(\mathbb{R}^n)} \leq C_{n,s,\alpha}[f]_{C^\alpha(\mathbb{R}^n)}$, as desired. □

13.3 REGULARITY ESTIMATES VIA THE METHOD OF SEMIGROUPS

In this section we apply the method of semigroups to prove Hölder, Schauder and Zygmund estimates for the fractional Laplacian in a unified way. In particular, we give a different proof of the results of Sections 13.1 and 13.2.

In addition, with the semigroup technique we reach the endpoint cases when $\alpha+2s$ and $\alpha - 2s$ are integers. At these endpoints, the natural spaces are not Lipschitz or C^k, but the Zygmund spaces.

Furthermore, we prove that if $f \in L^\infty$ with compact support then $u = (-\Delta)^{-s}f \in C^{2s}$ if $s \neq 1/2$, and $u = (-\Delta)^{-s}f \in \Lambda_*$ if $s = 1/2$. In other words, the inverse fractional Laplacian acts as *fractional integral operator of order* $2s$ on L^∞.

The reader should recall the definition and properties of the classical heat diffusion semigroup $\{e^{t\Delta}\}_{t\geq 0}$ in Chapter 3, and the semigroup formulas for the fractional Laplacian and the inverse fractional Laplacian of Chapter 12.

Let $\beta > 0$ and take any integer $k > \beta/2$. We define the following set of functions:

$$\Lambda^\beta = \Big\{u \in L^\infty(\mathbb{R}^n) : [u]_{\Lambda^\beta} := \sup_{x\in\mathbb{R}^n, t>0} |t^{k-\beta/2}\partial_t^k e^{t\Delta}u(x)| < \infty\Big\} \tag{13.5}$$

under the norm $\|u\|_{\Lambda^\beta} = \|u\|_{L^\infty(\mathbb{R}^n)} + [u]_{\Lambda^\beta}$. It can be seen that this definition is independent of the choice of $k > \beta/2$ and that the norms for different values of k are all equivalent. Moreover, if $0 < \beta_1 < \beta_2$ then $\Lambda^{\beta_2} \subset \Lambda^{\beta_1}$ with $\|u\|_{\Lambda^{\beta_1}} \leq \|u\|_{\Lambda^{\beta_2}}$.

The **Zygmund space** $\Lambda_* = \Lambda_*^1$ is the family of functions $u \in L^\infty(\mathbb{R}^n)$ such that

$$[u]_{\Lambda_*} = \sup_{x,h\in\mathbb{R}^n, h\neq 0} \frac{|u(x+h) + u(x-h) - 2u(x)|}{|h|} < \infty$$

under the norm $\|u\|_{\Lambda_*} = \|u\|_{L^\infty(\mathbb{R}^n)} + [u]_{\Lambda_*}$. It is clear that $C^{0,1}(\mathbb{R}^n) \subset \Lambda_*$ continuously and it is known that the inclusion is strict. For any integer $k \geq 2$ we denote

$$\Lambda_*^k = \{u \in C^{k-1}(\mathbb{R}^n) : D^\gamma u \in \Lambda_* \text{ for all } |\gamma| = k-1\}$$

under the norm $\|u\|_{\Lambda_*^k} = \|u\|_{C^{k-1}(\mathbb{R}^n)} + \max_{|\gamma|=k-1}[D^\gamma u]_{\Lambda_*}$. Then $C^{k-1,1}(\mathbb{R}^n) \subsetneq \Lambda_*^k$.

It turns out that the spaces Λ^β, which are given in terms of the rate of growth of the classical heat diffusion semigroup acting on u, coincide with the classical Hölder and Zygmund spaces. The following result is due to Taibleson.

Theorem 13.5 (Semigroup characterization of Hölder and Zygmund spaces). *If $\beta > 0$ then*

$$\Lambda^\beta = \begin{cases} C^{[\beta],\beta-[\beta]}(\mathbb{R}^n) & \text{if } \beta \text{ is not an integer} \\ \Lambda_*^k & \text{if } \beta = k \text{ is an integer} \end{cases}$$

with equivalent norms.

We use the characterization above and the semigroup formulas for $(-\Delta)^s$ and $(-\Delta)^{-s}$ to show that $(-\Delta)^{\pm s} : \Lambda^\beta \to \Lambda^{\beta\mp 2s}$, namely, Hölder, Zygmund and Schauder estimates, and all of them in a unified way.

Before we prove the following result, it is important to note that Theorem 12.3 says that, for sufficiently regular functions u on the whole space $\mathbb{R}^n$, the fractional Laplacian $(-\Delta)^s u$ of u is given in terms of the semigroup formula.

Theorem 13.6 (Unified Hölder–Zygmund estimates). *Let $0 < s < 1$ and $\beta > 0$ such that $2s < \beta$. If $u \in \Lambda^\beta$ then $(-\Delta)^s u \in \Lambda^{\beta-2s}$ with*

$$\|(-\Delta)^s u\|_{L^\infty(\mathbb{R}^n)} \leq C\|u\|_{\Lambda^\beta}$$

and

$$[(-\Delta)^s u]_{\Lambda^{\beta-2s}} \leq C[u]_{\Lambda^\beta}$$

where the constant $C > 0$ above depends only on s and β.

Proof. We begin by proving that $(-\Delta)^s u$ is bounded. Suppose that $2s < \beta < 2$. By the semigroup formula for the fractional Laplacian, the basic properties of $\{e^{t\Delta}\}_{t\geq 0}$ and the definition of the space Λ^β in (13.5) with $k = 1 > \beta/2$,

$$\begin{aligned}
|(-\Delta)^s u(x)| &= \left|\frac{1}{\Gamma(-s)}\int_0^\infty (e^{r\Delta}u(x) - u(x))\,\frac{dr}{r^{1+s}}\right| \\
&\leq C\int_0^1\int_0^r |\partial_\rho e^{\rho\Delta}u(x)|\,d\rho\,\frac{dr}{r^{1+s}} + 2C\|u\|_{L^\infty(\mathbb{R}^n)}\int_1^\infty \frac{dr}{r^{1+s}} \\
&\leq C[u]_{\Lambda^\beta}\int_0^1\int_0^r \rho^{\beta/2-1}\,d\rho\,\frac{dr}{r^{1+s}} + C\|u\|_{L^\infty(\mathbb{R}^n)} \\
&\leq C[u]_{\Lambda^\beta}\int_0^1 r^{\beta/2}\,\frac{dr}{r^{1+s}} + C\|u\|_{L^\infty(\mathbb{R}^n)} \\
&\leq C_{\beta,s}\|u\|_{\Lambda^\beta}
\end{aligned}$$

because $\beta > 2s$. For the case $\beta \geq 2$, choose $2s < \beta' = \frac{2s+2}{2} < 2 \leq \beta$, apply the estimate we just proved and the continuous inclusion $\Lambda^\beta \subset \Lambda^{\beta'}$ to get

$$\|(-\Delta)^s u\|_{L^\infty(\mathbb{R}^n)} \leq C_s\|u\|_{\Lambda^{\beta'}} \leq C_s\|u\|_{\Lambda^\beta}.$$

To control the seminorm $[(-\Delta)^s u]_{\Lambda^{\beta-2s}}$, fix an integer $k > \beta/2 > (\beta - 2s)/2 > 0$. We need to show that there is a constant $C > 0$ depending on s and β such that

$$|t^k\partial_t^k e^{t\Delta}[(-\Delta)^s u](x)| \leq C[u]_{\Lambda^\beta} t^{(\beta-2s)/2} \qquad \text{for all } x \in \mathbb{R}^n. \tag{13.6}$$

The semigroup formula for the fractional Laplacian gives

$$t^k\partial_t^k e^{t\Delta}[(-\Delta)^s u](x) = \frac{1}{\Gamma(-s)}\int_0^\infty t^k\partial_t^k e^{t\Delta}(e^{r\Delta}u(x) - u(x))\,\frac{dr}{r^{1+s}}.$$

On one hand, observe that, by the semigroup property of $\{e^{t\Delta}\}_{t\geq 0}$,

$$\partial_t^k e^{t\Delta}(\partial_\rho e^{\rho\Delta}u) = \partial_t^k\partial_\rho e^{(t+\rho)\Delta}u = \partial_\tau^{k+1}e^{\tau\Delta}u\big|_{\tau=t+r}.$$

With this and the hypothesis $u \in \Lambda^\beta$, that is, (13.5) with $k+1 > \beta/2$,

$$\begin{aligned}
\left|\int_0^t t^k\partial_t^k e^{t\Delta}(e^{r\Delta}u(x) - u(x))\,\frac{dr}{r^{1+s}}\right| &= \left|\int_0^t t^k\partial_t^k e^{t\Delta}\left[\int_0^r \partial_\rho e^{\rho\Delta}u(x)\,d\rho\right]\frac{dr}{r^{1+s}}\right| \\
&= t^k\left|\int_0^t\int_0^r \partial_\tau^{k+1}e^{\tau\Delta}u(x)\big|_{\tau=t+\rho}\,d\rho\,\frac{dr}{r^{1+s}}\right| \\
&\leq t^k[u]_{\Lambda^\beta}\int_0^t\int_0^r (t+\rho)^{\beta/2-k-1}\,d\rho\,\frac{dr}{r^{1+s}}.
\end{aligned}$$

For the latter integral, we change variables $w = \rho/t$ and use that $(1+w)^{\beta/2-k-1} \leq 1$:

$$\begin{aligned}
\int_0^t\int_0^r (t+\rho)^{\beta/2-k-1}\,d\rho\,\frac{dr}{r^{1+s}} &= t^{\beta/2-k-1}\int_0^t\int_0^r (1+\rho/t)^{\beta/2-k-1}\,d\rho\,\frac{dr}{r^{1+s}} \\
&= t^{\beta/2-k}\int_0^t\int_0^{r/t}(1+w)^{\beta/2-k-1}\,dw\,\frac{dr}{r^{1+s}} \\
&\leq t^{\beta/2-k}\int_0^t \frac{r}{t}\,\frac{dr}{r^{1+s}} = Ct^{\beta/2-k-s}.
\end{aligned}$$

Therefore,

$$\left|\int_0^t t^k \partial_t^k e^{t\Delta}(e^{r\Delta}u(x) - u(x))\,\frac{dr}{r^{1+s}}\right| \le C[u]_{\Lambda^\beta} t^{(\beta-2s)/2}.$$

On the other hand, by applying the semigroup property and the hypothesis $u \in \Lambda^\beta$,

$$\begin{aligned}
&\left|\int_t^\infty t^k \partial_t^k e^{t\Delta}(e^{r\Delta}u(x) - u(x))\,\frac{dr}{r^{1+s}}\right| \\
&\qquad \le t^k \int_t^\infty \left|\partial_\tau^k e^{\tau\Delta}u(x)\big|_{\tau=t+r}\right| \frac{dr}{r^{1+s}} + |t^k \partial_t^k e^{t\Delta}u(x)| \int_t^\infty \frac{dr}{r^{1+s}} \\
&\qquad \le C[u]_{\Lambda^\beta}\left[t^k \int_t^\infty (t+r)^{\beta/2-k}\,\frac{dr}{r^{1+s}} + t^{(\beta-2s)/2}\right].
\end{aligned}$$

By the change of variables $w = r/t$ and using that $\beta/2 - k - s < 0$,

$$\begin{aligned}
\int_t^\infty (t+r)^{\beta/2-k}\,\frac{dr}{r^{1+s}} &= t^{\beta/2-k}\int_t^\infty (1+r/t)^{\beta/2-k}\,\frac{dr}{r^{1+s}} \\
&= t^{\beta/2-k-s}\int_1^\infty (1+w)^{\beta/2-k}\,\frac{dw}{w^{1+s}} = Ct^{\beta/2-k-s}.
\end{aligned}$$

Hence,

$$\left|\int_t^\infty t^k \partial_t^k e^{t\Delta}(e^{r\Delta}u(x) - u(x))\,\frac{dr}{r^{1+s}}\right| \le C[u]_{\Lambda^\beta} t^{(\beta-2s)/2}.$$

We have shown (13.6), which means that $[(-\Delta)^s u]_{\Lambda^{\beta-2s}} \le C_{\beta,s}[u]_{\Lambda^\beta}$ as desired. □

Immediately from Theorem 13.6 and by writing $\Lambda^\beta = \Lambda^{k+\alpha}$ for some $k \ge 0$ and $0 < \alpha \le 1$ in Theorem 13.5, we obtain the Hölder estimates for the fractional Laplacian. Notice that if $0 < \alpha \le 1$ and $0 < s < 1$ then $-2 < \alpha - 2s < 1$.

Corollary 13.7 (Hölder estimates). *Let $u \in C^{k,\alpha}(\mathbb{R}^n)$ for some $k \ge 0$ and $0 < \alpha \le 1$.*

(1) *If $0 < \alpha - 2s < 1$ and $k \ge 0$, then $(-\Delta)^s u \in C^{k,\alpha-2s}(\mathbb{R}^n)$ and*

$$\|(-\Delta)^s u\|_{C^{k,\alpha-2s}(\mathbb{R}^n)} \le C\|u\|_{C^{k,\alpha}(\mathbb{R}^n)}.$$

(2) *If $0 < \alpha - 2s + 1 < 1$ and $k \ge 1$, then $(-\Delta)^s u \in C^{k-1,\alpha-2s+1}(\mathbb{R}^n)$ and*

$$\|(-\Delta)^s u\|_{C^{k-1,\alpha-2s+1}(\mathbb{R}^n)} \le C\|u\|_{C^{k,\alpha}(\mathbb{R}^n)}.$$

(3) *If $0 < \alpha - 2s + 2 < 1$ and $k \ge 2$, then $(-\Delta)^s u \in C^{k-2,\alpha-2s+2}(\mathbb{R}^n)$ and*

$$\|(-\Delta)^s u\|_{C^{k-2,\alpha-2s+2}(\mathbb{R}^n)} \le C\|u\|_{C^{k,\alpha}(\mathbb{R}^n)}.$$

The constants $C > 0$ above depend only on n, s, k and α.

Moreover, we get the endpoint cases when $\alpha - 2s = 0, -1$.

Corollary 13.8 (Zygmund estimates). *Let $u \in C^{k,\alpha}(\mathbb{R}^n)$ for some $k \ge 0$ and $0 < \alpha \le 1$.*

(1) *If* $0 < \alpha = 2s < 1$ *and* $k \geq 0$, *then* $(-\Delta)^s u \in \Lambda_*^k$ *and*

$$\|(-\Delta)^s u\|_{\Lambda_*^k} \leq C\|u\|_{C^{k,2s}(\mathbb{R}^n)}.$$

(2) *If* $0 < \alpha = 2s - 1 < 1$ *and* $k \geq 1$, *then* $(-\Delta)^s u \in \Lambda_*^{k-1}$ *and*

$$\|(-\Delta)^s u\|_{\Lambda_*^{k-1}} \leq C\|u\|_{C^{k,2s-1}(\mathbb{R}^n)}.$$

The constants $C > 0$ *above depend only on* n, s *and* k.

Now we move on to considering Schauder estimates. As with Theorem 13.6, we apply the method of semigroups in the spaces Λ^β to prove the following estimates in a unified way.

Theorem 13.9 (Unified Schauder–Zygmund estimates). *Let* $\beta > 0$ *and fix* $0 < s < \min\{1, n/2\}$. *Let* $f \in \Lambda^\beta$ *with compact support and define*

$$u(x) = (-\Delta)^{-s} f(x) = \frac{1}{\Gamma(s)} \int_0^\infty e^{t\Delta} f(x)\, \frac{dt}{t^{1-s}} \qquad \text{for } x \in \mathbb{R}^n.$$

Then $u \in \Lambda^{\beta+2s}$ *with*

$$\|u\|_{L^\infty(\mathbb{R}^n)} \leq C\|f\|_{L^\infty(\mathbb{R}^n)}$$

and

$$[u]_{\Lambda^{\beta+2s}} \leq C[f]_{\Lambda^\beta}$$

where the constant $C > 0$ *above depends only on* n, s *and* β.

Proof. We already proved in Theorem 12.10 that $\|u\|_{L^\infty(\mathbb{R}^n)} \leq C\|f\|_{L^\infty(\mathbb{R}^n)}$, where $C > 0$ depends only on n and s. To estimate $[u]_{\Lambda^{\beta+2s}}$, fix an integer $k > (\beta + 2s)/2$. Then, by using the semigroup property of $\{e^{t\Delta}\}_{t\geq 0}$, the hypothesis $f \in \Lambda^\beta$ (see (13.5) with $k > \beta/2$), the change of variables $w = r/t$ and the fact that $\beta/2 - k + s < 0$,

$$\begin{aligned}
|t^k \partial_t^k e^{t\Delta} u(x)| &= |t^k \partial_t^k e^{t\Delta}[(-\Delta)^{-s} f](x)| \\
&= \frac{t^k}{\Gamma(s)} \left| \int_0^\infty \partial_t^k e^{t\Delta}(e^{r\Delta} f(x))\, \frac{dr}{r^{1-s}} \right| \\
&\leq C t^k \int_0^\infty \left| \partial_\tau^k e^{\tau\Delta} f(x)\big|_{\tau=t+r} \right| \frac{dr}{r^{1-s}} \\
&\leq C[f]_{\Lambda^\beta} t^k \int_0^\infty (t+r)^{\beta/2-k}\, \frac{dr}{r^{1-s}} \\
&= C[f]_{\Lambda^\beta} t^{\beta/2} \int_0^\infty (1 + r/t)^{\beta/2-k}\, \frac{dr}{r^{1-s}} \\
&= C[f]_{\Lambda^\beta} t^{\beta/2+s} \int_0^\infty (1+w)^{\beta/2-k}\, \frac{dw}{w^{1-s}} \\
&= C[f]_{\Lambda^\beta} t^{(\beta+2s)/2}.
\end{aligned}$$

In other words, $[u]_{\Lambda^{\beta+2s}} \leq C_{\beta,s}[f]_{\Lambda^\beta}$. □

Directly from Theorems 13.9 and 13.5 we get the full range of Schauder estimates, including the cases when $\alpha + 2s$ is an integer.

Corollary 13.10 (Schauder and Zygmund estimates). *Let $f \in C^{0,\alpha}(\mathbb{R}^n)$, $0 < \alpha \leq 1$, with compact support. Let $0 < s < 1$ and define*

$$u(x) = (-\Delta)^{-s} f(x) = \frac{1}{\Gamma(s)} \int_0^\infty e^{t\Delta} f(x) \, \frac{dt}{t^{1-s}} \qquad \text{for } x \in \mathbb{R}^n.$$

(1) *If $0 < \alpha + 2s < 1$, then $u \in C^{0,\alpha+2s}(\mathbb{R}^n)$ and*

$$\|u\|_{C^{0,\alpha+2s}(\mathbb{R}^n)} \leq C\|f\|_{C^{0,\alpha}(\mathbb{R}^n)}.$$

(2) *If $1 < \alpha + 2s < 2$, then $u \in C^{1,\alpha+2s-1}(\mathbb{R}^n)$ and*

$$\|u\|_{C^{1,\alpha+2s-1}(\mathbb{R}^n)} \leq C\|f\|_{C^{0,\alpha}(\mathbb{R}^n)}.$$

(3) *If $2 < \alpha + 2s < 3$, then $u \in C^{2,\alpha+2s-2}(\mathbb{R}^n)$ and*

$$\|u\|_{C^{2,\alpha+2s-2}(\mathbb{R}^n)} \leq C\|f\|_{C^{0,\alpha}(\mathbb{R}^n)}.$$

(4) *If $\alpha + 2s = k$, $k = 1, 2$, then $u \in \Lambda_*^k$ and*

$$\|u\|_{\Lambda_*^k} \leq C\|f\|_{C^{0,\alpha}(\mathbb{R}^n)}.$$

The constant $C > 0$ above depends only on n, s and α. As a consequence, by Theorems 12.10 and 12.3, u satisfies $(-\Delta)^s u = f$ in $\mathbb{R}^n$ in the classical sense.

Next we consider the case of a bounded right hand side (not necessarily Hölder). We apply the method of semigroups to show that $(-\Delta)^{-s} : L^\infty \to \Lambda^{2s}$.

Theorem 13.11 (Unified estimates for bounded right hand side). *Let $f \in L^\infty(\mathbb{R}^n)$ with compact support. Fix $0 < s < \min\{1, n/2\}$ and define*

$$u(x) = (-\Delta)^{-s} f(x) = \frac{1}{\Gamma(s)} \int_0^\infty e^{t\Delta} f(x) \, \frac{dt}{t^{1-s}} \qquad \text{for } x \in \mathbb{R}^n.$$

Then $u \in \Lambda^{2s}$ with

$$\|u\|_{L^\infty(\mathbb{R}^n)} + [u]_{\Lambda^{2s}} \leq C\|f\|_{L^\infty(\mathbb{R}^n)}$$

where $C > 0$ is a constant depending only on n and s.

Proof. Theorem 12.10 gives that $\|u\|_{L^\infty(\mathbb{R}^n)} \leq C\|f\|_{L^\infty(\mathbb{R}^n)}$, where $C > 0$ depends only on n and s. Fix an integer $k > s$. We need to show that

$$[u]_{\Lambda^{2s}} = \sup_{x \in \mathbb{R}^n, t > 0} |t^k \partial_t^k e^{t\Delta}[(-\Delta)^{-s} f](x)| \leq C\|f\|_{L^\infty(\mathbb{R}^n)} t^s$$

for some constant $C > 0$ depending only on n and s. It is easy, but a bit tedious, to verify that the heat kernel $G_t(x) = \frac{1}{(4\pi t)^{n/2}} e^{-|x|^2/(4t)}$ satisfies the following estimate:

$$\int_{\mathbb{R}^n} |\partial_t^k G_t(x)| \, dx \leq \frac{c}{t^k} \qquad t > 0$$

where $c > 0$ depends only on n and k. In particular,

$$|\partial_t^k e^{t\Delta} f(x)| = |\partial_t^k G_t * f(x)| \le \frac{c}{t^k} \|f\|_{L^\infty(\mathbb{R}^n)} \qquad \text{for all } x \in \mathbb{R}^n,\ t > 0.$$

With this, the semigroup property of $\{e^{t\Delta}\}_{t\ge 0}$ and the change of variables $w = r/t$,

$$\begin{aligned}
|t^k \partial_t^k e^{t\Delta}[(-\Delta)^{-s} f](x)| &\le C t^k \int_0^\infty \left| \partial_\tau^k e^{\tau\Delta} f(x)\big|_{\tau=t+r} \right| \frac{dr}{r^{1-s}} \\
&\le C t^k \|f\|_{L^\infty(\mathbb{R}^n)} \int_0^\infty (t+r)^{-k} \frac{dr}{r^{1-s}} \\
&= C \|f\|_{L^\infty(\mathbb{R}^n)} \int_0^\infty (1+r/t)^{-k} \frac{dr}{r^{1-s}} \\
&= C \|f\|_{L^\infty(\mathbb{R}^n)} t^s \int_0^\infty (1+w)^{-k} \frac{dw}{w^{1-s}} \\
&= C_{n,s} \|f\|_{L^\infty(\mathbb{R}^n)} t^s
\end{aligned}$$

where in the last line we used that $s - k < 0$. □

In particular, from Theorems 13.11 and 13.5, we immediately obtain Schauder estimates in Hölder and Zygmund spaces for bounded right hand side, including the endpoint case $2s = 1$.

Corollary 13.12 (Schauder and Zygmund estimates for bounded right hand side). *Let $f \in L^\infty(\mathbb{R}^n)$ with compact support. Let $0 < s < 1$ and define*

$$u(x) = (-\Delta)^{-s} f(x) = \frac{1}{\Gamma(s)} \int_0^\infty e^{t\Delta} f(x) \frac{dt}{t^{1-s}} \qquad \textit{for } x \in \mathbb{R}^n.$$

(1) *If $0 < 2s < 1$, then $u \in C^{0,2s}(\mathbb{R}^n)$ and*

$$\|u\|_{C^{0,2s}(\mathbb{R}^n)} \le C \|f\|_{L^\infty(\mathbb{R}^n)}.$$

(2) *If $2s = 1$, then $u \in \Lambda_*$ and*

$$\|u\|_{\Lambda_*} \le C \|f\|_{L^\infty(\mathbb{R}^n)}.$$

(3) *If $1 < 2s < 2$, then $u \in C^{1,2s-1}(\mathbb{R}^n)$ and*

$$\|u\|_{C^{1,2s-1}(\mathbb{R}^n)} \le C \|f\|_{L^\infty(\mathbb{R}^n)}.$$

The constant $C > 0$ above depends only on n and s.

NOTES ON CHAPTER 13

The Hölder estimates of Section 13.1 were proved by Silvestre in [59, 61]. However, unlike in [59, 61], our proof of Theorem 13.2 does not use of the boundedness of the Riesz transforms on Hölder spaces.

The proof of Theorem 13.4 follows ideas of Gatto–Segovia–Vagi [41], where other, more general models of fractional Laplacians and inverse fractional Laplacians in metric measure spaces and spaces of homogeneous type are studied.

The Zygmund space Λ_*, which, as we have seen, in many instances turns out to be the natural endpoint space in the scale of Hölder spaces in place of $C^{0,1}$, was introduced by Zygmund in [76]. An example constructed by Hardy shows that $C^{0,1} \subsetneq \Lambda_*$, see Zygmund [76] or Stein [63].

The heat semigroup characterization of Hölder–Zygmund spaces of Theorem 13.5 is due to Taibleson [72], where a similar characterization using the Poisson semigroup was shown as well, see also Stein [63].

The proofs of the Hölder and Schauder estimates with the method of semigroups of Section 13.3 are from Stinga [68]. The proofs of Theorem 13.11 and Corollary 13.12 are due to Caffarelli–Stinga [15].

The method of semigroups has been developed for parabolic Hölder and Zygmund spaces and for Hölder, Schauder and Zygmund estimates for master equations driven by the fractional heat operator $(\partial_t - \Delta)^s u = f$ by Stinga–Torrea in [70].

CHAPTER 14

The Caffarelli–Silvestre extension problem

In this chapter we consider the Caffarelli–Silvestre extension problem, which is a way of characterizing the fractional Laplacian $(-\Delta)^s$, $0 < s < 1$, on $\mathbb{R}^n$ (a nonlocal operator of fractional order $2s$) as the Dirichlet-to-Neumann map for a second-order degenerate/singular elliptic local PDE problem. We develop the extension problem for the fractional Laplacian with the Stinga–Torrea method of semigroups. The characterization and semigroup methodology are very powerful for various reasons. In particular, local PDE methods can be applied in the extension to prove regularity estimates for solutions to fractional Laplacian equations. We show applications to interior Harnack inequalities, local Hölder estimates, the Liouville theorem and interior derivative estimates.

14.1 THE EXTENSION PROBLEM FOR $(-\Delta)^{1/2}$

To understand the ideas and possible applications of the extension problem, let us begin with an overview of the (perhaps more classical) case of the extension problem for the fractional Laplacian $(-\Delta)^{1/2}$ in $\mathbb{R}^n$, $n \geq 1$.

Given $u = u(x) \in \mathcal{S}$, let $U = U(x,y) : \mathbb{R}^n \times [0,\infty) \to \mathbb{R}$ denote the **harmonic extension** of u to the upper half space $\mathbb{R}^n \times (0,\infty)$. In other words, let U solve

$$\begin{cases} \Delta U + \partial_{yy} U = 0 & \text{in } \mathbb{R}^n \times (0,\infty) \\ U(x,0) = u(x) & \text{on } \mathbb{R}^n. \end{cases} \tag{14.1}$$

Here the Laplacian is taken with respect to the x variable: $\Delta = \Delta_x$. To solve the boundary value problem (14.1) and find a formula for U in terms of u, we apply Fourier's method. Let $\widehat{U}(\xi, y)$ denote the Fourier transform of U in the x variable, for each fixed $y \geq 0$:

$$\widehat{U}(\xi, y) = \frac{1}{(2\pi)^{n/2}} \int_{\mathbb{R}^n} U(x,y) e^{-i\xi \cdot x}\, dx.$$

DOI: 10.1201/9781032679853-14

Then, for each fixed $\xi \in \mathbb{R}^n$, the function $y \longmapsto f_\xi(y) := \widehat{U}(\xi, y)$ satisfies the second-order initial value ODE problem

$$\begin{cases} f_\xi'' = |\xi|^2 f_\xi & \text{for } y > 0 \\ f_\xi(0) = \widehat{u}(\xi). \end{cases}$$

Therefore,

$$f_\xi(y) = c_1 e^{-y|\xi|} + c_2 e^{y|\xi|}$$

for some constants c_1 and c_2. For uniqueness of solutions, we need a condition, say, at infinity. Since we are using the Fourier transform and $e^{y|\xi|}$ does not define a tempered distribution, we set $c_2 = 0$. Using the initial condition, we get

$$\widehat{U}(\xi, y) = e^{-y|\xi|}\widehat{u}(\xi). \tag{14.2}$$

Since, for each fixed $y > 0$, $e^{-y|\xi|} \in L^1(\mathbb{R}^n) \cap L^2(\mathbb{R}^n)$, the Fourier transform can be inverted and we find that

$$U(x, y) = P_y * u(x) = \int_{\mathbb{R}^n} P_y(x - z)u(z)\, dz \tag{14.3}$$

where $P_y(x)$ is the **Poisson kernel for the Laplacian in the upper-half space**, which is defined by $\widehat{P_y}(\xi) = e^{-y|\xi|}/(2\pi)^{n/2}$. It is known that

$$P_y(x) \equiv P_y^{1/2}(x) = d_n \frac{y}{(|x|^2 + y^2)^{(n+1)/2}} \qquad \text{for } x \in \mathbb{R}^n,\ y > 0$$

where $d_n = \frac{\Gamma((n+1)/2)}{\pi^{(n+1)/2}} > 0$. The constant d_n and the scaling $P_y(x) = \frac{1}{y^n}P_1(\frac{x}{y})$ give

$$\int_{\mathbb{R}^n} P_y(x)\, dx = 1 \qquad \text{for all } y > 0. \tag{14.4}$$

In particular, $U = P_y * u$ defines an approximation of the identity and $U(x, y) \to u(x)$ uniformly on compact subsets of $\mathbb{R}^n$ and in $L^p(\mathbb{R}^n)$ for $1 \leq p < \infty$, as $y \to 0$.

Noticing that $|\xi|$ in (14.2) is the Fourier multiplier of $(-\Delta)^{1/2}$, see (12.1), we can introduce the notation

$$e^{-y(-\Delta)^{1/2}}u(x) := U(x, y)$$

for (14.3). We have that $\{e^{-y(-\Delta)^{1/2}}\}_{y\geq 0}$ defines a semigroup of bounded linear operators (see Section 3.3 for the general definition of semigroup) on, say, $L^2(\mathbb{R}^n)$. Indeed, by using the Fourier transform definition in (14.2), we see that $e^{-0(-\Delta)^{1/2}}u = u$, $e^{-y(-\Delta)^{1/2}} : L^2(\mathbb{R}^n) \to L^2(\mathbb{R}^n)$ is a bounded linear operator for every $y \geq 0$, and $e^{-(y_1+y_2)(-\Delta)^{1/2}}u = e^{-y_1(-\Delta)^{1/2}}(e^{-y_2(-\Delta)^{1/2}}u)$, for every $y_1, y_2 \geq 0$ and $u \in L^2(\mathbb{R}^n)$. The semigroup $\{e^{-y(-\Delta)^{1/2}}\}_{y\geq 0}$ is called the **Poisson semigroup** generated by the Laplacian.

The semigroup notation $U(x, y) = e^{-y(-\Delta)^{1/2}}u(x)$ is consistent with the problem (14.1) in that

$$\begin{cases} \partial_{yy}(e^{-y(-\Delta)^{1/2}}u) = (-\Delta)(e^{-y(-\Delta)^{1/2}}u) & \text{for } y > 0 \\ e^{-0(-\Delta)^{1/2}}u = u. \end{cases}$$

We now claim that

$$-\partial_y U(x,y)\big|_{y=0} = (-\Delta)^{1/2}u(x) \qquad \text{for } x \in \mathbb{R}^n. \tag{14.5}$$

This can be shown in at least two different ways. One is with the Fourier transform. Indeed, by (14.2),

$$\begin{aligned}-\partial_y U(x,y) &= -\mathcal{F}^{-1}(\partial_y e^{-y|\xi|}\widehat{u}(\xi))(x) \\ &= \mathcal{F}^{-1}(e^{-y|\xi|}|\xi|\widehat{u}(\xi))(x) \\ &\to \mathcal{F}^{-1}(|\xi|\widehat{u}(\xi))(x) = (-\Delta)^{1/2}u(x)\end{aligned}$$

as $y \to 0$. Another way of proving (14.5) is by means of the Poisson formula (14.3). Notice that, by changing variables and using that $P_y(-z) = P_y(z)$, we can write

$$U(x,y) = \frac{1}{2}\int_{\mathbb{R}^n} P_y(z)(u(x+z) + u(x-z))\,dz.$$

Since (14.4) holds, we have $\int_{\mathbb{R}^n} \partial_y P_y(x)\,dx = 0$, so that we can introduce the term $-2u(x)$ inside the following integral and get

$$\begin{aligned}\partial_y U(x,y) &= \frac{1}{2}\int_{\mathbb{R}^n} \partial_y P_y(z)(u(x+z) + u(x-z) - 2u(x))\,dz \\ &= \frac{d_n}{2}\int_{\mathbb{R}^n} \left[1 - \frac{(n+1)y^2}{|z|^2+y^2}\right]\frac{u(x+z)+u(x-z)-2u(x)}{(|z|^2+y^2)^{(n+1)/2}}\,dz \\ &\to \frac{d_n}{2}\int_{\mathbb{R}^n} \frac{u(x+z)+u(x-z)-2u(x)}{|z|^{n+1}}\,dz = -(-\Delta)^{1/2}u(x)\end{aligned}$$

as $y \to 0$, see the pointwise formula for the fractional Laplacian in Theorem 12.1(3) and notice that, since $\Gamma(1/2) = \pi$, the constant d_n here coincides with the constant $c_{n,s}$ in (12.9) for the fractional Laplacian when $s = 1/2$. There is a third way of proving (14.5) by means of the method of semigroups that will be shown in Section 14.2. Let us also observe here that (14.5) is also consistent with the Poisson semigroup notation because

$$-\partial_y(e^{-y(-\Delta)^{1/2}}u) = (-\Delta)^{1/2}e^{-y(-\Delta)^{1/2}}u \to (-\Delta)^{1/2}u$$

as $y \to 0$.

The characterization of the fractional Laplacian given in (14.5) says that $(-\Delta)^{1/2}$ is the **Dirichlet-to-Neumann map** for the harmonic extension of u to the upper half space. Indeed, we extended $u = u(x)$ to the upper half space by adding a new variable $y > 0$ as a harmonic function $U(x,y)$ such that $U(x,0) = u(x)$ (a Dirichlet boundary condition for U at $y = 0$), see (14.1), and showed that $-\partial_y U(x,0) = (-\Delta)^{1/2}u(x)$, which is a Neumann boundary condition for U at $y = 0$. This is the **extension problem** characterization of the fractional Laplacian of order $s = 1/2$.

To begin understanding the significance and power of the extension problem, let us consider the simplest situation. As we know, the Poisson equation for the fractional Laplacian

$$(-\Delta)^{1/2}u = f \qquad \text{in } \mathbb{R}^n \tag{14.6}$$

is a nonlocal equation of fractional order $2s$. Since the problem is nonlocal, many of the local PDE regularity techniques that we have seen in this book do not directly apply to the analysis of regularity of u. For example, one cannot use compactly supported test functions to perform localized estimates, because if φ has compact support then, in general, $(-\Delta)^{1/2}\varphi$ has noncompact support. We also saw in Section 12.7 that the viscosity solutions approach needs to be modified as to take into account the values of u at infinity. However, with the extension problem characterization, we can convert the problem into an equivalent local PDE problem.

Let u be a solution to the Poisson equation (14.6). If U is the harmonic extension of u to the upper half space, see (14.1), then, by (14.5), U satisfies

$$\begin{cases} \Delta U + \partial_{yy} U = 0 & \text{in } \mathbb{R}^n \times (0,\infty) \\ -\partial_y U(x,0) = f(x) & \text{on } \mathbb{R}^n. \end{cases}$$

This is now a local Neumann problem for the Laplacian for which many of the regularity techniques from this book are available. Now, if we are able to prove regularity estimates for U all the way up to the boundary $y = 0$ then those estimates automatically transfer back to u simply because

$$U(x,0) = u(x).$$

As a matter of fact, this means that $U(x,0) = (-\Delta)^{-1/2} f(x)$, so U also characterizes the inverse fractional Laplacian.

The tradeoff between nonlocality and locality with this extension technique is the appearance of the new variable $y > 0$ which, in some sense, encodes the values of u at infinity, see (14.3).

To illustrate some of the many possible applications of the extension problem, let us consider a bounded solution u to

$$(-\Delta)^{1/2} u = 0 \qquad \text{in } B_1.$$

Since the fractional Laplacian commutes with derivatives, we expect u to be smooth in, say, $B_{1/8}$. Unfortunately, we cannot differentiate the equation directly, as that would require differentiating u outside of B_1, where it may not be smooth.

Consider now the extension U of u that solves (14.1) and is given by convolution with the Poisson kernel, see (14.3). By (14.4), U is bounded with

$$\|U\|_{L^\infty(\mathbb{R}^n \times [0,\infty))} \le \|u\|_{L^\infty(\mathbb{R}^n)}.$$

One can, however, obtain a local bound: for any $|x| < 1/4$ and $y < 1$, by (14.4) and using the explicit formula for P_y,

$$\begin{aligned} |U(x,y)| &\le \int_{B_1} P_y(x-z)|u(z)|\,dz + \int_{\mathbb{R}^n \setminus B_1} P_y(x-z)|u(z)|\,dz \\ &\le \|u\|_{L^\infty(B_1)} + d_n \int_{\mathbb{R}^n \setminus B_1} \frac{|u(z)|}{|x-z|^{n+1}}\,dz \\ &\le \|u\|_{L^\infty(B_1)} + C \int_{\mathbb{R}^n \setminus B_1} \frac{|u(z)|}{1+|z|^{n+1}}\,dz \end{aligned} \tag{14.7}$$

where in the last line we used that if $|x| < 1/4$ and $|z| > 1$ then $1+|z| \le 2|z| \le 2|x-z| + 2|x| < 2|x-z| + 1/2$, so that $1/2 + |z| \le 2|x-z|$ and, consequently, $1+|z| \le 1+2|z| \le 4|x-z|$. Moreover, by (14.5),

$$-\partial_y U(x,0) = (-\Delta)^{1/2} u = 0 \qquad \text{for } x \in B_1.$$

Next, one can check (see Section 14.2) that the even reflection $\tilde{U}(x,y) := U(x,|y|)$ of U to $y<0$ is harmonic across $y=0$ in B_1:

$$\Delta_{x,y}\tilde{U} = 0 \qquad \text{in } B_1 \times (-1,1). \tag{14.8}$$

Now we can apply the regularity results for harmonic functions of Chapter 2 to $\tilde{U}$ and deduce estimates for u.

As a first example, recall that the interior derivative estimates for harmonic functions in Corollary 2.15(1) give that, for any multi-index $\tilde{\gamma} = (\gamma, \gamma_{n+1}) \in \mathbb{N}_0^{n+1}$,

$$\|D^{\tilde{\gamma}}_{x,y}\tilde{U}\|_{L^\infty(B_{1/8}\times(-1/8,1/8))} \le C_{n,\tilde{\gamma}} \|\tilde{U}\|_{L^\infty(B_{1/4}\times(-1/4,1/4))}.$$

In particular, since $\tilde{U}(x,0) = u(x)$, we find that u is smooth in $B_{1/8}$ and, by (14.7),

$$\begin{aligned}
\|D^\gamma u\|_{L^\infty(B_{1/8})} &= \|D^\gamma_x U\|_{L^\infty(B_{1/8})} \\
&\le \|D^{\tilde{\gamma}}_{x,y}\tilde{U}\|_{L^\infty(B_{1/8}\times(-1/8,1/8))} \\
&\le C_{n,\tilde{\gamma}} \|\tilde{U}\|_{L^\infty(B_{1/4}\times(-1/4,1/4))} \\
&\le 2C \|U\|_{L^\infty(B_{1/4}\times[0,1/4))} \\
&\le C\left(\|u\|_{L^\infty(B_1)} + \int_{\mathbb{R}^n\setminus B_1} \frac{|u(z)|}{1+|z|^{n+1}}\,dz \right)
\end{aligned}$$

where $C>0$ depends only on n and γ. These are the interior derivative estimates for u.

A second example is the Harnack inequality for u. Since the Poisson kernel P_y is positive, if $u \ge 0$ in $\mathbb{R}^n$ then its extension $U(x,y) \ge 0$ and, therefore, $\tilde{U}$ is a nonnegative harmonic function in $B_1 \times (-1,1)$. The Harnack inequality of Theorem 2.16 then gives that there exists a constant $C_H > 0$ depending only on n such that

$$\sup_{B_{1/2}\times(-1/2,1/2)} \tilde{U} \le C_H \inf_{B_{1/2}\times(-1/2,1/2)} \tilde{U}.$$

Hence, $u(x) = \tilde{U}(x,0)$ satisfies the Harnack inequality in $B_{1/2}$:

$$\begin{aligned}
\sup_{B_{1/2}} u = \sup_{x\in B_{1/2}} \tilde{U}(x,0) &\le \sup_{B_{1/2}\times(-1/2,1/2)} \tilde{U} \\
&\le C_H \inf_{B_{1/2}\times(-1/2,1/2)} \tilde{U} \\
&\le C_H \inf_{x\in B_{1/2}} \tilde{U}(x,0) = C_H \inf_{B_{1/2}} u.
\end{aligned}$$

A third example is the Liouville theorem. Suppose now that u is a globally bounded solution to

$$(-\Delta)^{1/2}u = 0 \qquad \text{in } \mathbb{R}^n.$$

Then, since the extension U of u satisfies

$$-\partial_y U(x,0) = (-\Delta)^{1/2}u = 0 \qquad \text{in } \mathbb{R}^n$$

the even reflection $\tilde{U}$ of U defined as above is a bounded harmonic function in $\mathbb{R}^{n+1}$. Therefore, by the Liouville theorem for harmonic functions (see Theorem 2.14), $\tilde{U}$ is constant in $\mathbb{R}^{n+1}$ and, in particular, u is constant on $\mathbb{R}^n$.

The striking fact is that other powers of the Laplacian different than $s = 1/2$ can be characterized in a similar way with a local PDE extension problem. This is the **Caffarelli–Silvestre extension problem**. In fact, the extension problem characterization was generalized by means of the method of semigroups to hold for more general operators on Hilbert spaces, and this is the **Stinga–Torrea extension problem**. We next present the extension problem for the fractional Laplacian using the method of semigroups, and applications to regularity estimates.

14.2 THE EXTENSION PROBLEM FOR $(-\Delta)^s$

Throughout the rest of this chapter, we take $0 < s < 1$ and set

$$a = 1 - 2s \in (-1,1).$$

The extension problem for the fractional Laplacian $(-\Delta)^s$ is driven by a degenerate/singular modification of the Laplacian. As we will see in a moment, the new extension equation can be written in divergence form as

$$\operatorname{div}_{x,y}(y^a \nabla_{x,y} U) = 0.$$

By passing the derivatives through,

$$y^{-a} \operatorname{div}_{x,y}(y^a \nabla_{x,y} U) = \Delta U + \tfrac{a}{y}\partial_y U + \partial_{yy} U.$$

Thus, the extension equation can also be written as

$$\Delta U + \tfrac{a}{y}\partial_y U + \partial_{yy} U = 0$$

where Δ is the Laplacian in the x variable. If $s = 1/2$, then $a = 0$ and we are back to the harmonic extension equation (14.1).

The coefficient y^a is degenerate at $y = 0$ when $0 < a < 1$ (that corresponds to $s < 1/2$) and singular when $-1 < a < 0$ (corresponding to $s > 1/2$). By multiplying the equation in divergence form by a test function and integrating by parts, we find that the conormal derivative at $y = 0$ associated to this new PDE is

$$-y^a \partial_y U(x,y)\big|_{y=0}.$$

We first present the extension problem characterization for the case when $u \in \mathcal{S}$, where all computations are easily justified. For this, we use the method of semigroups. Then we show how to write down the problem for the cases when $u \in L_s$ (distributional solutions) and $u \in H^s(\mathbb{R}^n)$ (weak solutions).

The reader should recall the definition and properties of the Fourier transform and the classical heat diffusion semigroup $\{e^{t\Delta}\}_{t\geq 0}$, and the semigroup formulas for the fractional Laplacian, see Chapters 3 and 12, respectively.

Theorem 14.1 (Extension problem for the fractional Laplacian). *Let $u \in \mathcal{S}$. Define $U = U(x,y) : \mathbb{R}^n \times [0,\infty) \to \mathbb{R}$ by*

$$\begin{aligned} U(x,y) &= \frac{y^{2s}}{4^s\Gamma(s)} \int_0^\infty e^{-y^2/(4t)} e^{t\Delta} u(x) \, \frac{dt}{t^{1+s}} \\ &= \frac{1}{\Gamma(s)} \int_0^\infty e^{-t} e^{\frac{y^2}{4t}\Delta} u(x) \, \frac{dt}{t^{1-s}} \\ &= \frac{1}{\Gamma(s)} \int_0^\infty e^{-y^2/(4t)} e^{t\Delta} [(-\Delta)^s u](x) \, \frac{dt}{t^{1-s}} \\ &= \frac{\Gamma(n/2+s)}{\Gamma(s)\pi^{n/2}} \int_{\mathbb{R}^n} \frac{y^{2s}}{(|x-z|^2+y^2)^{(n+2s)/2}} u(z)\,dz. \end{aligned} \tag{14.9}$$

Then $U \in C^\infty(\mathbb{R}^n \times (0,\infty)) \cap C(\mathbb{R}^n \times [0,\infty))$ and

$$\|U\|_{L^\infty(\mathbb{R}^n\times(0,\infty))} \leq \|u\|_{L^\infty(\mathbb{R}^n)}.$$

Moreover, U is the unique classical solution to

$$\begin{cases} \Delta U + \frac{a}{y}\partial_y U + \partial_{yy} U = 0 & \text{in } \mathbb{R}^n \times (0,\infty) \\ U(x,0) = u(x) & \text{on } \mathbb{R}^n \end{cases} \tag{14.10}$$

that weakly vanishes as $y \to \infty$. Furthermore, $y^a \partial_y U \in C(\mathbb{R}^n \times [0,\infty))$ with

$$-\lim_{y\to 0} y^a \partial_y U(x,y) = \frac{\Gamma(1-s)}{4^{s-1/2}\Gamma(s)} (-\Delta)^s u(x) \tag{14.11}$$

and, also,

$$-\lim_{y\to 0} \frac{U(x,y)-U(x,0)}{y^{2s}} = \frac{\Gamma(1-s)}{4^s\Gamma(1+s)} (-\Delta)^s u(x) \tag{14.12}$$

for all $x \in \mathbb{R}^n$.

Before we present the proof of Theorem 14.1, let us understand where the semigroup formulas (14.9) for the solution U to the extension problem (14.10) come from. As we did in Section 14.1, we apply the Fourier transform in the variable x for each $y \geq 0$ fixed, that is, we look at $U(x,y) \longmapsto \widehat{U}(\xi,y)$. We have

$$\begin{cases} -|\xi|^2\widehat{U} + \frac{a}{y}\partial_y \widehat{U} + \partial_{yy}\widehat{U} = 0 & \text{in } \mathbb{R}^n \times (0,\infty) \\ \widehat{U}(\xi,0) = \widehat{u}(\xi) & \text{on } \mathbb{R}^n. \end{cases}$$

Thus, for each $\xi \in \mathbb{R}^n$ fixed, one finds that $y \longmapsto f_\xi(y) = \widehat{U}(\xi, y)$ solves the second-order ODE problem

$$\begin{cases} f_\xi''(y) + \frac{a}{y} f_\xi'(y) = \lambda f_\xi(y) & \text{for } y > 0 \\ f_\xi(0) = \widehat{u}(\xi) \end{cases} \tag{14.13}$$

where $\lambda = |\xi|^2$. The ODE is a **Bessel differential equation**. There are several types of Bessel functions that can be used to build up two independent solutions for (14.13). We will not go through all the analysis here (the whole derivation can be found in [67, 69]). It turns out that, by using the condition that $U(x, y) \to 0$ weakly in $L^2(\mathbb{R}^n)$ as $y \to \infty$, we find that the solution to (14.13) is

$$\widehat{U}(\xi, y) = \frac{2^{1-s}}{\Gamma(s)} (y|\xi|)^s K_s(y|\xi|) \widehat{u}(\xi). \tag{14.14}$$

The function $K_s(z)$, $z > 0$, is the **modified Bessel function of the second kind** or **Macdonald's function**, which is defined, for $z > 0$, as

$$K_{1/2}(z) = \left(\frac{\pi}{2z}\right)^{1/2} e^{-z}$$

and, for $0 < s < 1$,

$$K_s(z) = \frac{1}{2}\left(\frac{z}{2}\right)^s \int_0^\infty e^{-t} e^{-z^2/(4t)} \frac{dt}{t^{1+s}}. \tag{14.15}$$

We have that $K_s(z) = K_{-s}(z)$,

$$K_s(z) \sim \frac{2^{s-1}\Gamma(s)}{z^s} \text{ as } z \to 0 \qquad \text{and} \qquad K_s(z) \sim \left(\frac{\pi}{2z}\right)^{1/2} e^{-z} \text{ as } z \to \infty \tag{14.16}$$

see [51]. Then, by (14.14) and (14.16),

$$\widehat{U}(\xi, y) \to \widehat{u}(\xi) \text{ as } y \to 0 \qquad \text{and} \qquad \widehat{U}(\xi, y) \to 0 \text{ as } y \to \infty.$$

By noticing that $|\xi|$ is the Fourier multiplier of $(-\Delta)^{1/2}$, we introduce the notation

$$U(x, y) = \frac{2^{1-s}}{\Gamma(s)} [(y(-\Delta)^{1/2})^s K_s(y(-\Delta)^{1/2})] u(x)$$

to denote (14.14). In particular, when $s = 1/2$,

$$\widehat{U}(\xi, y) = \frac{2^{1-1/2}}{\Gamma(1/2)} (y|\xi|)^{1/2} \left(\frac{\pi}{2y|\xi|}\right)^{1/2} e^{-y|\xi|} \widehat{u}(\xi) = e^{-y|\xi|} \widehat{u}(\xi)$$

that is,

$$U(x, y) = \frac{2^{1-1/2}}{\Gamma(1/2)} [(y(-\Delta)^{1/2})^{1/2} K_{1/2}(y(-\Delta)^{1/2})] u(x) = e^{-y(-\Delta)^{1/2}} u(x)$$

as we found in Section 14.1. To obtain the integral representation of the solution U in (14.9) in terms of the semigroup $\{e^{t\Delta}\}_{t\geq 0}$ for any $0<s<1$, we choose $z=y|\xi|$ in (14.15) and apply the change of variables $r=y^2/(4t)$ to get

$$\begin{aligned}\widehat{U}(\xi,y) &= \frac{2^{1-s}}{\Gamma(s)}(y|\xi|)^s\frac{1}{2}\left(\frac{y|\xi|}{2}\right)^s\int_0^\infty e^{-t}e^{-(y|\xi|)^2/(4t)}\widehat{u}(\xi)\,\frac{dt}{t^{1+s}}\\ &= \frac{y^{2s}}{4^s\Gamma(s)}\int_0^\infty e^{-t}e^{-\frac{y^2}{4t}|\xi|^2}\left(|\xi|^{2s}\widehat{u}(\xi)\right)\frac{dt}{t^{1+s}} \\ &= \frac{1}{\Gamma(s)}\int_0^\infty e^{-y^2/(4r)}e^{-r|\xi|^2}\left(|\xi|^{2s}\widehat{u}(\xi)\right)\frac{dr}{r^{1-s}}.\end{aligned}\tag{14.17}$$

By taking the inverse Fourier transform, we arrive at

$$U(x,y)=\frac{1}{\Gamma(s)}\int_0^\infty e^{-y^2/(4r)}e^{r\Delta}[(-\Delta)^s u](x)\,\frac{dr}{r^{1-s}}$$

which is the third formula in (14.9). If in (14.17) we now perform the change of variables $t=r|\xi|^2$, we find that

$$\begin{aligned}\widehat{U}(\xi,y) &= \frac{y^{2s}}{4^s\Gamma(s)}\int_0^\infty e^{-t}e^{-\frac{y^2}{4t}|\xi|^2}\left(|\xi|^{2s}\widehat{u}(\xi)\right)\frac{dt}{t^{1+s}}\\ &= \frac{y^{2s}}{4^s\Gamma(s)}\int_0^\infty e^{-r|\xi|^2}e^{-y^2/(4r)}\widehat{u}(\xi)\,\frac{dr}{r^{1+s}}\end{aligned}$$

which, after inverting the Fourier transform, gives

$$U(x,y)=\frac{y^{2s}}{\Gamma(s)}\int_0^\infty e^{-y^2/(4r)}e^{r\Delta}u(x)\,\frac{dr}{r^{1+s}}$$

and this is the first formula in (14.9). Finally, the second semigroup formula in (14.9) follows directly from the first one by the change of variables $r=y^2/(4t)$.

Proof of Theorem 14.1. We have just seen that the first three formulas for U in (14.9) coincide. For the fourth one, we use the explicit expression for $e^{t\Delta}u(x)$ in terms of the heat kernel, Fubini's theorem, the change of variables $r=\frac{|x-z|^2+y^2}{4t}$ and the definition of the Gamma function:

$$\begin{aligned}U(x,y) &= \frac{y^{2s}}{4^s\Gamma(s)}\int_0^\infty e^{-y^2/(4t)}e^{t\Delta}u(x)\,\frac{dt}{t^{1+s}}\\ &= \frac{y^{2s}}{\Gamma(s)\pi^{n/2}}\int_0^\infty\int_{\mathbb{R}^n}\frac{e^{-\frac{|x-z|^2+y^2}{4t}}}{(4t)^{n/2+s}}u(z)\,dz\,\frac{dt}{t}\\ &= \int_{\mathbb{R}^n}u(z)\left[\frac{y^{2s}}{\Gamma(s)\pi^{n/2}}\int_0^\infty\frac{e^{-\frac{|x-z|^2+y^2}{4t}}}{(4t)^{n/2+s}}\frac{dt}{t}\right]dz\\ &= \int_{\mathbb{R}^n}u(z)\left[\frac{y^{2s}}{\Gamma(s)\pi^{n/2}}\int_0^\infty e^{-r}\left(\frac{r}{|x-z|^2+y^2}\right)^{n/2+s}\frac{dr}{r}\right]dz\\ &= \frac{\Gamma(n/2+s)}{\Gamma(s)\pi^{n/2}}\int_{\mathbb{R}^n}\frac{y^{2s}}{(|x-z|^2+y^2)^{(n+2s)/2}}u(z)\,dz.\end{aligned}\tag{14.18}$$

By differentiating under the integral sign in, for instance, this last formula, it can be readily seen that $U \in C^\infty(\mathbb{R}^n \times (0,\infty))$. Also, by using the second formula for U in (14.9), the dominated convergence theorem (recall that $\|e^{t\Delta}u\|_{L^\infty(\mathbb{R}^n)} \le \|u\|_{L^\infty(\mathbb{R}^n)}$ for all $t \ge 0$) and the definition of the Gamma function,

$$\lim_{y\to 0} U(x,y) = \frac{1}{\Gamma(s)}\int_0^\infty e^{-t}\Big(\lim_{y\to 0} e^{-\frac{y^2}{4t}\Delta}u(x)\Big)\,\frac{dt}{t^{1-s}} = u(x)$$

which shows that $U \in C(\mathbb{R}^n \times [0,\infty))$.

Next, by the change of variables $y^2/(4t) = r$ and the definition of the Gamma function, we find that

$$\frac{y^{2s}}{4^s\Gamma(s)}\int_0^\infty e^{-y^2/(4t)}\,\frac{dt}{t^{1+s}} = 1 \qquad \text{for all } y > 0. \tag{14.19}$$

Therefore, by the first formula in (14.9) and using that $|e^{t\Delta}u(x)| \le \|u\|_{L^\infty(\mathbb{R}^n)}$,

$$|U(x,y)| \le \frac{y^{2s}}{4^s\Gamma(s)}\int_0^\infty e^{-y^2/(4t)}|e^{t\Delta}u(x)|\,\frac{dt}{t^{1+s}} \le \|u\|_{L^\infty(\mathbb{R}^n)}$$

for all $x \in \mathbb{R}^n$ and $y > 0$.

The analysis performed after the statement of the theorem showed that U is the unique solution to (14.10) that weakly vanishes at infinity. However, we can verify this again directly by using the first formula in (14.9). Indeed, by differentiating inside the integral sign and integrating by parts,

$$\begin{aligned}
\Delta U(x,y) &= \frac{y^{2s}}{4^s\Gamma(s)}\int_0^\infty e^{-y^2/(4t)}\Delta e^{t\Delta}u(x)\,\frac{dt}{t^{1+s}} \\
&= \frac{y^{2s}}{4^s\Gamma(s)}\int_0^\infty e^{-y^2/(4t)}\partial_t e^{t\Delta}u(x)\,\frac{dt}{t^{1+s}} \\
&= -\frac{y^{2s}}{4^s\Gamma(s)}\int_0^\infty \partial_t\Big(\frac{e^{-y^2/(4t)}}{t^{1+s}}\Big)e^{t\Delta}u(x)\,dt \\
&= -\frac{1}{4^s\Gamma(s)}\int_0^\infty \big(\tfrac{a}{y}\partial_y + \partial_{yy}\big)\big(y^{2s}e^{-y^2/(4t)}\big)e^{t\Delta}u(x)\,\frac{dt}{t^{1+s}} \\
&= -\big(\tfrac{a}{y}\partial_y + \partial_{yy}\big)\Big[\frac{y^{2s}}{4^s\Gamma(s)}\int_0^\infty e^{-y^2/(4t)}e^{t\Delta}u(x)\,\frac{dt}{t^{1+s}}\Big] \\
&= -\tfrac{a}{y}\partial_y U(x,y) - \partial_{yy}U(x,y).
\end{aligned} \tag{14.20}$$

To see that U weakly vanishes as $y \to \infty$, let $\varphi \in \mathcal{S}$ and use the Fourier transform formula for U in (14.14) to get

$$\begin{aligned}
\Big|\int_{\mathbb{R}^n} U(x,y)\varphi(x)\,dx\Big| &= \Big|\int_{\mathbb{R}^n} \widehat{U}(\xi,y)\overline{\widehat{\varphi}(\xi)}\,d\xi\Big| \\
&\le C\int_{\mathbb{R}^n} (y|\xi|)^s K_s(y|\xi|)|\widehat{u}(\xi)||\widehat{\varphi}(\xi)|\,d\xi.
\end{aligned}$$

By the asymptotic properties of the Bessel function $K_s(z)$ in (14.16), there is a constant $C > 0$ depending only on s such that

$$0 \leq (y|\xi|)^s K_s(y|\xi|) \leq C \qquad \text{for all } y > 0,\ \xi \in \mathbb{R}^n.$$

Moreover, for all $|\xi| \neq 0$, as $y \to \infty$,

$$(y|\xi|)^s K_s(y|\xi|) \leq C(y|\xi|)^{s-1/2} e^{-y|\xi|} \to 0.$$

Therefore, since $\widehat{u}\widehat{\varphi} \in L^1(\mathbb{R}^n)$, by the dominated convergence theorem,

$$\int_{\mathbb{R}^n} (y|\xi|)^s K_s(y|\xi|) |\widehat{u}(\xi)||\widehat{\varphi}(\xi)|\, d\xi \to 0$$

as $y \to \infty$ and $U(x.y) \to 0$ weakly in $L^2(\mathbb{R}^n)$ as $y \to \infty$.

Finally, we check the limiting identities (14.11) and (14.12) with the fractional Laplacian. By the first formula for U in (14.9) and recalling that (14.19) holds, $a = 1 - 2s$ and the semigroup formula for the fractional Laplacian (see (12.6)),

$$\begin{aligned}
-y^a \partial_y U(x,y) &= -\frac{y^{1-2s}}{4^s\Gamma(s)} \int_0^\infty \partial_y \big(y^{2s} e^{-y^2/(4t)}\big) e^{t\Delta} u(x) \, \frac{dt}{t^{1+s}} \\
&= -\frac{y^{1-2s}}{4^s\Gamma(s)} \int_0^\infty \partial_y \big(y^{2s} e^{-y^2/(4t)}\big) \big(e^{t\Delta} u(x) - u(x)\big) \frac{dt}{t^{1+s}} \\
&= -\frac{1}{4^s\Gamma(s)} \int_0^\infty \left(2s - \frac{y^2}{2t}\right) e^{-y^2/(4t)} \big(e^{t\Delta} u(x) - u(x)\big) \frac{dt}{t^{1+s}} \\
&\to -\frac{2s}{4^s\Gamma(s)} \int_0^\infty \big(e^{t\Delta} u(x) - u(x)\big) \frac{dt}{t^{1+s}} = \frac{\Gamma(1-s)}{4^{s-1/2}\Gamma(s)} (-\Delta)^s u(x).
\end{aligned}$$

Thus, (14.11) is proved. To prove (14.12) we begin by observing that, by the semigroup formula for $(-\Delta)^{-s}$, see (12.24), we can write

$$U(x,0) = u(x) = (-\Delta)^{-s}[(-\Delta)^s u](x) = \frac{1}{\Gamma(s)} \int_0^\infty e^{t\Delta}[(-\Delta)^s u](x) \, \frac{dt}{t^{1-s}}.$$

With this, from the third formula for U in (14.9), by the change of variables $r = y^2/(4t)$, the fact that $e^{t\Delta} u \to u$ as $t \to 0$ and the definition of the Gamma function,

$$\begin{aligned}
-\frac{U(x,y) - U(x,0)}{y^{2s}} &= -\frac{1}{\Gamma(s)} \int_0^\infty \frac{e^{-y^2/(4t)} - 1}{y^{2s}} e^{t\Delta}[(-\Delta)^s u](x) \, \frac{dt}{t^{1-s}} \\
&= -\frac{1}{4^s\Gamma(s)} \int_0^\infty (e^{-r} - 1) e^{\frac{y^2}{4r}\Delta}[(-\Delta)^s u](x) \, \frac{dr}{r^{1+s}} \\
&\to -\frac{1}{4^s\Gamma(s)} (-\Delta)^s u(x) \int_0^\infty (e^{-t} - 1) \, \frac{dt}{t^{1+s}} \\
&= \frac{\Gamma(1-s)}{4^s\Gamma(1+s)} (-\Delta)^s u(x)
\end{aligned}$$

as $y \to 0$, and (14.12) is proved. □

As we saw in (14.18), the last convolution formula for the solution U in (14.9) was obtained by means of the first semigroup formula and the explicit form of the heat kernel. In other words,

$$U(x,y) = P_y^s * u(x) \qquad \text{for } x \in \mathbb{R}^n,\ y > 0$$

where

$$\begin{aligned} P_y^s(x) &= \frac{y^{2s}}{4^s\Gamma(s)} \int_0^\infty e^{-y^2/(4t)} G_t(x) \frac{dt}{t^{1+s}} \\ &= d_{n,s} \frac{y^{2s}}{(|x|^2 + y^2)^{(n+2s)/2}} \end{aligned} \tag{14.21}$$

and $d_{n,s} = \frac{\Gamma(n/2+s)}{\Gamma(s)\pi^{n/2}} > 0$. We call this the **Poisson kernel for the extension problem**.

It is clear that the rescaling $P_y^s(x) = \frac{1}{y^n} P_1^s(\frac{x}{y})$ holds and that $P_1^s \in L^p(\mathbb{R}^n)$ for all $1 \le p \le \infty$. Moreover, since $\int_{\mathbb{R}^n} G_t(x)\,dx = 1$ and (14.19) holds, it is clear that

$$\int_{\mathbb{R}^n} P_y^s(x)\,dx = \frac{y^{2s}}{4^s\Gamma(s)} \int_0^\infty e^{-y^2/(4t)} \int_{\mathbb{R}^n} G_t(x)\,dx \frac{dt}{t^{1+s}} = 1 \tag{14.22}$$

for all $y > 0$. As a consequence, $U(x,y) = P_y^s * u(x)$ defines an approximation of the identity and $U(x,y) \to u(x)$ as $y \to 0$ uniformly on compact subsets of $\mathbb{R}^n$ if u is continuous and bounded, and in $L^p(\mathbb{R}^n)$ and a.e. if $u \in L^p(\mathbb{R}^n)$, for $1 \le p < \infty$. However, $U(x,y)$ does not define a semigroup unless $s = 1/2$.

As we did in Section 14.1, we can again prove (14.11) with the Fourier transform or with the explicit formula for the Poisson kernel for the extension problem.

For the Fourier transform argument, we use that

$$\partial_z[z^s K_s(z)] = -z^s K_{s-1}(z) = -z^s K_{1-s}(z)$$

(see [51]), the chain rule in (14.14), and (14.16) to get that, as $y \to 0$,

$$\begin{aligned} -y^a \partial_y \widehat{U}(\xi,y) &= y^{1-2s} \frac{2^{1-s}}{\Gamma(s)} (y|\xi|)^s K_{1-s}(y|\xi|) |\xi| \widehat{u}(\xi) \\ &\sim y^{1-2s} \frac{2^{1-s}}{\Gamma(s)} (y|\xi|)^s \frac{2^{-s}\Gamma(1-s)}{(y|\xi|)^{1-s}} |\xi| \widehat{u}(\xi) \\ &= \frac{\Gamma(1-s)}{4^{s-1/2}\Gamma(s)} |\xi|^{2s} \widehat{u}(\xi). \end{aligned}$$

After inverting the Fourier transform, we arrive at (14.11).

Now for the proof using the Poisson kernel for the extension problem P_y^s, by changing variables and the symmetry property $P_y^s(-x) = P_y^s(x)$, we can write

$$U(x,y) = \frac{d_{n,s}}{2} \int_{\mathbb{R}^n} P_y^s(z)(u(x+z) + u(x-z))\,dz.$$

Hence, since (14.22) holds, we can introduce the term $-2u(x)$ inside the integral formula for $\partial_y U(x,y)$, recall that $a = 1-2s$ and apply the dominated convergence theorem to obtain

$$\begin{aligned} y^a \partial_y U(x,y) &= \frac{d_{n,s}}{2} \int_{\mathbb{R}^n} y^a \partial_y P_y^s(z)\big(u(x+z)+u(x-z)-2u(x)\big)\,dz \\ &= \frac{d_{n,s}}{2} \int_{\mathbb{R}^n} \left(2s - \frac{(n+2s)y^2}{|z|^2+y^2}\right) \frac{u(x+z)+u(x-z)-2u(x)}{(|z|^2+y^2)^{(n+2s)/2}}\,dz \\ &\to (2s)\frac{d_{n,s}}{2} \int_{\mathbb{R}^n} \frac{u(x+z)+u(x-z)-2u(x)}{|z|^{n+2s}}\,dz \\ &= -\frac{\Gamma(1-s)}{4^{s-1/2}\Gamma(s)}(-\Delta)^s u(x) \end{aligned} \tag{14.23}$$

where in the last line we used the pointwise formula in Theorem 12.1(3) and the explicit value of the constant $c_{n,s}$ in (12.9).

We next give some observations about the extension problem of Theorem 14.1. The reader should verify the consistency of all these remarks and comments with the case $s = 1/2$ we analyzed in Section 14.1.

Remark 14.2 (Harmonic functions in $n+1+a = n+2-2s$ fractional dimensions). The extension equation in (14.10) was interpreted by Caffarelli and Silvestre as an equation for a harmonic function in "fractional" dimensions. Indeed, if a is a nonnegative integer and $U(x,z) : \mathbb{R}^n \times \mathbb{R}^{1+a} \to \mathbb{R}$ is a radially symmetric function in the z variable, that is, $U(x,z) = U(x,z')$ whenever $|z| = |z'| = y$, then one can think of U as a function of the space variables $x \in \mathbb{R}^n$ and the radial variable $y = |z|$. In these variables, the Laplacian of U is

$$\Delta_x U + \tfrac{a}{y}\partial_y U + \partial_{yy} U.$$

This is exactly the extension equation in (14.10). At this point, there is no need to consider only integer values for a in this equation. The solution to (14.10) can then be thought of as the *harmonic extension of u to $1+a$ dimensions more.* Of course, $\mathbb{R}^{n+1+a}$ has no meaning when a is not an integer. However, with this interpretation one can, for example, guess the form of the Poisson kernel P_y^s.

Remark 14.3 (Extension problem in nondivergence form). As we mentioned at the beginning of this section, the extension equation in (14.10) can be written in divergence form as

$$\operatorname{div}_{x,y}(y^a \nabla_{x,y} U) = 0.$$

The problem (14.10) can also be transformed into a nondivergence form problem via a convenient change of variables. Consider the new variable $z = (y/(2s))^{2s} > 0$ so that $y = (2s)z^{1/(2s)}$, and define

$$V(x,z) = U(x,y).$$

Then

$$\tfrac{a}{y}\partial_y U + \partial_{yy} U = z^{2-1/s} V_{zz}$$

and

$$y^a \partial_y U(x,z) = (2s)^a V_z(x,z).$$

Therefore, by Theorem 14.1, V solves the nondivergence form elliptic problem

$$\begin{cases} \Delta V + z^{2-1/s} V_{zz} = 0 & \text{in } \mathbb{R}^n \times (0,\infty) \\ V(x,0) = u(x) & \text{on } \mathbb{R}^n \end{cases}$$

and, for its (classical) Neumann condition, we get

$$-\lim_{z\to 0} V_z(x,z) = \frac{s^{2s}\Gamma(1-s)}{\Gamma(1+s)}(-\Delta)^s u(x).$$

Remark 14.4 (Extension problem for the inverse fractional Laplacian)**.** Let $u \in \mathcal{S}$ and f such that

$$(-\Delta)^s u = f \qquad \text{in } \mathbb{R}^n.$$

Equivalently (see Section 12.4),

$$u = (-\Delta)^{-s} f \qquad \text{in } \mathbb{R}^n.$$

With this, the third semigroup formula for U in Theorem 14.1, see (14.9), gives that

$$U(x,y) = \frac{1}{\Gamma(s)} \int_0^\infty e^{-y^2/(4t)} e^{t\Delta} f(x) \, \frac{dt}{t^{1-s}}.$$

Therefore, the results of Theorem 14.1 say that U is a classical solution to the Neumann problem

$$\begin{cases} \Delta U + \frac{a}{y}\partial_y U + \partial_{yy} U = 0 & \text{in } \mathbb{R}^n \times (0,\infty) \\ -\lim_{y\to 0} y^a \partial_y U(x,y) = \frac{\Gamma(1-s)}{4^{s-1/2}\Gamma(s)} f(x) & \text{on } \mathbb{R}^n \end{cases} \tag{14.24}$$

that weakly vanishes at infinity and, in addition,

$$U(x,0) = (-\Delta)^{-s} f(x)$$

for all $x \in \mathbb{R}^n$. In other words, (14.24) is the extension problem characterization of the inverse fractional Laplacian $(-\Delta)^{-s}$.

Next, we move on to generalizing the extension problem characterization of $(-\Delta)^s u$ to other natural classes of functions u for the fractional Laplacian. We first consider functions u in L_s, for which the fractional Laplacian is defined in the sense of distributions, see Sections 12.1 and 12.4. Second, we look at functions u in the fractional Sobolev space $H^s(\mathbb{R}^n)$, where the fractional Laplacian is defined in the sense of $L^2(\mathbb{R}^n)$, see Section 12.5.

We begin with the case of $u \in L_s$.

Theorem 14.5 (Extension problem for the distributional fractional Laplacian). *Let $u \in L_s$ and define $U = U(x,y)$ by*

$$U(x,y) = P_y^s * u(x) = d_{n,s} \int_{\mathbb{R}^n} \frac{y^{2s}}{(|x-z|^2 + y^2)^{(n+2s)/2}} u(z)\, dz \tag{14.25}$$

where $d_{n,s} = \frac{\Gamma(n/2+s)}{\Gamma(s)\pi^{n/2}} > 0$. Then U is well defined and coincides with the first two semigroup formulas in (14.9). Moreover, $U \in C^\infty(\mathbb{R}^n \times (0,\infty))$ and satisfies

$$\begin{cases} \Delta U + \frac{a}{y}\partial_y U + \partial_{yy} U = 0 & \text{in } \mathbb{R}^n \times (0,\infty) \\ \lim_{y\to 0} U = u & \text{in the norm of } L_s. \end{cases} \tag{14.26}$$

Furthermore,

$$-\lim_{y\to 0} y^a \partial_y U = \frac{\Gamma(1-s)}{4^{s-1/2}\Gamma(s)} (-\Delta)^s u \tag{14.27}$$

in the sense of distributions, that is, for any $\varphi \in \mathcal{S}$,

$$-\lim_{y\to 0} \int_{\mathbb{R}^n} y^a \partial_y U(x,y)\varphi(x)\, dx = \frac{\Gamma(1-s)}{4^{s-1/2}\Gamma(s)} \int_{\mathbb{R}^n} u(x)(-\Delta)^s \varphi(x)\, dx.$$

Proof. To see that U is well defined, fix $x \in \mathbb{R}^n$ and $y > 0$. Then

$$\begin{aligned} |U(x,y)| &\le C \int_{\mathbb{R}^n} \frac{y^{2s}}{(|x-z|^2+y^2)^{(n+2s)/2}} |u(z)|\, dz \\ &\le \frac{C}{y^n} \int_{B_{2(|x|+1)}} |u(z)|\, dz + C y^{2s} \int_{\mathbb{R}^n \setminus B_{2(|x|+1)}} \frac{|u(z)|}{|x-z|^{n+2s}}\, dz \\ &\le \frac{C}{y^n} \int_{B_{2(|x|+1)}} |u(z)|\, dz + C y^{2s} \int_{\mathbb{R}^n \setminus B_{2(|x|+1)}} \frac{|u(z)|}{1+|z|^{n+2s}}\, dz < \infty \end{aligned}$$

where in the last line we used that if $|z|/2 > |x|+1$, then $|z| \le |z-x| + |x| \le |z-x| + |z|/2 - 1$ which gives that $1 + |z|/2 \le |x-z|$, together with the fact that $u \in L_s$. Then, Fubini's theorem can be applied as in (14.18) and we obtain that U coincides with the first two semigroup formulas in (14.9). Moreover, by a similar reasoning as above, we see that one can differentiate inside the integral sign in (14.25) to get that $U \in C^\infty(\mathbb{R}^n \times (0,\infty))$. Since the Poisson kernel for the extension problem P_y^s satisfies the extension equation, it follows that U satisfies the equation in (14.26).

To prove the validity of the second condition in (14.26), that is, $U(x,y) \to u(x)$ as $y \to 0$ in the L_s norm, notice that, by (14.22),

$$|U(x,y) - u(x)| \le \int_{\mathbb{R}^n} P_y^s(z) |u(x-z) - u(x)|\, dz.$$

Then, by Tonelli's theorem and changing variables,

$$\begin{aligned} \int_{\mathbb{R}^n} \frac{|U(x,y)-u(x)|}{1+|x|^{n+2s}}\, dx &\le \int_{\mathbb{R}^n} y^{-n} P_1^s(z/y) \int_{\mathbb{R}^n} \frac{|u(x-z)-u(x)|}{1+|x|^{n+2s}}\, dx\, dz \\ &= \int_{\mathbb{R}^n} P_1^s(z) \int_{\mathbb{R}^n} \frac{|u(x-yz)-u(x)|}{1+|x|^{n+2s}}\, dx\, dz. \end{aligned}$$

Now, exactly as we did in the proof of Theorem 12.3, see (12.18), the generalized dominated convergence theorem can be applied to deduce that the last term above vanishes in the limit $y \to 0$, so we obtain that

$$\|U(\cdot, y) - u(\cdot)\|_{L_s} = \int_{\mathbb{R}^n} \frac{|U(x,y) - u(x)|}{1 + |x|^{n+2s}}\, dx \to 0 \qquad \text{as } y \to 0.$$

To show that (14.27) holds in the sense of distributions, fix $\varphi \in \mathcal{S}$. Since we can differentiate inside the integral in (14.25) and $P_y^s(-x) = P_y^s(x)$,

$$\begin{aligned}\int_{\mathbb{R}^n} y^a \partial_y U(x,y)\varphi(x)\, dx &= \int_{\mathbb{R}^n}\int_{\mathbb{R}^n} y^a \partial_y P_y^s(x-z)u(z)\, dz\, \varphi(x)\, dx \\ &= \int_{\mathbb{R}^n} u(z)\left[\int_{\mathbb{R}^n} y^a \partial_y P_y^s(z-x)\varphi(x)\, dx\right] dz.\end{aligned} \tag{14.28}$$

For each fixed $y > 0$, the interchange of integrals in (14.28) is justified because $u \in L_s$ and $\varphi \in \mathcal{S}$. By Theorem 14.1, the integral in brackets converges in the pointwise sense to $-\frac{\Gamma(1-s)}{4^{s-1/2}\Gamma(s)}(-\Delta)^s\varphi(z)$ as $y \to 0$. Moreover, by the second identity in (14.23) and Lemma 12.2,

$$\begin{aligned}\left|\int_{\mathbb{R}^n} y^a \partial_y P_y(z-x)\varphi(x)\, dx\right| &\leq C\int_{\mathbb{R}^n} \frac{|\varphi(z+x) + \varphi(z-x) - 2\varphi(z)|}{|x|^{n+2s}}\, dx \\ &\leq \frac{C}{1 + |z|^{n+2s}}\end{aligned}$$

for some constant $C > 0$ depending on n, s and Schwartz seminorms of φ. Therefore, since $u \in L_s$, we can apply the dominated convergence theorem in (14.28) to get

$$-\int_{\mathbb{R}^n} y^a \partial_y U(x,y)\varphi(x)\, dx \to \frac{\Gamma(1-s)}{4^{s-1/2}\Gamma(s)}\int_{\mathbb{R}^n} u(z)(-\Delta)^s\varphi(z)\, dz$$

as $y \to 0$, as desired. □

Now we give the extension problem characterization of the fractional Laplacian for the case when $u \in H^s(\mathbb{R}^n)$.

To introduce the definition of weak solution to the extension problem, first observe that if $u \in L^2(\mathbb{R}^n)$ and U is defined as

$$U(x,y) = \frac{y^{2s}}{4^s\Gamma(s)}\int_0^\infty e^{-y^2/(4t)} e^{t\Delta}u(x)\, \frac{dt}{t^{1+s}}, \tag{14.29}$$

then we can carry out the computation in (14.18) to get that $U(x,y) = P_y^s * u(x)$. By Minkowski's integral inequality, the fact that $\|e^{t\Delta}u\|_{L^2(\mathbb{R}^n)} \leq \|u\|_{L^2(\mathbb{R}^n)}$ and (14.19), for each $y > 0$,

$$\|U(\cdot, y)\|_{L^2(\mathbb{R}^n)} \leq \frac{y^{2s}}{4^s\Gamma(s)}\int_0^\infty e^{-y^2/(4t)}\|e^{t\Delta}u\|_{L^2(\mathbb{R}^n)}\, \frac{dt}{t^{1+s}} \leq \|u\|_{L^2(\mathbb{R}^n)}.$$

Moreover, since $P_y^s * u$ defines an approximation of the identity, we get $U(x,y) \to u(x)$ in $L^2(\mathbb{R}^n)$ as $y \to 0$. This can also be seen from (14.29). Indeed, since $e^{t\Delta}u \to u$ in $L^2(\mathbb{R}^n)$ as $t \to 0$, by (14.19), Minkowski's integral inequality, the change of variables $r = y^2/(4t)$ and the dominated convergence theorem,

$$\begin{aligned}\|U(\cdot,y) - u(\cdot)\|_{L^2(\mathbb{R}^n)} &\leq \frac{y^{2s}}{4^s\Gamma(s)}\int_0^\infty e^{-y^2/(4t)}\|e^{t\Delta}u - u\|_{L^2(\mathbb{R}^n)}\,\frac{dt}{t^{1+s}} \\ &= \frac{1}{\Gamma(s)}\int_0^\infty e^{-r}\|e^{-\frac{y^2}{4t}\Delta}u - u\|_{L^2(\mathbb{R}^n)}\,\frac{dr}{r^{1-s}} \to 0\end{aligned}$$

as $y \to 0$.

On the other hand, let us recall that the extension equation can be written in divergence form as

$$\operatorname{div}_{x,y}(y^a\nabla_{x,y}U) = 0.$$

If U is a sufficiently smooth solution then, by multiplying the equation by a smooth function $V = V(x,y)$ with compact support in $\mathbb{R}^n \times [0,\infty)$, integrating over $\mathbb{R}^n \times (0,\infty)$ and integrating by parts, we find that

$$\int_0^\infty y^a \int_{\mathbb{R}^n} \nabla_{x,y}U\nabla_{x,y}V\,dx\,dy = -\lim_{y\to 0}\int_{\mathbb{R}^n} y^a\partial_y U(x,y)V(x,y)\,dx. \tag{14.30}$$

By Hölder's inequality, the left hand side in (14.30) is finite if

$$\int_0^\infty y^a \int_{\mathbb{R}^n} |\nabla_{x,y}U|^2\,dx\,dy < \infty.$$

As for the right hand side in (14.30), we saw in Theorem 14.1 that if $u \in \mathcal{S}$ then U given as in (14.29) is a smooth solution to the extension equation and, by calling $v(x) = V(x,0)$ in (14.30) and using the dominated convergence theorem, we get

$$\begin{aligned}\int_0^\infty \int_{\mathbb{R}^n} y^a\nabla_{x,y}U\nabla_{x,y}V\,dx\,dy &= \frac{\Gamma(1-s)}{4^{s-1/2}\Gamma(s)}\int_{\mathbb{R}^n}(-\Delta)^s u(x)v(x)\,dx \\ &= \frac{\Gamma(1-s)}{4^{s-1/2}\Gamma(s)}\int_{\mathbb{R}^n}(-\Delta)^{s/2}u(x)(-\Delta)^{s/2}v\,dx.\end{aligned}$$

In particular, if a weak solution U to the extension problem should exist, then we must have $u \in H^s(\mathbb{R}^n)$. Motivated by these observations, we give the following definition.

Given $u \in L^2(\mathbb{R}^n)$, we say that a function $U = U(x,y)$ such that

$$\int_0^\infty y^a \int_{\mathbb{R}^n} |\nabla_{x,y}U|^2\,dx\,dy < \infty.$$

is a **weak solution to the extension problem**

$$\begin{cases}\operatorname{div}_{x,y}(y^a\nabla_{x,y}U) = 0 & \text{in } \mathbb{R}^n \times (0,\infty) \\ U(x,0) = u(x) & \text{on } \mathbb{R}^n\end{cases} \tag{14.31}$$

if

$$\lim_{y\to 0} U(x,y) = u(x) \qquad \text{in } L^2(\mathbb{R}^n)$$

and, for every $V \in C_c^\infty(\mathbb{R}^n \times [0,\infty))$,

$$\int_0^\infty y^a \int_{\mathbb{R}^n} \nabla_{x,y} U \nabla_{x,y} V \, dx \, dy = -\lim_{y\to 0} \int_{\mathbb{R}^n} y^a \partial_y U(x,y) V(x,y) \, dx.$$

where the limit on the right hand side exists.

Theorem 14.6 (Extension problem for $u \in H^s(\mathbb{R}^n)$). *Let $u \in L^2(\mathbb{R}^n)$ and define $U = U(x,y)$ by*

$$\begin{aligned} U(x,y) &= \frac{y^{2s}}{4^s\Gamma(s)} \int_0^\infty e^{-y^2/(4t)} e^{t\Delta} u(x) \frac{dt}{t^{1+s}} \\ &= \frac{\Gamma(n/2+s)}{\Gamma(s)\pi^{n/2}} \int_{\mathbb{R}^n} \frac{y^{2s}}{(|x-z|^2+y^2)^{(n+2s)/2}} u(z) \, dz \end{aligned}$$

or, equivalently, as in (14.14)*:*

$$\widehat{U}(\xi,y) = \frac{2^{1-s}}{\Gamma(s)} (y|\xi|)^s K_s(y|\xi|) \widehat{u}(\xi).$$

Then $U \in C^\infty(\mathbb{R}^n \times (0,\infty)) \cap C([0,\infty); L^2(\mathbb{R}^n))$ with

$$\|U(\cdot,y)\|_{L^2(\mathbb{R}^n)} \le \|u\|_{L^2(\mathbb{R}^n)} \qquad \text{for all } y > 0$$

and

$$\begin{cases} \operatorname{div}_{x,y}(y^a \nabla_{x,y} U) = 0 & \text{in the classical sense in } \mathbb{R}^n \times (0,\infty) \\ \lim_{y\to 0} U(x,y) = u(x) & \text{in } L^2(\mathbb{R}^n). \end{cases}$$

Moreover,

$$\int_0^\infty y^a \int_{\mathbb{R}^n} |\nabla_{x,y} U|^2 \, dx \, dy < \infty \tag{14.32}$$

if and only if $u \in H^s(\mathbb{R}^n)$ and, in this case,

$$\int_0^\infty y^a \int_{\mathbb{R}^n} |\nabla_{x,y} U|^2 \, dx \, dy = c_s \int_{\mathbb{R}^n} |(-\Delta)^{s/2} u|^2 \, dx$$

where $c_s > 0$ is a constant depending only on s. Furthermore, U is a weak solution to the extension problem (14.31) *and, for any $V \in C_c^\infty(\mathbb{R}^n \times [0,\infty))$, if we denote $v(x) = V(x,0)$, then*

$$\begin{aligned} \int_0^\infty y^a \int_{\mathbb{R}^n} \nabla_{x,y} U \nabla_{x,y} V \, dx \, dy &= = -\lim_{y\to 0} \int_{\mathbb{R}^n} y^a \partial_y U(x,y) V(x,y) \, dx \\ &= \frac{\Gamma(1-s)}{4^{s-1/2}\Gamma(s)} \int_{\mathbb{R}^n} (-\Delta)^{s/2} u (-\Delta)^{s/2} v \, dx. \end{aligned} \tag{14.33}$$

Proof. As we mentioned above, $u \in L^2(\mathbb{R}^n)$ is sufficient to show that the formulas for U in the statement coincide. Since the Poisson kernel P_y^s is smooth and satisfies the extension equation in the classical sense, differentiation under the integral sign shows that U verifies the same properties. Furthermore, we already proved the $L^2(\mathbb{R}^n)$ bound in U and saw that $U(x,y) \to u(x)$ in $L^2(\mathbb{R}^n)$ as $y \to 0$.

Next, for (14.32), on one hand,

$$-\partial_y \widehat{U}(\xi, y) = C(y|\xi|)^s K_{1-s}(y|\xi|)|\xi|\widehat{u}(\xi).$$

Hence, by the change of variables $r = y|\xi|$ and the asymptotic formulas in (14.16),

$$\begin{aligned}\int_0^\infty y^a \int_{\mathbb{R}^n} |\partial_y U|^2 \,dx\,dy &= \int_0^\infty y^a \int_{\mathbb{R}^n} |\partial_y \widehat{U}(\xi, y)|^2 \,d\xi\,dy \\ &= C \int_{\mathbb{R}^n} |\xi|^2 |\widehat{u}(\xi)|^2 \int_0^\infty y^a |(y|\xi|)^s K_{1-s}(y|\xi|)|^2 \,dy\,d\xi \\ &= C \int_{\mathbb{R}^n} |\xi|^{2s} |\widehat{u}(\xi)|^2 \int_0^\infty r|K_{1-s}(r)|^2 \,dr\,d\xi \\ &= C \int_{\mathbb{R}^n} ||\xi|^s \widehat{u}(\xi)|^2 \,d\xi\end{aligned}$$

which is finite if and only if $u \in H^s(\mathbb{R}^n)$. Similarly,

$$\begin{aligned}\int_0^\infty y^a \int_{\mathbb{R}^n} |\nabla U|^2 \,dx\,dy &= \int_0^\infty y^a \int_{\mathbb{R}^n} ||\xi|\widehat{U}(\xi, y)|^2 \,d\xi\,dy \\ &= C \int_{\mathbb{R}^n} |\xi|^2 |\widehat{u}(\xi)|^2 \int_0^\infty y^a |(y|\xi|)^s K_s(y|\xi|)|^2 \,dy\,d\xi \\ &= C \int_{\mathbb{R}^n} |\xi|^{2s} |\widehat{u}(\xi)|^2 \int_0^\infty r|K_s(r)|^2 \,dr\,d\xi \\ &= C \int_{\mathbb{R}^n} ||\xi|^s \widehat{u}(\xi)|^2 \,d\xi\end{aligned}$$

and this is finite if and only if $u \in H^s(\mathbb{R}^n)$. In other words, (14.32) holds if and only if $u \in H^s(\mathbb{R}^n)$ and, by the two identities above,

$$\int_0^\infty y^a \int_{\mathbb{R}^n} |\nabla_{x,y} U|^2 \,dx\,dy = c_s \int_{\mathbb{R}^n} |(-\Delta)^{s/2} u|^2 \,dx$$

where $c_s > 0$ depends only on s.

Finally, by multiplying the equation for U by a test function $V \in C_c^\infty(\mathbb{R}^n \times [0, \infty))$ and integrating by parts,

$$\int_0^\infty y^a \int_{\mathbb{R}^n} \nabla_{x,y} U \nabla_{x,y} V \,dx\,dy = -\lim_{y \to 0} \int_{\mathbb{R}^n} y^a \partial_y U(x, y) V(x, y)\,dx.$$

To compute the limit in the right hand side, we apply Parseval's identity and the asymptotic property of the Bessel function in (14.16) to obtain that

$$\begin{aligned}-\int_{\mathbb{R}^n} y^a \partial_y U(x, y) V(x, y)\,dx &= -\int_{\mathbb{R}^n} y^a \partial_y \widehat{U}(\xi, y) \overline{\widehat{V}(\xi, y)}\,d\xi \\ &= -\frac{2^{1-s}}{\Gamma(s)} \int_0^\infty y^{1-2s} (y|\xi|)^s K_{1-s}(y|\xi|)|\xi| \widehat{u}(\xi) \overline{\widehat{V}(\xi, y)}\,d\xi \\ &\to \frac{\Gamma(1-s)}{4^{s-1/2}\Gamma(s)} \int_0^\infty |\xi|^{2s} \widehat{u}(\xi) \overline{\widehat{v}(\xi)}\,d\xi \\ &= \frac{\Gamma(1-s)}{4^{s-1/2}\Gamma(s)} \int_{\mathbb{R}^n} (-\Delta)^{s/2} u (-\Delta)^{s/2} v\,dx\end{aligned}$$

as $y \to 0$. Thus, (14.33) holds. □

We finish this section with the following reflection lemma that, together with local regularity estimates for the extension problem, will allow us to prove regularity results for fractional Laplacian equations.

Lemma 14.7 (Reflection lemma). *Let $U : B_1 \times (0,1) \to \mathbb{R}$ be a weak solution to*

$$\begin{cases} \operatorname{div}_{x,y}(y^a \nabla_{x,y} U) = 0 & \text{in } B_1 \times (0,1) \\ -y^a \partial_y U(x,y)\big|_{y=0} = 0 & \text{on } B_1. \end{cases}$$

In other words, suppose that $y^a |\nabla_{x,y} U|^2 \in L^1(B_1 \times (0,1))$ and that, for every smooth test function $V \in C_c^\infty(B_1 \times [0,1))$,

$$\int_0^1 y^a \int_{B_1} \nabla_{x,y} U \nabla_{x,y} V \, dx \, dy = 0$$

and

$$-\lim_{y \to 0} \int_{B_1} y^a \partial_y U(x,y) V(x,y) \, dx = 0.$$

Then, the even reflection of U in the variable y defined as

$$\tilde{U}(x,y) = U(x,|y|) \qquad \text{for } y \in (-1,1)$$

is a weak solution to

$$\operatorname{div}(|y|^a \nabla_{x,y} \tilde{U}) = 0 \qquad \text{in } B_1 \times (-1,1)$$

namely, $|y|^a |\nabla_{x,y} \tilde{U}|^2 \in L^1(B_1 \times (-1,1))$ and, for any $W \in C_c^\infty(B_1 \times (-1,1))$,

$$\int_{-1}^1 |y|^a \int_{B_1} \nabla_{x,y} \tilde{U} \nabla_{x,y} W \, dx \, dy = 0.$$

Proof. Since, for $y < 0$, $\partial_y \tilde{U}(x,y) = -\partial_y U(x,-y)$ in the weak sense, we get that $|y|^a |\nabla_{x,y} \tilde{U}|^2 \in L^1(B_1 \times (-1,1))$. Observe that U is smooth in the interior of $B_1 \times (0,1)$ because the coefficient y^a is smooth, bounded and uniformly elliptic on any compact subdomain of $B_1 \times (0,1)$. Hence, the equation $\operatorname{div}_{x,y}(y^a \nabla_{x,y} U) = 0$ is satisfied in the classical sense in $B_1 \times (0,1)$. Let $W \in C_c^\infty(B_1 \times (-1,1))$. For any $\delta > 0$, by using the equation and applying the divergence theorem,

$$\begin{aligned} &\int_{-1}^1 |y|^a \int_{B_1} \nabla_{x,y} \tilde{U} \nabla_{x,y} W \, dx \, dy \\ &\quad = \int_{B_1} \int_{\delta < |y| < 1} |y|^a \nabla_{x,y} \tilde{U} \nabla_{x,y} W \, dy \, dx + \int_{B_1} \int_{|y| \le \delta} |y|^a \nabla_{x,y} \tilde{U} \nabla_{x,y} W \, dy \, dx \\ &\quad = \int_{B_1} \int_{\delta < |y| < 1} \operatorname{div}(|y|^a W \nabla_{x,y} \tilde{U}) \, dy \, dx + \int_{B_1} \int_{|y| \le \delta} |y|^a \nabla_{x,y} \tilde{U} \nabla_{x,y} W \, dy \, dx \\ &\quad = -\int_{B_1} \delta^a \partial_y U(x,\delta) [W(x,\delta) + W(x,-\delta)] \, dx \\ &\qquad + \int_{B_1} \int_{|y| \le \delta} |y|^a \nabla_{x,y} \tilde{U} \nabla_{x,y} W \, dy \, dx. \end{aligned}$$

To conclude we take the limit as $\delta \to 0$ in the last identity: the first term goes to zero by hypothesis and the second one also converges to zero because, by Hölder's inequality, its integrand is locally integrable. □

Immediately from Theorem 14.6, Lemma 14.7 and the density of $C_c^\infty(B_1)$ in $H_0^s(B_1)$ we obtain the following.

Corollary 14.8. *Let $u \in H^s(\mathbb{R}^n)$ be a weak solution to*

$$(-\Delta)^s u = 0 \qquad \text{in } B_1 \tag{14.34}$$

that is,

$$\int_{\mathbb{R}^n} (-\Delta)^{s/2} u (-\Delta)^{s/2} v \, dx = 0 \qquad \text{for all } v \in H_0^s(B_1).$$

Let $U = U(x, y)$ be the extension of u given as in Theorem 14.6. Then U satisfies the hypotheses of Lemma 14.7 and, consequently, its even reflection $\tilde{U} = U(x, |y|)$ is a weak solution to

$$\operatorname{div}(|y|^a \nabla_{x,y} \tilde{U}) = 0 \qquad \text{in } B_1 \times (-1, 1).$$

14.3 A DETOUR TO DEGENERATE ELLIPTIC EQUATIONS

Corollary 14.8 says that if $u \in H^s(\mathbb{R}^n)$ is a weak solution to

$$(-\Delta)^s u = 0 \qquad \text{in } B_1$$

then the even reflection $\tilde{U}$ of the extension U of u in the variable y satisfies

$$\int_0^1 y^a \int_{B_1} |\nabla_{x,y} \tilde{U}|^2 \, dx \, dy < \infty$$

and is a weak solution to the degenerate/singular elliptic equation

$$\operatorname{div}_{x,y}(|y|^a \nabla_{x,y} \tilde{U}) = 0 \qquad \text{for } x \in B_1, \ y \in (-1, 1). \tag{14.35}$$

Therefore, it is important to understand the regularity theory of elliptic equations in divergence form with degenerate/singular coefficients. In this section we summarize some known facts about them.

In order not to confuse the reader with the rest of the results in this chapter, and only for this section, we change the notation. Let $B_1 \subset \mathbb{R}^N$, $N \geq 1$, be the unit ball and denote its points by $X \in B_1$. Consider equations of the form

$$\operatorname{div}(A(X) \nabla U) = 0 \tag{14.36}$$

in B_1, where the matrix of coefficients $A(X) = (A^{ij}(X))_{i,j=1}^N$ is **degenerate elliptic** in the sense that there exist ellipticity constants $0 < \lambda \leq \Lambda$ such that

$$\lambda \omega(X) |\xi|^2 \leq A^{ij}(X) \xi_i \xi_j \leq \Lambda \omega(X) |\xi|^2$$

for all $\xi \in \mathbb{R}^N$ and a.e. $X \in B_1$. Here $\omega = \omega(X)$ is a **weight**, that is, an a.e. positive measurable function on $\mathbb{R}^N$. The equation is degenerate in the sense that $\omega(X)$ may vanish or become infinity at some points.

The extension equation for the even reflection (14.35) is degenerate elliptic as in (14.36) in $X = (x, y)$ with the weight $\omega(x, y) = |y|^a$, which vanishes at $y = 0$ if $a > 0$

and blows up at $y = 0$ if $a < 0$. The only case when (14.35) is uniformly elliptic is when $a = 0$.

We denote by $L^2(B_1, \omega)$ the **weighted L^2 space** of all measurable functions U in B_1 such that

$$\|U\|^2_{L^2(B_1,\omega)} = \int_{B_1} |U(X)|^2 \omega(X)\, dX < \infty.$$

Then $H^1(B_1, \omega)$ denotes the **weighted Sobolev space** of functions $U \in L^2(B_1, \omega)$ for which $\nabla U \in L^2(B_1, \omega)$. We say that $U \in H^1(B_1, \omega)$ is a **weak solution** to (14.36) if, for all $\Psi \in C_c^\infty(B_1)$,

$$\int_{B_1} A(X) \nabla U \nabla \Psi \, dX = 0$$

The main question is to find suitable conditions on the weight ω in order to be able to prove the Harnack inequality for nonnegative weak solutions to (14.36). Let us recall Moser's proof of the Harnack inequality in Chapter 6. It is clear that the following ingredients are needed to implement the Moser iteration technique.

Energy estimates. If $k \neq 1/2$ and $V = U^k$ then

$$\int_{B_1} \Psi^2 |\nabla V|^2 \omega \, dX \leq C \int_{B_1} |\nabla \Psi|^2 V^2 \omega \, dX$$

and if $V = \log U$ then

$$\int_{B_1} \Psi^2 |\nabla V|^2 \omega \, dX \leq C \int_{B_1} |\nabla \Psi|^2 \omega \, dX$$

for all test functions Ψ. These can be achieved as we did in Chapter 6.

Sobolev embedding. For any $1 < p < \infty$, there exist $k > 1$ and $C > 0$ such that for any ball $B \subset \mathbb{R}^N$ and any $\Psi \in C_c^\infty(B)$ we have

$$\left(\frac{1}{\omega(B)} \int_B |\Psi|^{pk} \omega \, dX \right)^{1/(pk)} \leq C \operatorname{diam}(B) \left(\frac{1}{\omega(B)} \int_B |\nabla \Psi|^p \omega \, dX \right)^{1/p}$$

where

$$\omega(B) = \int_B \omega \, dX.$$

With these two ingredients, as we did in Sections 6.1 and 6.2, the local improvement of integrability can be shown and, by applying Moser's iteration, the upper and lower bounds follow. In particular, one obtains the local boundedness of solutions.

To close the gap and finish the proof of the Harnack inequality as we did in Section 6.3, one needs two more ingredients.

Poincaré inequality. For any $1 \leq p < \infty$, there exists $C > 0$ such that for any ball $B \subset \mathbb{R}^N$ and $\Psi \in C^1(B)$ we have

$$\left(\frac{1}{\omega(B)} \int_B |\Psi - \Psi_B|^p \omega \, dX \right)^{1/p} \leq C \operatorname{diam}(B) \left(\frac{1}{\omega(B)} \int_B |\nabla \Psi|^p \omega \, dX \right)^{1/p}$$

where

$$\Psi_B = \frac{1}{\omega(B)} \int_B \Psi\omega \, dX.$$

John–Nirenberg inequality. Suppose that for every ball B,

$$\frac{1}{\omega(B)} \int_B |U - U_B|\omega \, dX \leq C.$$

Then, for all $t > 0$,

$$\omega(\{X \in B : |U(X) - U_B| > t\}) \leq c_1 e^{-c_2 t}\omega(B).$$

It turns out that the Sobolev, Poincaré and John–Nirenberg inequalities can be established whenever ω is a **Muckenhoupt A_2 weight**, namely, for weights ω for which there is a constant $C_\omega > 0$ such that for any ball $B \subset \mathbb{R}^N$

$$\left(\frac{1}{|B|} \int_B \omega \, dX\right)\left(\frac{1}{|B|} \int_B \omega^{-1} \, dX\right) \leq C_\omega.$$

Thus, the following result is valid.

Theorem 14.9 (Fabes–Kenig–Serapioni Harnack inequality). *Let $U \in H^1(B_1, \omega)$ be a weak solution to* (14.36) *such that $U \geq 0$ in B_1. There exists a constant $C_H > 0$ depending only on N and ω such that*

$$\sup_{B_{1/2}} U \leq C_H \inf_{B_{1/2}} U.$$

As a consequence, weak solutions U to (14.36) *are locally Hölder continuous of order $0 < \alpha < 1$ in B_1 with*

$$\|U\|_{C^{0,\alpha}(B_{1/2})} \leq C_H \|U\|_{L^2(B_1,\omega)}$$

where the constant $C > 0$ depends only on N and ω. The Liouville theorem also follows: if U is a weak solution to (14.36) *in $\mathbb{R}^N$ such that $U \in L^\infty(\mathbb{R}^N)$ then U is constant.*

It is easy to check that $\omega(x, y) = |y|^a$ is a Muckenhoupt A_2 weight in $\mathbb{R}^{n+1}$ whenever $-1 < a < 1$. Then, by Theorem 14.9, we obtain the Harnack inequality, local Hölder continuity and the Liouville theorem for weak solutions to (14.35).

14.4 APPLICATIONS TO REGULARITY ESTIMATES

In this section we show how to use the regularity estimates for degenerate elliptic equations of Theorem 14.9 to obtain regularity results for solutions to fractional Laplacian equations. The reader should compare the results of the following subsections with those for the Laplacian in Chapter 2.

14.4.1 Hölder regularity and Harnack inequality

Theorem 14.10 (Hölder regularity and Harnack inequality). *Let $u \in H^s(\mathbb{R}^n)$ be a weak solution to*

$$(-\Delta)^s u = 0 \qquad \text{in } B_1. \tag{14.37}$$

Then u is locally Hölder continuous in B_1: there exist $0 < \alpha < 1$ and $C > 0$ depending only on n and s such that

$$\|u\|_{C^{0,\alpha}(B_{1/2})} \le C\|u\|_{L^2(\mathbb{R}^n)}.$$

If, in addition,

$$u \ge 0 \qquad \text{in } \mathbb{R}^n$$

then u satisfies the interior Harnack inequality: there exists a constant $C_H > 0$ depending only on n and s such that

$$\sup_{B_{1/2}} u \le C_H \inf_{B_{1/2}} u.$$

Proof. By Corollary 14.8, the reflection extension $\tilde{U}$ of the extension U of u is a weak solution to

$$\operatorname{div}_{x,y}(|y|^a \nabla \tilde{U}) = 0 \qquad \text{in } B_1 \times (-1,1).$$

By Theorem 14.9, $\tilde{U}$ is locally Hölder continuous in B_1 and, therefore, since $\|U(\cdot,y)\|_{L^2(\mathbb{R}^n)} \le \|u\|_{L^2(\mathbb{R}^n)}$ and $-1 < a < 1$,

$$\begin{aligned}
\|u\|_{C^{0,\alpha}(B_{1/2})} &= \|U(\cdot,0)\|_{C^{0,\alpha}(B_{1/2})} \\
&\le \|\tilde{U}\|_{C^{0,\alpha}(B_{1/2}\times(-1/2,1/2))} \\
&\le C\|\tilde{U}\|_{L^2(B_1\times(-1,1),|y|^a)} \\
&\le 2C\int_0^1 y^a \int_{B_1} |U(x,y)|^2\,dx\,dy \le C\|u\|_{L^2(\mathbb{R}^n)}.
\end{aligned}$$

If, in addition, $u \ge 0$ then $U(x,y) = P_y^s * u(x) \ge 0$ and $\tilde{U}$ is nonnegative. Hence, by Theorem 14.9, $\tilde{U}$ satisfies the Harnack inequality and, therefore,

$$\begin{aligned}
\sup_{B_{1/2}} u &\le \sup_{B_{1/2}\times(-1/2,1/2)} \tilde{U} \\
&\le C_H \inf_{B_{1/2}\times(-1/2,1/2)} \tilde{U} \le C_H \inf_{B_{1/2}} u.
\end{aligned}$$

□

Remark 14.11. Observe that the Harnack inequality in Theorem 14.10 requires that $u \ge 0$ in the whole space $\mathbb{R}^n$, not just in B_1. This is to ensure that its extension U is also nonnegative. Moreover, the proof of the local Hölder continuity of u was not done using the Harnack inequality satisfied by u, but was a consequence of the local Hölder continuity of U (and this one can be proved using the Harnack inequality, see Section 6.4). In fact, if one tries to mimic the proof in Section 6.4 for u, the method immediately breaks down because if $M = \sup_{B_1} u$, then $M - u$ is not necessarily nonnegative in $\mathbb{R}^n$ and is not even in $H^s(\mathbb{R}^n)$.

14.4.2 Liouville theorem

Theorem 14.12 (Liouville theorem). *If $u \in H^s(\mathbb{R}^n) \cap L^\infty(\mathbb{R}^n)$ is a weak solution to*

$$(-\Delta)^s u = 0 \qquad \text{in } \mathbb{R}^n$$

then u is constant.

Proof. Let U be the extension of u given by Theorem 14.6. Since $\int_{\mathbb{R}^n} P_y^s(x)\,dx = 1$ for all $y > 0$, we get that $|U(x,y)| \le \|u\|_{L^\infty(\mathbb{R}^n)}$. Therefore, by Corollary 14.8, $\tilde{U}(x,y) = U(x,|y|)$ is a bounded weak solution to

$$\operatorname{div}_{x,y}(|y|^a \nabla \tilde{U}) = 0 \qquad \text{in } \mathbb{R}^n \times (-\infty, \infty).$$

By Theorem 14.9, $\tilde{U}$ is constant, and so is u. □

14.4.3 Interior derivative estimates

We next prove that weak solutions to $(-\Delta)^s u = 0$ in B_4 are smooth in the interior.

Theorem 14.13 (Interior derivative estimates). *Let $u \in H^s(\mathbb{R}^n)$ be a weak solution to*

$$(-\Delta)^s u = 0 \qquad \text{in } B_4.$$

Then u is smooth in the interior of B_4: for any multi-index $\gamma \in \mathbb{N}_0^n$, there is a constant $C > 0$ depending only on n, s and $|\gamma|$ such that

$$\sup_{B_1} |D^\gamma u(x)| \le C\bigg(\|u\|_{L^\infty(B_4)} + \int_{\mathbb{R}^n \setminus B_4} \frac{|u(x)|}{1+|x|^{n+2s}}\,dx \bigg).$$

The passage from B_4 to B_1 in Theorem 14.13 is only for convenience. In fact, these balls can be replaced by bounded domains Ω and $\Omega' \subset\subset \Omega$, respectively, and the constant in the result will also depend on $\operatorname{dist}(\Omega', \partial\Omega)$.

For the proof of Theorem 14.13 we apply the following, very useful incremental quotient technique devised by Caffarelli.

Lemma 14.14 (Incremental quotient approach to Hölder regularity). *Fix constants $0 < \alpha < 1$, $0 < \beta \le 1$ and $K > 0$. Let u be a bounded function defined on the interval $[-1,1]$ with $\|u\|_{L^\infty([-1,1])} \le K$. Define, for any $h \in \mathbb{R}$ with $0 < |h| \le 1$, the incremental quotient of order β as*

$$u_{\beta,h}(x) = \frac{u(x+h) - u(x)}{|h|^\beta}$$

for $x \in I_h$, where $I_h = [-1, 1-h]$ if $h > 0$ and $I_h = [-1-h, 1]$ if $h < 0$. Suppose that $u_{\beta,h} \in C^\alpha(I_h)$ with

$$\|u_{\beta,h}\|_{C^\alpha(I_h)} \le K \qquad \text{for any } 0 < |h| \le 1.$$

Then there exists a constant $C > 0$ depending only on α and β such that

(1) *if $\alpha + \beta < 1$, then $u \in C^{\alpha+\beta}([-1,1])$ and $[u]_{C^{\alpha+\beta}([-1,1])} \leq CK$;*

(2) *if $\alpha + \beta > 1$, then $u \in C^{0,1}([-1,1])$ and $[u]_{C^{0,1}([-1,1])} \leq CK$.*

Proof of Theorem 14.13. By Corollary 14.8, the reflection $\tilde{U}$ of the extension U of u is a weak solution to

$$\operatorname{div}(|y|^a \nabla_{x,y} \tilde{U}) = 0 \qquad \text{in } B_4 \times (-4,4).$$

Observe that, for any $|x| < 2$ and $|y| < 2$

$$\begin{aligned}
|\tilde{U}(x,y)| &\leq 2 \int_{\mathbb{R}^n} P_y^s(x-z)|u(z)|\,dz \\
&\leq 2\|u\|_{L^\infty(B_4)} \int_{B_4} P_y^s(x-z)\,dz + C \int_{\mathbb{R}^n \setminus B_4} \frac{|u(z)|}{|x-z|^{n+2s}}\,dz \\
&\leq C\Big(\|u\|_{L^\infty(B_4)} + \int_{\mathbb{R}^n \setminus B_4} \frac{|u(z)|}{1+|z|^{n+2s}}\,dz\Big)
\end{aligned}$$

where in the last integral we used that for $|z| > 4$ and $|x| < 1$ we have $1+|z| \leq 4|x-z|$. Therefore, since $\tilde{U}(x,0) = u(x)$, to prove the regularity of u on B_1, it is enough to prove that $\tilde{U}$ is differentiable with respect to the horizontal variable x with

$$\sup_{(x,y)\in B_1\times(-1,1)} |D_x^\gamma \tilde{U}(x,y)| \leq C\|\tilde{U}\|_{L^\infty(B_2\times(-2,2))}$$

and then restrict back to $y = 0$. From the Harnack inequality in Theorem 14.9, we know that there exists $0 < \alpha < 1$ depending only on n and s such that

$$\|\tilde{U}\|_{C^\alpha(B_{3/2}\times(-3/2,3/2))} \leq C\|\tilde{U}\|_{L^\infty(B_2\times(-2,2))} = K.$$

Let e be a unit vector in $\mathbb{R}^n$. Then, by this estimate, the horizontal incremental quotient of order α of $\tilde{U}$ given by

$$\tilde{U}_{\alpha,h}(x,y) = \frac{\tilde{U}(x+he,y) - \tilde{U}(x,y)}{|h|^\alpha}$$

is bounded in $B_{3/2} \times (-3/2, 3/2)$ independently of h. The extension equation is a linear equation that is translation invariant with respect to the horizontal variable x, so this incremental quotient $\tilde{U}_{\alpha,h}$ is also a weak solution to the extension equation and, as a consequence, it is C^α. Therefore, Lemma 14.14 implies that $U \in C^{2\alpha}$ in x in the interior. By iteration of this argument a finite number of steps (that is, by considering incremental quotients of order 2α, 3α and so on), we obtain that $\tilde{U}$ is Lipschitz in the variable x. Finally, by applying again Lemma 14.14 with $\beta = 1$, $\tilde{U}$ is $C^{1,\alpha}$ in x. At this point we can differentiate the equation with respect to x and bootstrap the reasoning above to obtain that $\tilde{U}$ is smooth with respect to x, with the corresponding estimates for the derivatives. □

NOTES ON CHAPTER 14

The extension problem characterization for the fractional Laplacian was first introduced in the PDE context by Caffarelli–Silvestre [11]. They found the formula for the Poisson kernel for the extension, proved the reflection lemma and developed applications to interior regularity for solutions to fractional Laplacian equations. The case $s = 1/2$ was more or less known in the harmonic analysis literature. However, it was not until the appearance of [11] that it began to be explicitly used to prove regularity estimates.

The extension problem characterization and method of semigroups for fractional powers of nonnegative, normal linear operators in Hilbert spaces is due to Stinga–Torrea [67,69]. In particular, their result includes the case of the fractional Laplacian. The deduction of the semigroup formulas for U by using Bessel functions and the uniqueness result can be found in these references. We also based our presentation on Stinga [68].

Bessel functions can be seen in Lebedev [51].

The Harnack inequality for degenerate elliptic equations was proved by Fabes–Kenig–Serapioni in [34]. Muckenhoupt weights are, by now, a classical topic in harmonic analysis, see [30].

The proof of Caffarelli's incremental quotients lemma, Lemma 14.14, can be found in Caffarelli–Cabré [7]. In fact, in [7], the lemma is applied to prove $C^{1,\alpha}$ estimates for fully nonlinear elliptic equations. The proof of Theorem 14.13 using the extension equation is originally due to Caffarelli–Salsa–Silvestre [10], where the obstacle problem for the fractional Laplacian is studied by means of the extension problem.

The most general extension problem for fractional operators was proved by Galé–Miana–Stinga [40]. The extension characterization of [40] holds for fractional powers of generators of integrated semigroups on Banach space, of which generators of uniformly bounded C_0-semigroups are a particular case.

The above-mentioned generalizations of the extension problem have important applications to many other contexts. For example, Hölder and Schauder estimates for solutions to fractional nonlocal elliptic equations in divergence form $(-\operatorname{div}(a(x)\nabla))^s u = f$ were proved by Caffarelli–Stinga in [15]; the Harnack inequality and local Hölder continuity of solutions to fractional nonlocal elliptic equations in nondivergence form $(-a^{ij}(x)\partial_{ij})^s u = f$ was proved by Stinga–Vaughan in [71]; a complete regularity theory for fractional nonlocal space-time master equations driven by the fractional heat operator $(\partial_t - \Delta)^s u = f$ in parabolic Hölder spaces was established by Stinga–Torrea in [70].

Bibliography

[1] R. A. Adams. *Sobolev Spaces.* Pure and Applied Mathematics, Vol. 65. Academic Press [Harcourt Brace Jovanovich, Publishers], New York-London, 1975.

[2] A. Benedek, A. P. Calderón, and R. Panzone. Convolution operators on Banach space valued functions. *Proc. Nat. Acad. Sci. U.S.A.*, 48:356–365, 1962.

[3] L. A. Caffarelli. Elliptic second order equations. *Rend. Sem. Mat. Fis. Milano*, 58:253–284 (1990), 1988.

[4] L. A. Caffarelli. Interior a priori estimates for solutions of fully nonlinear equations. *Ann. of Math. (2)*, 130(1):189–213, 1989.

[5] L. A. Caffarelli. *The Obstacle Problem.* Lezioni Fermiane. [Fermi Lectures]. Accademia Nazionale dei Lincei, Rome; Scuola Normale Superiore, Pisa, 1998.

[6] L. A. Caffarelli. Lecture notes. Nonlocal School on Fractional Equations NSFE 2017, Iowa State University, August 2017.

[7] L. A. Caffarelli and X. Cabré. *Fully Nonlinear Elliptic Equations*, volume 43 of *American Mathematical Society Colloquium Publications.* American Mathematical Society, Providence, RI, 1995.

[8] L. A. Caffarelli and I. Peral. On $W^{1,p}$ estimates for elliptic equations in divergence form. *Comm. Pure Appl. Math.*, 51(1):1–21, 1998.

[9] L. A Caffarelli and S. Salsa. *A Geometric Approach to Free Boundary Problems*, volume 68 of *Graduate Studies in Mathematics.* American Mathematical Society, Providence, RI, 2005.

[10] L. A. Caffarelli, S. Salsa, and L. Silvestre. Regularity estimates for the solution and the free boundary of the obstacle problem for the fractional Laplacian. *Invent. Math.*, 171(2):425–461, 2008.

[11] L. A. Caffarelli and L. Silvestre. An extension problem related to the fractional Laplacian. *Comm. Partial Differential Equations*, 32(7-9):1245–1260, 2007.

[12] L. A. Caffarelli and L. Silvestre. Regularity theory for fully nonlinear integro-differential equations. *Comm. Pure Appl. Math.*, 62(5):597–638, 2009.

[13] L. A. Caffarelli and L. Silvestre. The Evans-Krylov theorem for nonlocal fully nonlinear equations. *Ann. of Math. (2)*, 174(2):1163–1187, 2011.

[14] L. A. Caffarelli and L. Silvestre. Regularity results for nonlocal equations by approximation. *Arch. Ration. Mech. Anal.*, 200(1):59–88, 2011.

[15] L. A. Caffarelli and P. R. Stinga. Fractional elliptic equations, Caccioppoli estimates and regularity. *Ann. Inst. H. Poincaré C Anal. Non Linéaire*, 33(3):767–807, 2016.

[16] L. A. Caffarelli and A. F. Vasseur. The De Giorgi method for regularity of solutions of elliptic equations and its applications to fluid dynamics. *Discrete Contin. Dyn. Syst. Ser. S*, 3(3):409–427, 2010.

[17] A. P. Calderón. Inequalities for the maximal function relative to a metric. *Studia Math.*, 57(3):297–306, 1976.

[18] A. P. Calderón and A. Zygmund. On the existence of certain singular integrals. *Acta Math.*, 88:85–139, 1952.

[19] S. Campanato. Proprietà di hölderianità di alcune classi di funzioni. *Ann. Scuola Norm. Sup. Pisa Cl. Sci. (3)*, 17:175–188, 1963.

[20] S. Campanato. Proprietà di una famiglia di spazi funzionali. *Ann. Scuola Norm. Sup. Pisa Cl. Sci. (3)*, 18:137–160, 1964.

[21] F. Chiarenza, M. Frasca, and P. Longo. Interior $W^{2,p}$ estimates for nondivergence elliptic equations with discontinuous coefficients. *Ricerche Mat.*, 40(1):149–168, 1991.

[22] M. G. Crandall, H. Ishii, and P.-L. Lions. User's guide to viscosity solutions of second order partial differential equations. *Bull. Amer. Math. Soc. (N.S.)*, 27(1):1–67, 1992.

[23] M. G. Crandall and P.-L. Lions. Viscosity solutions of Hamilton-Jacobi equations. *Trans. Amer. Math. Soc.*, 277(1):1–42, 1983.

[24] E. De Giorgi. Sulla differenziabilità e l'analiticità delle estremali degli integrali multipli regolari. *Mem. Accad. Sci. Torino. Cl. Sci. Fis. Mat. Nat. (3)*, 3:25–43, 1957.

[25] E. De Giorgi. *Selected Papers.* Springer-Verlag, Berlin, 2006. Edited by L. Ambrosio, G. Dal Maso, M. Forti, M. Miranda and S. Spagnolo.

[26] E. Di Benedetto and N. S. Trudinger. Harnack inequalities for quasiminima of variational integrals. *Ann. Inst. H. Poincaré Anal. Non Linéaire*, 1(4):295–308, 1984.

[27] G. Di Fazio. L^p estimates for divergence form elliptic equations with discontinuous coefficients. *Boll. Un. Mat. Ital. A (7)*, 10(2):409–420, 1996.

[28] E. Di Nezza, G. Palatucci, and E. Valdinoci. Hitchhiker's guide to the fractional Sobolev spaces. *Bull. Sci. Math.*, 136(5):521–573, 2012.

[29] E. DiBenedetto. *Partial Differential Equations.* Cornerstones. Birkhäuser Boston, Ltd., Boston, MA, second edition, 2010.

[30] J. Duoandikoetxea. *Fourier Analysis*, volume 29 of *Graduate Studies in Mathematics.* American Mathematical Society, Providence, RI, 2001. Translated and revised from the 1995 Spanish original by David Cruz-Uribe.

[31] A. Erdélyi, W. Magnus, F. Oberhettinger, and F. G. Tricomi. *Tables of Integral Transforms. Vol. II.* McGraw-Hill Book Co., Inc., New York-Toronto-London, 1954. Based, in part, on notes left by Harry Bateman.

[32] L. C. Evans. *Partial Differential Equations*, volume 19 of *Graduate Studies in Mathematics.* American Mathematical Society, Providence, RI, second edition, 2010.

[33] L. C. Evans and R. F. Gariepy. *Measure Theory and Fine Properties of Functions.* Textbooks in Mathematics. CRC Press, Boca Raton, FL, revised edition, 2015.

[34] E. B. Fabes, C. E. Kenig, and R. P. Serapioni. The local regularity of solutions of degenerate elliptic equations. *Comm. Partial Differential Equations*, 7(1):77–116, 1982.

[35] N. Fava and F. Zó. *Medida e Integral de Lebesgue*, volume 1 of *Colección Textos Universitarios.* Red Olímpica, Olimpíada Matemática Argentina, Buenos Aires, Argentina, 1996.

[36] A. Fiscella, R. Servadei, and E. Valdinoci. Density properties for fractional Sobolev spaces. *Ann. Acad. Sci. Fenn. Math.*, 40(1):235–253, 2015.

[37] G. B. Folland. *Fourier Analysis and its Applications.* The Wadsworth & Brooks/Cole Mathematics Series. Wadsworth & Brooks/Cole Advanced Books & Software, Pacific Grove, CA, 1992.

[38] G. B. Folland. *Introduction to Partial Differential Equations.* Princeton University Press, Princeton, NJ, second edition, 1995.

[39] G. B. Folland. *Real Analysis.* Pure and Applied Mathematics (New York). John Wiley & Sons, Inc., New York, second edition, 1999. Modern techniques and their applications, A Wiley-Interscience Publication.

[40] J. E. Galé, P. J. Miana, and P. R. Stinga. Extension problem and fractional operators: semigroups and wave equations. *J. Evol. Equ.*, 13(2):343–368, 2013.

[41] A. E. Gatto, C. Segovia, and S. Vági. On fractional differentiation and integration on spaces of homogeneous type. *Rev. Mat. Iberoamericana*, 12(1):111–145, 1996.

[42] R. K. Getoor. First passage times for symmetric stable processes in space. *Trans. Amer. Math. Soc.*, 101:75–90, 1961.

[43] D. Gilbarg and N. S. Trudinger. *Elliptic Partial Differential Equations of Second Order.* Classics in Mathematics. Springer-Verlag, Berlin, 2001. Reprint of the 1998 edition.

[44] Q. Han and F.-H. Lin. *Elliptic Partial Differential Equations*, volume 1 of *Courant Lecture Notes in Mathematics.* Courant Institute of Mathematical Sciences, New York; American Mathematical Society, Providence, RI, second edition, 2011.

[45] R. Jensen. The maximum principle for viscosity solutions of fully nonlinear second order partial differential equations. *Arch. Rational Mech. Anal.*, 101(1):1–27, 1988.

[46] F. John and L. Nirenberg. On functions of bounded mean oscillation. *Comm. Pure Appl. Math.*, 14:415–426, 1961.

[47] M. Kassmann. Harnack inequalities: an introduction. *Bound. Value Probl.*, pages Art. ID 81415, 21, 2007.

[48] N. V. Krylov and M. V. Safonov. An estimate for the probability of a diffusion process hitting a set of positive measure. *Dokl. Akad. Nauk SSSR*, 245(1):18–20, 1979.

[49] N. V. Krylov and M. V. Safonov. A property of the solutions of parabolic equations with measurable coefficients. *Izv. Akad. Nauk SSSR Ser. Mat.*, 44(1):161–175, 239, 1980.

[50] N. S. Landkof. *Foundations of Modern Potential Theory.* Die Grundlehren der mathematischen Wissenschaften, Band 180. Springer-Verlag, New York-Heidelberg, 1972. Translated from the Russian by A. P. Doohovskoy.

[51] N. N. Lebedev. *Special Functions and their Applications.* Dover Publications, Inc., New York, 1972. Revised edition, translated from the Russian and edited by Richard A. Silverman, Unabridged and corrected republication.

[52] J. Moser. A new proof of De Giorgi's theorem concerning the regularity problem for elliptic differential equations. *Comm. Pure Appl. Math.*, 13:457–468, 1960.

[53] J. Moser. On Harnack's theorem for elliptic differential equations. *Comm. Pure Appl. Math.*, 14:577–591, 1961.

[54] J. L. Rubio de Francia, F. J. Ruiz, and J. L. Torrea. Calderón-Zygmund theory for operator-valued kernels. *Adv. in Math.*, 62(1):7–48, 1986.

[55] W. Rudin. *Functional Analysis.* McGraw-Hill Series in Higher Mathematics. McGraw-Hill Book Co., New York-Düsseldorf-Johannesburg, 1973.

[56] W. Rudin. *Real and Complex Analysis.* McGraw-Hill Book Co., New York, third edition, 1987.

[57] S. Salsa. *Partial Differential Equations in Action*, volume 99 of *Unitext*. Springer, [Cham], third edition, 2016. From modelling to theory, La Matematica per il 3+2.

[58] O. Savin. Small perturbation solutions for elliptic equations. *Comm. Partial Differential Equations*, 32(4-6):557–578, 2007.

[59] L. Silvestre. *Regularity of the obstacle problem for a fractional power of the Laplace operator*. ProQuest LLC, Ann Arbor, MI, 2005. Thesis (Ph.D.)–The University of Texas at Austin.

[60] L. Silvestre. Hölder estimates for solutions of integro-differential equations like the fractional Laplace. *Indiana Univ. Math. J.*, 55(3):1155–1174, 2006.

[61] L. Silvestre. Regularity of the obstacle problem for a fractional power of the Laplace operator. *Comm. Pure Appl. Math.*, 60(1):67–112, 2007.

[62] G. Stampacchia. Le problème de Dirichlet pour les équations elliptiques du second ordre à coefficients discontinus. *Ann. Inst. Fourier (Grenoble)*, 15(fasc. 1):189–258, 1965.

[63] E. M. Stein. *Singular Integrals and Differentiability Properties of Functions*. Princeton Mathematical Series, No. 30. Princeton University Press, Princeton, NJ, 1970.

[64] E. M. Stein. *Topics in Harmonic Analysis Related to the Littlewood-Paley Theory*. Annals of Mathematics Studies, No. 63. Princeton University Press, Princeton, NJ; University of Tokyo Press, Tokyo, 1970.

[65] E. M. Stein. *Harmonic Analysis: Real-variable Methods, Orthogonality, and Oscillatory Integrals*, volume 43 of *Princeton Mathematical Series*. Princeton University Press, Princeton, NJ, 1993. With the assistance of Timothy S. Murphy, Monographs in Harmonic Analysis, III.

[66] E. M. Stein and G. Weiss. *Introduction to Fourier Analysis on Euclidean Spaces*. Princeton Mathematical Series, No. 32. Princeton University Press, Princeton, NJ, 1971.

[67] P. R. Stinga. *Fractional powers of second order partial differential operators: extension problem and regularity theory*. Universidad Autónoma de Madrid, 2010. Thesis (Ph.D.)–Universidad Autónoma de Madrid.

[68] P. R. Stinga. User's guide to the fractional Laplacian and the method of semigroups. In *Handbook of fractional calculus with applications. Vol. 2*, pages 235–265. De Gruyter, Berlin, 2019.

[69] P. R. Stinga and J. L. Torrea. Extension problem and Harnack's inequality for some fractional operators. *Comm. Partial Differential Equations*, 35(11):2092–2122, 2010.

[70] P. R. Stinga and J. L. Torrea. Regularity theory and extension problem for fractional nonlocal parabolic equations and the master equation. *SIAM J. Math. Anal.*, 49(5):3893–3924, 2017.

[71] P. R. Stinga and M. Vaughan. Fractional elliptic equations in nondivergence form: definition, applications and Harnack inequality. *J. Math. Pures Appl. (9)*, 156:245–306, 2021.

[72] M. H. Taibleson. On the theory of Lipschitz spaces of distributions on Euclidean n-space. I. Principal properties. *J. Math. Mech.*, 13:407–479, 1964.

[73] J. L. Torrea. *Integrales Singulares Vectoriales*, volume 12 of *Notas de Álgebra y Análisis [Notes on Algebra and Analysis]*. Universidad Nacional del Sur, Instituto de Matemática, Bahía Blanca, 1984. Algunas aplicaciones de una versión actualizada de un resultado de A. Benedek, A. P. Calderón y R. Panzone. [Some applications of a modern version of a result of A. Benedek, A. P. Calderón and R. Panzone], With an English summary.

[74] R. L. Wheeden and A. Zygmund. *Measure and Integral.* Pure and Applied Mathematics (Boca Raton). CRC Press, Boca Raton, FL, second edition, 2015. An introduction to real analysis.

[75] F. Zó. A note on approximation of the identity. *Studia Math.*, 55(2):111–122, 1976.

[76] A. Zygmund. Smooth functions. *Duke Math. J.*, 12:47–76, 1945.

Index

Regularity Techniques for Elliptic PDEs and the Fractional Laplacian

Regularity Techniques for Elliptic PDEs and the Fractional Laplacian presents important analytic and geometric techniques to prove regularity estimates for solutions to second order elliptic equations, both in divergence and nondivergence form, and to nonlocal equations driven by the fractional Laplacian. The emphasis is placed on ideas and the development of intuition, while at the same time being completely rigorous. The reader should keep in mind that this text is about how analysis can be applied to regularity estimates. Many methods are nonlinear in nature, but the focus is on linear equations without lower order terms, thus avoiding bulky computations. The philosophy underpinning the book is that ideas must be flushed out in the cleanest and simplest ways, showing all the details and always maintaining rigor.

Features

- Self-contained treatment of the topic.
- Bridges the gap between upper undergraduate textbooks and advanced monographs to offer a useful, accessible reference for students and researchers.
- Replete with useful references.